Handbook of Genomics, Health and Society

T0362225

The *Handbook* provides an essential resource at the interface of Genomics, Health and Society, and forms a crucial research tool for both new students and established scholars across biomedicine and social sciences. Building from and extending the first *Routledge Handbook of Genetics and Society*, the book offers a comprehensive introduction to pivotal themes within the field, an overview of the current state of the art knowledge on genomics, science and society, and an outline of emerging areas of research.

Key themes addressed include the way genomic based DNA technologies have become incorporated into diverse arenas of clinical practice and research whilst also extending beyond the clinic; the role of genomics in contemporary 'bioeconomies'; how challenges in the governance of medical genomics can both reconfigure and stabilise regulatory processes and jurisdictional boundaries; how questions of diversity and justice are situated across different national and transnational terrains of genomic research; and how genomics informs – and is shaped by – developments in fields such as epigenetics, synthetic biology, stem cell, microbial and animal model research.

Presenting cutting edge research from leading social science scholars, the *Handbook* provides a unique and important contribution to the field. It brings a rich and varied cross disciplinary social science perspective that engages with both the history and contemporary context of genomics and 'post-genomics', and considers the now global and transnational terrain in which these developments are unfolding.

Sahra Gibbon is Reader in Medical Anthropology in the Anthropology Department at University College London, UK.

Barbara Prainsack is a Professor at the Department of Political Science at the University of Vienna, Austria, and at the Department of Global Health & Social Medicine at King's College London, UK.

Stephen Hilgartner is Professor of Science & Technology Studies at Cornell University, USA.

Janelle Lamoreaux is Assistant Professor of Anthropology at University of Arizona, USA.

Handbook of Genomics, Health and Society

Second edition

Edited by Sahra Gibbon, Barbara Prainsack, Stephen Hilgartner and Janelle Lamoreaux

Routledge
Taylor & Francis Group

LONDON AND NEW YORK

Second edition published 2018
by Routledge
2 Park Square, Milton Park, Abingdon, Oxon OX14 4RN

and by Routledge
52 Vanderbilt Avenue, New York, NY 10017

First issued in paperback 2020

Routledge is an imprint of the Taylor & Francis Group, an informa business

Second edition published 2018

First edition published by Routledge 2012

British Library Cataloguing in Publication Data
A catalogue record for this book is available from the British Library

Library of Congress Cataloging in Publication Data
Names: Gibbon, Sahra, editor.
Title: Handbook of genomics, health & society / edited by Sahra Gibbon, Barbara Prainsack, Stephen Hilgartner and Janelle Lamoreaux.
Other titles: Handbook of genomics, health and society
Description: Milton Park, Abingdon, Oxon ; New York, NY : Routledge, 2018. | Includes bibliographical references.
Identifiers: LCCN 2017043979 | ISBN 9781138211957 (hbk) | ISBN 9781315451695 (ebk)
Subjects: LCSH: Medical genetics–Handbooks, manuals, etc. | Genomics–Social aspects–Handbooks, manuals, etc.
Classification: LCC RB155 .H355 2018 | DDC 616/.042–dc23
LC record available at https://lccn.loc.gov/2017043979

ISBN 13: 978–0–367–65994–3 (pbk)
ISBN 13: 978–1–138–21195–7 (hbk)

Typeset in Bembo
by Taylor & Francis Books

Contents

Contents

Contents

Contributors

Editors:

Sahra Gibbon is Reader in Medical Anthropology in the Department of Anthropology at University College, London. She has carried out research related to genomics, health and society in the UK, Cuba and Brazil and has a particular interest in cancer genetics, identity, activism and public health. Some of her major publications have focused on developments in breast cancer genetics. They include *Breast Cancer Genes and the Gendering of Knowledge* (Palgrave Macmillan, 2007) as well as the co-edited volume *Breast Cancer Gene Research and Medical Practices: Transnational Perspectives in the Time of BRCA* (Routledge, 2014). A monograph (under contract with UCL Press) on her most recent research in Brazil is currently being finalised, provisionally entitled 'Making Biological Difference: cancer genetics, local biologies and the politics of public health in Brazil'.

Barbara Prainsack is Professor at the Department of Political Science at the University of Vienna, with a joint appointment as Professor at the Department of Global Health & Social Medicine at King's College, London. Her work explores the societal, regulatory and ethical aspects of biomedicine and bioscience, with special focus on DNA technologies and personalised medicine. Her latest books are: *Personalized Medicine: Empowered Patients in the 21st Century?* (New York University Press, 2017), and *Solidarity in Biomedicine and Beyond* (with Alena Buyx, Cambridge University Press, 2017). Barbara is also intensely involved in policy-related work. She led the European Science Foundation's (ESF) Forward Look on Personalised Medicine (with Stephen Holgate and Aarno Palotie, 2012) and she is a member of the Austrian National Bioethics Committee and the European Group on Ethics and New Technologies advising the European Commission.

Stephen Hilgartner is Professor of Science & Technology Studies at Cornell University. He specialises in the social dimensions and politics of contemporary and emerging science and technology, especially in the life sciences. His research focuses on situations in which scientific knowledge is implicated in establishing, contesting, and maintaining social order, a theme he has explored in studies of expertise, property formation, risk disputes, and biotechnology. His most recent book, *Reordering Life: Knowledge and Control in the Genomics Revolution* (MIT Press, 2017), examines how new forms of knowledge and new law-like regimes of governance took shape as the field of genomics emerged. His other books include *Science on Stage: Expert Advice as Public Drama* (Stanford University Press, 2000), which won the Carson Prize from the Society for Social Studies of Science, and *Science & Democracy: Making Knowledge and Making Power in the Biosciences and Beyond* (Routledge, 2015). Hilgartner is a Fellow of the American Association for the Advancement of Science.

Janelle Lamoreaux is Assistant Professor of Anthropology at the University of Arizona. She conducts research on gender, genomics and environmental change. Her book manuscript, provisionally entitled *Infertile Futures: Reproducing Chinese Environments*, is currently being finalised. It is an ethnographic exploration of epigenetic toxicology and the intersection of environmental and reproductive health in China.

Contributors:

Amber Benezra is a sociocultural anthropologist researching how studies of human microbiota intersect with biomedical ethics and public health/technological infrastructures. She is developing an 'anthropology of microbes' to address global health problems across disciplines, and will begin as an industry Assistant Professor of Science and Technology Studies at NYU in the fall of 2017. Her forthcoming book from University of Minnesota Press, *Anthrobiota*, is the first ethnography of the microbiome, and seeks to develop terms of ethical co-engagement for social and biological scientists. The book follows the development of Benezra's collaborative partnership with a top-tier US scientific lab researching gut microbes, as well as local scientists at the field site in Bangladesh. As scientists study the interrelationships between microbes and malnutrition, Benezra explores ways to reconcile the scale and speed differences between the lab, the intimate biosocial practices of Bangladeshi mothers and children, the looming structural violence of poverty and profound inequality.

Dominic Berry is a historian and philosopher of science integrating methods and analysis from science and technology studies. He is a Research Fellow at the University of Edinburgh and a member of the ERC funded Engineering Life project, focusing on the history of biological engineering. His research interests include biotechnology, agriculture, intellectual property, and the history and philosophy of biological experimentation.

Kean Birch is Associate Professor in the Department of Geography and member of the geography, sociology, and science and technology studies Graduate Programs at York University, Canada. He is also Senior Associate at the Innovation Policy Lab, University of Toronto, Canada and Associate Editor on the STS journal *Science as Culture*. His recent books include *We Have Never Been Neoliberal* (Zero Books, 2015); *The Handbook of Neoliberalism* (Routledge, 2016 – co-edited with Simon Springer and Julie MacLeavy); *Innovation, Regional Development and the Life Sciences: Beyond Clusters* (Routledge, 2016); *Business and Society: A Critical Introduction* (Zed Books, 2017 – co-authored with Mark Peacock, Richard Wellen, Caroline Shenaz Hossein, Sonya Scott, and Alberto Salazar); and *A Research Agenda for Neoliberalism* (Edward Elgar, 2017). He is currently working on a book called 'Neoliberal Bio-economies? The Co-construction of Markets and Natures' for Palgrave Macmillan.

Catherine Bliss is Associate Professor of Sociology at the University of California, San Francisco. Her research explores the sociology of race, gender and sexuality in science and society. Bliss's award-winning book *Race Decoded: The Genomic Fight for Social Justice* (Stanford University Press, 2012) examines how genomics became today's new science of race. Her forthcoming book *Social by Nature: How Sociogenomics is Redefining What it Means to be Human* (Stanford University Press, 2018) examines postgenomic convergences in social and genetic science led by new sociogenomic sciences, including their implications for equality, identity, and belonging. Her latest book project, 'Our Genomic World', introduces social debates around genomics to academic and non-academic audiences.

Pascale Bourret is Associate Professor at Aix-Marseille Université where she teaches sociology. She is also a researcher at SESSTIM (Economy and Social Sciences, Health Care Systems and Societies) an INSERM-IRD-Aix-Marseille Université UMR. At the crossroads of science studies and the sociology of medicine, her work focuses on the transformation of biomedical practices in connection to the development of genomic tools, with a focus on biology/clinic interface, the transformation of clinical work, the production of clinical judgement and clinical decision-making. She has published articles on bio-clinical collectives in the domain of BRCA testing, on regulation issues linked to new genomics tools, and on the emergence of genomic-driven clinical trials. She is presently coordinating a project investigating the implementation of precision medicine in oncology, which explores the conditions surrounding the development of targeted therapies in the context of clinical and translational research by focusing on a series of on-going biomarker-driven trials.

Jane Calvert is a Reader in Science, Technology and Innovation Studies at the University of Edinburgh. Her current research, funded by a European Research Council Consolidator grant, focuses on attempts to engineer living things in the emerging field of synthetic biology, which raises intriguing questions about design, evolution and 'life'. She is also interested in the governance of emerging technologies and interdisciplinary collaborations of all sorts. She is a co-author of the interdisciplinary book *Synthetic Aesthetics: Investigating Synthetic Biology's Designs on Nature*, published by MIT Press in 2014.

Alberto Cambrosio is Professor at McGill University's Department of Social Studies of Medicine. His recent work examines 'genomics in action', i.e., as applied to concrete instances of medical work, by investigating public, academic and commercial programmes that capitalise on the therapeutic insights offered by the new molecular genetics of cancer. His most recent book (*Cancer on Trial: Oncology as a New Style of Practice*, University of Chicago Press, 2012, co-authored with Peter Keating) argues that, contrary to common assumptions, clinical trials do not boil down to a mere 'technology' or a few methodological principles: rather, they are an institution that corresponds to a profound transformation of biomedical activities, and rise to the level of a 'new style of practice'. This work builds on a previous book (*Biomedical Platforms*, MIT Press, 2003, also co-authored with Peter Keating) that analysed the transformation of medicine into biomedicine.

Silvia Camporesi (PhD, PhD) is a bioethicist with an interdisciplinary background in biotechnology, and philosophy of medicine. She began her career as a biotechnologist working in the gene therapy laboratory at the International Centre for Genetic Engineering and Biotechnology in Triest, but subsequently left the bench to study the ethical and social implications of gene therapy applied for enhancement purposes. She is currently Assistant Professor in the Department of Global Health & Social Medicine and Director of the MSc programme in Bioethics & Society at King's College London. Over the past decade, Silvia has been writing about the ethics of emerging technologies, and is author of *From Bench to Bedside to Track and Field: The Context of Enhancement and its Ethical Relevance* (UC Medical Humanities Press, 2014) and (forthcoming for Routledge, 2018) of *Bioethics, Genetics and Sport*, co-authored with Mike McNamee. She tweets at @silviacamporesi, and info about her research is available on her personal webpage: https://silviacamporesiresearch.org.

Giulia Cavaliere is a PhD candidate in Bioethics and Society at the Department of Global Health and Social Medicine at King's College, London. She has a background in Philosophy, Ethics and Politics (MA) and Bioethics (MSc). Currently, she is working on a Wellcome Trust funded project on the ethics of new reproductive technologies and eugenics. Her research interests lie primarily in bioethics, meta-ethics, moral theory and democratic theories. She has published on various topics in bioethics, including embryo research, surrogacy, eugenics, genome editing technologies and mitochondrial replacement techniques.

Margaret Chiappetta is currently a PhD candidate in Science and Technology Studies at York University, Canada. Her doctoral research is supported by the Social Sciences and Humanities Research Council of Canada and is focused on open science and intellectual property as it relates to pharmaceutical research and innovation. Previously, she completed an MA in Science and Technology Studies at the University of British Columbia, Canada.

Dong Dong is Research Assistant Professor at the David C. Lam Institute for East-West Studies, Hong Kong Baptist University. She obtained her PhD in Mass Communication (with a minor in Epidemiology) from the University of Minnesota, Twin Cities. Specialised in media sociology, communication studies, and public health, Dong's research has been published in numerous journals and presented at more than twenty international conferences in the field of communication, sociology, anthropology, and medical sciences. Her current research interest focuses on issues related to rare diseases in China, especially the socio-economic status of people living with rare disorders, the health advocacy movement led by Chinese patient organisations, and the interpretation and practice of prenatal genetic testing by families affected by rare diseases. In addition to academic work, Dong also serves as a consultant to several patient organisations in China, helping them with policy research and social advocacy.

Edward Dove is Lecturer in Risk and Regulation at the School of Law, University of Edinburgh, and Deputy Director of the J Kenyon Mason Institute for Medicine, Life Sciences and Law. From 2011 through 2014, he was an Academic Associate at the Centre of Genomics and Policy at McGill University. Edward is the Case Comments Editor for *Medical Law International*; he has also served as Editor-in-Chief of *SCRIPTed*, an online, peer-reviewed open access journal dedicated to law, medicine, science and technology. From 2013 to 2015, he served as Coordinator for the Regulatory and Ethics Working Group of the Global Alliance for Genomics and Health (GA4GH). Currently, Edward is a member of the Data Access Committee of METADAC (Managing Ethico-social, Technical and Administrative issues in Data ACcess), which is a multi-agency, multi-study data access structure that services several of the UK's major longitudinal studies.

Ulrike Felt is Professor of Science and Technology Studies (STS) and Head of the interfaculty research platform 'Responsible research and innovation in academic practice'. Her research focuses on governance, democracy and public participation as well as on shifting research cultures. Her areas of study cover life science/(bio)medicine, nanotechnology, nuclear energy and sustainability research. From 2002 to 2007 she was editor-in-chief of *Science, Technology, & Human Values*. More recently she led the editorial team of the most recent *Handbook of Science and Technology Studies* (MIT Press, 2017). Since 2017 she is president of the European Association for the Study of Science and Technology (EASST).

Carrie Friese is Associate Professor of Sociology at the London School of Economics and Political Science. Her initial research focused on assisted reproductive technologies for humans and endangered species, including the development of interspecies nuclear transfer for species preservation in zoos. Her book *Cloning Wild Life: Zoos, Captivity and the Future of Endangered*

Animals (NYU Press) appeared in 2013. Friese's new research project explores animal husbandry and care in scientific knowledge production, including comparisons of care practices and their regulation in the United Kingdom.

Duana Fullwiley is Associate Professor in the Department of Anthropology at Stanford University. She also teaches in the interdepartmental programme in Comparative Studies of Race and Ethnicity as well as the programme in Science, Technology and Society. Dr Fullwiley is an anthropologist of science and medicine interested in how social identities, health outcomes, and molecular genetic findings increasingly intersect. Her first book, *The Enculturated Gene: Sickle Cell Health Politics and Biological Difference in West Africa* (Princeton, 2011), is a detailed ethnography of sickle cell population genetics and economies of care in postcolonial Senegal. She is currently finishing her second book entitled *Tabula Raza: Mapping Race and Human Diversity in American Genome Science*. Her current work focuses on ancestry genetics and how scientists use this data to create new technologies for personalised medicine, genealogical ancestry tracing, and DNA forensics. She is also interested in questions of privacy, informed consent, information ownership and ethics.

Anna Harris is an anthropologist studying medical and other craft practices. Her work concerns issues of sensorality, embodiment and learning (see more on her website www.annaroseharris.wordpress.com). She works as Assistant Professor in the Department of Technology and Society Studies at Maastricht University, in the Netherlands. Currently she is leading an ERC-funded project which examines the role of digital and other technologies in the training of doctors' skills (www.makingclinicalsense.com). She also blogs about pneumatic tube technologies at www.pneumaticpost.blogspot.com.

Stuart Hogarth's work focuses on biomedical innovation and his research has investigated a diverse range of emergent biotechnologies, such as stem cell therapies and synthetic biology. His primary interest is political economy of diagnostic innovation, with a particular focus on regulatory governance, intellectual property rights and the impact of genomic science on the diagnostics sector. Dr Hogarth uses an international comparative methodology to explore the continued salience of national institutions such as regulatory regimes and healthcare systems, in a bioeconomy which is increasingly characterised by global governance structures, international scientific collaborations and transnational flows of capital and scientific labour. His work combines empirical research with normative analysis of public policy and commercial strategy. He has produced policy reports for the European Commission and Health Canada, and the UK Human Genetics Commission.

Ine Van Hoyweghen is Professor of Sociology of Biomedicine at the Centre for Sociological Research of the University of Leuven. Her research interests are the sociology of biomedicine, science and technology studies and sociology of health care markets. Her recent work centres on the empirical investigation of the role of biomedicine in reconfiguring identities, responsibilities and solidarity. She is the PI of the project Postgenomic Solidarity: European Life Insurance in the Era of Personalised Medicine, for which she received an Odysseus grant by the Research Foundation Flanders (FWO). She is the director of the Life Sciences & Society Lab and founding board member of the Leuven Institute for Genomics and Society (LIGAS). She is also a member of the Young Academy (JA) of the Royal Flemish Academy of Belgium for Science and the Arts (KVAB) and the founding chair of the Belgian Science, Technology & Society (B.STS) Network.

J. Benjamin Hurlbut, PhD, is Associate Professor of Biology and Society in the School of Life Sciences at Arizona State University. His research lies at the intersection of science and technology studies, bioethics and political theory. Hurlbut studies relationships between science, politics and law in the governance of biomedical research and innovation in the twentieth and twenty-first centuries, examining the interplay of science and technology with shifting notions of democracy, of religious and moral pluralism, and of public reason. He is the author of *Experiments in Democracy: Human Embryo Research and the Politics of Bioethics* (Columbia University Press, 2017). He holds an A.B. from Stanford University, and a PhD in the History of Science from Harvard University.

Susan Kelly received her PhD in 1994 from the University of California, and continued her training at Stanford University in the History and Philosophy of Science, Genetics and Bioethics. She is currently Professor of Sociology at the University of Exeter.

Peter Keating is Associate Professor in the Department of History at the University of Quebec at Montreal. He pursues research in the history and sociology of medicine. His last book (with Alberto Cambrosio) was *Cancer on Trial: Oncology as a New Style of Practice* (University of Chicago Press, 2012).

Jackie Leach Scully is Professor of Social Ethics and Bioethics, and Executive Director of the Policy, Ethics and Life Sciences Research Centre, Newcastle University, UK. Her research interests include disability bioethics, feminist approaches to bioethics, narrative and identity in bioethical contexts, empirical methodologies, and public bioethics. She is the author of *Disability Bioethics: Moral Bodies, Moral Difference* (Rowman and Littlefield Publishers, 2008) and co-editor of *Feminist Bioethics: At the Center, On the Margins* (John Hopkins University Press, 2010), and is currently editor of the *International Journal of Feminist Approaches to Bioethics* (IJFAB).

Sandra Soo-Jin Lee, PhD, is Senior Research Scholar at the Stanford Center for Biomedical Ethics and teaches in the Program in Science, Technology and Society (STS) at Stanford University. She is a medical anthropologist whose research focuses on the ethics of inclusion, equity and social difference in biomedicine. Building on her interests in the sociocultural and ethical dimensions of emerging genomic technologies, Dr. Lee leads studies on the governance and use of biospecimens and data in research and the ethics of identifying race, ethnicity and ancestry in human genetic variation. Her work also includes ethnographic research of entrepreneurship and the social and cultural dimensions of innovation. Dr. Lee serves on the Scientific Advisory Board for the Kaiser Permanente National Biobank, the NIH Genomics and Society Working Group at the National Human Genome Research Institute, and on the editorial board of *Narrative Inquiry in Bioethics*.

Sabina Leonelli is Professor in philosophy and history of science at the University of Exeter, UK, where she co-directs the Exeter Centre for the Study of the Life Sciences and leads the 'Data Studies' research strand. Her research focuses on the philosophy of data-intensive science, especially the methods and assumptions involved in the production, dissemination and use of data in biology and biomedicine; and the ways in which the open science movement is redefining what counts as research and knowledge across different research environments. She is also interested in the epistemic status of experimental organisms as models and data sources, which she investigates jointly with Rachel Ankeny. She has published widely within the philosophy of science as well as biology and science & technology studies, and authored *Data-Centric Biology: A Philosophical Study* (Chicago University Press, 2016).

Jennifer Liu is Associate Professor in the Department of Anthropology at the University of Waterloo, and cross-appointed in the School of Public Health and Health Systems. She is a Fellow at the Balsillie School of International Affairs. Liu earned her PhD in medical anthropology from the University of California, Berkeley and San Francisco, and was a Fulbright Fellow in Taiwan.

Stephanie Lloyd is a medical anthropologist at Université Laval. Her ongoing research examines emerging studies in environmental epigenetics that attempts to link early life experiences to behaviours later in life. She is particularly interested in forms of reasoning specific to the postgenomic era and how stability and change are modelled in epigenetics and neuroscience research. Her emerging work focuses on models of sensory development, most specifically, how scientists and clinicians explain the effects of 'sensory deprivation' and 'sensory restoration' in the case of deaf people who use cochlear implants to hear. In continuity with her ongoing research, this study explores how environmental factors are studied as signals in scientific research and considers more broadly how we understand the interactions between body and society. She is co-editor of *The Handbook of Biology and Society* (with Meloni, Cromby, and Fitzgerald; Palgrave, 2017).

Margaret Lock is Marjorie Bronfman Professor Emerita in the Department of Social Studies of Medicine and the Department of Anthropology, McGill University. Her research focuses on embodiment, comparative epistemologies of medical knowledge, and the global impact of biomedical technologies. She is the author and/or co-editor of 18 books and 220 articles. Her five monographs have won major awards. Lock is a Fellow of the Royal Society of Canada, Officier de L'Ordre national du Québec, Officer of the Order of Canada, and an elected Member of the American Academy of Arts and Sciences. She has been awarded the Canada Council for the Arts Molson Prize, the Canada Council for the Arts Killam Prize, a Trudeau Foundation Fellowship, a Gold Medal for Research by the Social Sciences and Humanities Research Council of Canada, an SMA Career Achievement Award, and the McGill Medal for Exceptional Academic Achievement.

Ilana Löwy is Senior Researcher at INSERM (Institut National de la Santé et de la Recherche Scientifique), Paris. Trained as a biologist with a PhD in immunology, she then retrained as a historian of science and medicine. Her main research interests are relationships between laboratory sciences, clinical medicine and public health, intersections between gender studies and biomedicine, and material practices of biomedicine. She studied the history of bacteriology and immunology, tropical medicine, oncology and hereditary pathologies. Another area of interest is present-time ramification of the epistemology of Ludwik Fleck. She is currently studying the Zika epidemic in Brazil, and its consequences for woman's reproductive rights. Her recent books are: *Preventive Strikes: Women, Precancer and Prophylactic Surgery* (Johns Hopkins University Press, 2009); *A Woman's Disease: A History of Cervical Cancer* (Oxford University Press, 2011), and *Imperfect Pregnancies: A History of Birth Defects and Prenatal Diagnosis* (Johns Hopkins, forthcoming fall 2017).

Claire Marris is Reader at the Centre for Food Policy, City, University of London. Her research is in the field of Social Studies of Science, with a focus on the use of modern biotechnologies and genetic modification in food and agriculture. Her work explores links between science and democracy, and the relationship between scientific evidence and policy making. From 2009 to 2015 she was involved in a collaborative programme on synthetic biology involving engineers, natural scientists, and social scientists, that was initially framed as

ELSI (Ethical, Legal, Social Implications) and later as RRI (Responsible, Research and Innovation).

Paul Martin is Professor of Sociology of Science and Technology at the Department of Sociological Studies, University of Sheffield. He has a first training in molecular biology and works at the interface of STS and medical sociology. His current research interests include: the clinical and commercial development of genome editing; the development of epigenetics and the role of science in (social) policy; novel biosocial concepts and methods in the social sciences; and Responsible Research and Innovation. He is Co-Director of a recently established research centre – the Institute for the Study of the Human (iHuman), which seeks to explore what it means to be human in the twenty-first century.

Nicole Nelson is Assistant Professor in the Department of History and the Department of Medical History and Bioethics at the University of Wisconsin–Madison. She is the author of *Model Behavior: Animal Experiments, Complexity, and the Genetics of Psychiatric Disorders*, and is a Collaborating Editor for the journal *Social Studies of Science*.

Edward Nik-Khah is Associate Professor of Economics at Roanoke College. He has completed research on interactions between the Chicago School of Economics, the pharmaceutical industry, and pharmaceutical science; the neoliberal origins of economics imperialism; the distinctive role of George Stigler as architect of the Chicago School; and the tensions emerging from economists' assumption of a professional identity as designers of markets, for which he won the K. William Kapp Prize from the European Association for Evolutionary Political Economy. He is co-editor (with Robert Van Horn) of *The Contributions of Business to Economics* (History of Political Economy, Special Issue, Duke University Press 2017). His book (co-authored with Philip Mirowski) *The Knowledge We Have Lost in Information* (Oxford University Press, 2017) examines the history of information in modern economics.

Aaron Panofsky is Associate Professor in the Institute for Society and Genetics and Departments of Public Policy and Sociology at the University of California, Los Angeles. His research focuses on the sociology of science and knowledge, public participation in science, and Bourdieu's field theory. His award winning book *Misbehaving Science* (Chicago University Press, 2014) is about how controversies have successively reshaped the field of behaviour genetics socially and intellectually. Current research considers genetics' impact on knowledge of race, white nationalists' use of genetic ancestry tests, controversies over the statistical evaluation of teachers, and the sociology of the reproducibility crisis.

Shobita Parthasarathy is Associate Professor of Public Policy and Women's Studies, and Director of the Science, Technology, and Public Policy Program, at University of Michigan. Her research focuses on the governance of ethically and socially controversial science and technology, particularly in comparative perspective. She is the author of numerous articles and two books: *Building Genetic Medicine: Breast Cancer, Technology, and the Comparative Politics of Health Care* (MIT Press, 2007) and *Patent Politics: Life Forms, Markets, and the Public Interest in the United States and Europe* (University of Chicago Press, 2017). Findings from *Building Genetic Medicine* influenced the 2013 US Supreme Court decision prohibiting patents on isolated human genes. She holds a Bachelor degree in Biology from the University of Chicago and Master and PhD degrees in Science and Technology Studies from Cornell University.

Martyn Pickersgill is Wellcome Trust Reader in Social Studies of Biomedicine, Edinburgh Medical School, and Associate Director of the University of Edinburgh Centre for Science, Knowledge, and Policy (SKAPE). A sociologist of bioscience and the health professions, Martyn's work has primarily focussed on the historical, legal, and social dimensions of neuroscience, psychiatry, and psychology. He has received grants and fellowships from bodies such as the British Academy, Leverhulme Trust, and Wellcome Trust to support research and engagement in these and other areas. Martyn is a member of the BBSRC Bioscience for Society Strategy Panel, as well as various editorial boards. In 2015, he was awarded the Royal Society of Edinburgh Henry Duncan Medal.

Eugene Raikhel is Associate Professor in the Department of Comparative Human Development at the University of Chicago. A cultural and medical anthropologist by training, he studies the circulation of new forms of knowledge and clinical intervention produced by biomedicine, neuroscience and psychiatry. For the past several years, he has been collaborating with Stephanie Lloyd (U Laval) on a project examining the emerging field of 'behavioral epigenetics', with a particular focus on research about suicidal risk. He is the author of *Governing Habits: Treating Alcoholism in the Post-Soviet Clinic* (Cornell, 2016), co-editor of *Addiction Trajectories* (Duke, 2013), and co-founder and editor of *Somatosphere*, an online forum for medical anthropology.

Deborah Scott is a Research Fellow at the University of Edinburgh in Science, Technology, and Innovation Studies. She is a human geographer and lawyer. Her research explores decision-making processes around the governance of new and emerging technologies and sciences.

Niccolò Tempini is Research Fellow in the Department of Sociology, Philosophy and Anthropology at the University of Exeter, UK. Following studies in Philosophy (BA, MA) and Information Systems (MSc), he gained a PhD in Information Systems from LSE in 2015, with a thesis on the organization of medical research through patient-centred social media networks. Part of the 'Epistemology of data-intensive science' project led by Sabina Leonelli, his current research continues to focus on the development of infrastructures for the collection and re-use of biomedical data, but with a broader range and scope. Working at the intersection between information systems, STS and philosophy of science, he approaches the field with an interest in the practices of data scientists, software developers, and medical researchers, to understand how these intersect with each other as they come to grips with new methods and new forms of data, information technology and organisation of work.

Etienne Vignola-Gagné is a Postdoctoral Fellow at the Department of Social Studies of Medicine, McGill University. He works on change in biomedical innovation practices and policies, examining among others, iterations of genomic cancer medicine or reform movements that have defined and advocated 'translational research' agendas. He alternatively deploys approaches from interpretative policy analysis, organisational theory and science and technology studies. Findings from these projects have been published in journals such as *Science and Public Policy* (with Alberto Cambrosio and Peter Keating), *History and Philosophy of the Life Sciences* and (with George Weisz) *Population and Development Review*.

Sally Wyatt is Professor of 'digital cultures', Maastricht University, and academic director of the Netherlands Graduate Research School for Science, Technology and Modern Culture (WTMC). She originally studied economics (BA McGill, 1976; MA Sussex, 1979), but later did a PhD in science and technology studies (Maastricht, 1998), which focused on different ways of transmitting data over networks. Her current research interests include digital media in the production of knowledge in the humanities and the social sciences, and the ways in which people incorporate the internet into their practices for finding health information. On the latter, together with Anna Harris and Susan Kelly, she published a book called *CyberGenetics: Health Genetics and New Media* (Routledge, 2016). Together with Andrew Webster (York University, UK), she co-edits the book series, *Health, Technology and Society*, published by Palgrave Macmillan.

Jianfeng Zhu is an Associate Professor in Fudan-Harvard Medical Anthropology Collaborative Research Center, Fudan University, Shanghai. She received her BA and MA in law in China and PhD in anthropology from University of Minnesota. Her academic interests focus on reproductive medicine and biotechnologies, eugenic practices, population control policies as well as the broader issues of gender, modernity and globalisation. Her book, *Winning the Competition at the Start Line: Chinese Modernity, Reproduction and the Desire for a 'High Quality' Population* (East China Normal University Press, 2014) provides a nuanced understanding of the relations among Chinese population quality control policy, its health care system, various prenatal testing technologies as well as the everyday practices of childbirth and childrearing in Chinese ordinary families. Her current project is examining the blooming genetic testing industry in China under the framework of 'precision medicine' propagated by the Chinese state.

Acknowledgements

A handbook such as this is always the result of collective concerted efforts on the part of multiple individuals. We extend our thanks therefore in the first instance to our contributors and section editors who have taken the time and effort to help craft the content of this *Handbook*. The editors are particularly grateful to Aaron Parkhurst whose unwavering commitment, professional attitude and skill in managing the work required to bring a project of this scale to fruition has been invaluable. At Routledge, we would also like to thank Gerhard Boomgaarden and his colleagues, Alyson Claffey and Mihaela Diana Ciobotea for their support and assistance. We would finally also like to thank the anonymous reviewers who provided feedback and commentary on the volume in the early stages of development and peer reviewed several of the chapters.

Introduction to *Handbook of Genomics, Health and Society*

Sahra Gibbon, Barbara Prainsack, Stephen Hilgartner and
Janelle Lamoreaux

As we reflect on the period between the publication of the first edition of this *Handbook* in 2009 and the second edition, a great deal seems to have changed. In terms of scope, scale and speed, genomic technologies have become increasingly embedded within different health care and research arenas. In the process, the 'new genetics' seems to have seamlessly segued into genomics, even post-genomics including epigenetics. At the same time, while some of the core themes raised in the first edition, including questions of ethics, regulation and commercialisation, remain vital to current social science engagements with the evolving terrain of genomic science and medicine, these are increasingly seen through the lenses of justice, governance and the bioeconomy. Such shifts are in part reflected in the thematic focus (and renamed title) of the current Handbook that places Genomics, Health and Society centre stage. Whilst recognising that what constitutes 'health' in an era of genomics remains contested, inequitably distributed and not always easily defined, the renamed title reflects how 'health', broadly construed, has been and continues to be a vital resource, a site of transformation and a tool in the reshaping of genomics and society. In this sense, the new title points to the focus of the volume on genomics in human health-related contexts, and not, for example, forensics or environmental genomics. While the discussions in this volume do touch upon many areas beyond health, given the expansion and growth of genomic technologies in fields outside of health, we have had nevertheless to limit the area which we can claim to cover systematically.

We argue that the 'novelty' of the present moment in genomic research related to 'Health and Society' might be characterised in terms of a series of tensions, contradictions and paradoxes. Whilst these can propel different fields of research and medical care at particular intersections, they can also at times pull against each other. Some of these dynamics reflect themes that have long been entangled with historical and contemporary change in genetic science and medicine, such as individual rights versus societal obligations. Yet there are new dimensions at stake also, particularly when as Sabina Leonelli succinctly highlights, genomic practices are 'caught in a web of technical acceleration, societal changes and logistical chaos' (this volume). We identify four cross-cutting themes in the *Handbook* that reveal the porous and fluid boundaries of topics and themes related to Genomics, Health and Society.

'Genetics for the world?';[1] globalising genomics, national histories and inequities

One of the novel themes illuminated in this edition of the *Handbook* is the global terrain on which genomic research, technologies and medical interventions are unfolding and the way that this process constitutes dense and complex intersections between the so-called global 'north' and 'south'. National governments throughout the world, multinational corporations that operate throughout it, and transnational scientific communities are all engaged in building genomic medicine. Interest and investment in genetics by international organisations such as the WHO has a long history. But the increased use of genomic tools and techniques in low income countries for addressing health care challenges and the widening global market for genomic information delineates a current moment of transnational expansion.[2] In this expansion, compliance of practices and tools with international standards is an important consideration. Yet standardisation is not the only outcome with national or local regional contexts, practices and concerns continuing to shape developments in genomic science and medicine. This is the case within as well as between North America and Europe and also in and at the interface with other regions of the world (Sleeboom-Faulkner, 2011, Fullwiley, 2011, Wade et al., 2014). Consider how differently prenatal testing is configured in, say, North America than in China, where an emphasis on 'population quality' and 'good births' informs how 'choice' is configured in the context of prenatal testing (Zhu, this volume). Or consider the diversity of stem cell research, which is simultaneously a 'global biological' (Franklin, 2005) and a locally shaped practice (Thompson, 2010, Bharadwaj, 2013). As Jennifer Liu puts it, the often used label of 'Asian Tiger' belies a resistance to standardisation when it comes to stem cell research in certain national and international contexts (this volume).

The increased transnational expansion of genomic research and technologies is often rhetorically and practically tied to 'humanitarian' efforts to widen inclusion, access and participation (within) and far beyond the 'global north'. However, this very process can at the same time also reveal the stark inequities that have and continue to shape these developments (Prainsack; Fullwiley and Gibbon, this volume). This therefore constitutes another central tension that cuts across the chapters in this *Handbook*. We see how both colonial and postcolonial histories of scientific research and entrenched 'power asymmetries' continue to have a central place in the way genomic research and medicine is being extended, even if improving the health of the poor is often the laudable, if still somewhat elusive, aim of such actions.

In this sense, the comparative and transnational perspectives offered by many chapters in the *Handbook* reveal the 'frictions' and 'zones of awkward engagement' (Tsing, 2005) that characterise genomics as a global enterprise and which remain a key and ongoing focus of concern and analysis for social science research.

Stability and instability in 'post-genomics'

Another particularly striking theme in the new edition of this volume is the frequency with which stability and instability now enter into conversations of (post-) genomic research. Social scientists of science who previously challenged scientific black boxing, reductionism, and determinism, now often find themselves studying biological fields that embrace theories of complex systems, multifactorial causalities, and intricate interaction between genes and what stands outside them. While there is widespread acknowledgement that the genetic inheritance of disease is, with few single-gene exceptions, a process involving multiple factors, biomedical professionals, patients and other users of genetic data often seek stability in messy genetic information. (Post-) genomic instability, then, does not replace genetic stability, but conscripts a history of presumptive solidity into the assumed variability of the present.

Contributions in this volume point to a variety of sites where tensions between new and old, stability and instability, impact scientific practices in genomic research and treatment settings. Martyn Pickersgill's contribution, for instance, discusses the way epigenetic research in brain science does not necessarily move health policy in new directions; instead 'novel articulations of the imagined biological potentially reify policy paths already mapped or trod' (this volume). Other examples include the ways that the growing field of genetic counseling (Löwy, this volume) and the increase in personal genetic testing (Kelly, Harris and Wyatt, this volume) raise complicated issues about how genetic instability is potentially translated into what is often interpreted as stable information when shared with patients and consumers. Genetic instability, and the increased orientation of the social and natural sciences toward complexity, interaction and temporal change, leaves those striving to improve health outcomes with a multiplicity of paths forward.

Today instability comes not only in the form of new genomic theories, where genes exist and express only in relation to a variety of environments across time and space. Instability also comes in the ways that disciplinary boundaries, and lines between the social and biological sciences, are being transgressed. While such genomic and disciplinary instabilities do not replace stable genetic theories or institutional mechanisms, the flux characterising the contemporary moment does create a propensity for many social scientists to experiment with what have been called 'biosocial', 'biocultural' or 'bioethnographic' approaches (Ingold and Palsson, 2013; Roberts, 2015; Callard and Fitzgerald, 2015). Such interdisciplinary research often attempts to bring experts from the social and natural sciences together, resulting in collaborations with mixed degrees of success (see Scott, Berry, and Calvert, this volume). Bringing genomic and social scientists together potentially creates concomitant research (see Benezra, this volume) or para-ethnography (Nading, 2016) that hopes to improve health outcomes across geographic and species boundaries – a multispecies co-flourishing (Haraway, 2008). But some fear that the disciplinary instability on which such theories of health rest also risks the loss of distance necessary to enable critical perspectives (Helmreich, 2015). As with the present moment of (post) genomic instability, disciplinary flux inspires both optimistic and pessimistic forecasting of futures.

Big data as 'biofuel'

An important new actor that has entered the stage since the first edition of the *Handbook* is 'big data'. There are various definitions of the concept, ranging from references to work with very large datasets, to new epistemologies that focus hypothesis-free data mining, to the view that only comprehensive datasets (N=all) deserve this name. But like DNA, the concept of big data has also developed a mystique, becoming a new cultural icon (Nelkin and Lindee, 1995). One might say that big data are the genes of the twenty-first century: they fuel imaginations about what genomics can and should do, they determine the status and power of research groups, shaping access to funding and other resources. *The Economist* called data the new fuel of our economies (N.A., 2017). Data have certainly already become the fuel of bioeconomies – the system of processing, selling and consuming biological resources, as discussed below.

Social scientists have criticised aspects of these trends. Media studies scholar Gina Neff famously warned of the temptation to invest more money merely in collecting more data; to create benefits for patients we need to do more than collect and integrate data. We need to understand what they mean. We need to enhance not only the technical but also the social interoperability of our systems. In Neff's words, 'big data won't cure us' (Neff, 2013). We also need to avoid the danger of treating unstructured information such as patients' stories, values, and practices – information that is meaning-full – as worth-less in this context. If the focus on information that is easily quantifiable and/or readily available in digital formats comes at the cost of what is meaningful for patients, then this would also have a negative effect on current efforts to personalise medicine. What kind of person would this type of medicine serve? (Reardon, 2011; Prainsack, 2017)

The trend toward personalised medicine also highlights another concern, namely control over data use. The data used to tailor prevention, diagnosis and treatment more closely to the individual characteristics of patients have 'escaped the clinic' (Leonelli, this volume; Nettleton 2004); they are coming from a range of contexts including people's digital devices, their homes, and public archives (Weber, 2014). This means that what is at stake here is no longer merely individual privacy as we know it; it is the ability of people – both personally and as a collective of citizens – to have a say over what information about them is collected, used, shared, and for what purpose.

Innovation and value

Another cross-cutting theme in the collection is the increasing prominence of imaginaries of genomics and post-genomics as a source of innovation and value. Since the Human Genome Project (HGP), visions of high-throughput analysis of biological data leading to dramatic improvements in health and significant economic opportunities have inspired many actors, stimulating investments of money, talent, and hope in the possibilities of the 'bioeconomy'. The emergence of this term as prominent policy buzzword in the latter half of the 2000s was not merely a matter of attaching a new label to the biotechnology industry. It also marked the growing salience of new visions of the *value* of genome-based knowledge and technology, expressed by authorities in government, finance, and biomedicine, as well as by patient advocacy groups and many others. The bioeconomic futures that they promoted were not always neatly aligned, but contradictions and slippages did not prevent the formation of a 'discourse coalition' (Hajer, 2009) that made the bioeconomy into an object for policymakers, investors, and others to care about and seek to nurture.

A macro-level conception of 'the' bioeconomy as a policy object cannot be taken at face value, however. For one thing, as genomic technologies have grown increasingly entrenched in biomedical research and practices, the transformation of medicine and health care has proceeded more slowly than anticipated, and investment has been sustained more by the perpetual renewal of its future promise than by actual delivery of new cures or spectacular profits (Martin, this volume). For another, specific bioeconomies – implicating particular materials, forms of knowledge, actors, and social relations – are taking shape, as new patterns of circulation (often global ones) are transforming political economies of health (Sunder Rajan, 2017). Contributors to this volume raise pressing questions about emerging bioeconomies and the political economies (Birch, this volume) and gender politics (Lamoreaux, this volume) that they instantiate and reflect.

The challenges of governing innovation in genomics has been a salient concern of science policymakers and social analysts alike. Since its earliest days, genomics has been a site where new forms of knowledge and new regimes of control have been co-produced (Hilgartner, 2017). The contributors to this volume explore how governance of medical genomics has become a complex, distributed process involving a variety of regulatory mechanisms, both formal and informal, that are often contested (Cambrosio et al. this volume; Hogarth, this volume). Extant formal regulatory frameworks and concepts are often challenged by developments, and jurisdictional boundaries, both formal and informal, are being reconfigured (Dove, this volume). At the same time, at a deep, quasi-constitutional level, sociotechnical imaginaries (Jasanoff and Kim, 2015) help to stabilise allocations of authority grounded in the (intellectually indefensible) ontology of the linear model of innovation (Hurlbut, this volume).

With these overarching themes in mind we have organised the *Handbook* in relation to five key areas of cross disciplinary social science inquiry and investigation. The first section, edited by Sabina Leonelli, entitled 'Genomic Based DNA Technologies in the Clinic and Beyond', presents a series of chapters which explore how (both historically and in the contemporary moment) genomic knowledge, technologies, and medicine have become central to certain fields of clinical practice, whilst also exceeding and extending beyond the confines of the clinic. Chapters in this section reflect on the ongoing scope and limits of biomedicalisation in the context of a shifting terrain of genomic medicine (Bliss); the consequences of 'scaling' up technologies and incorporating 'big data' in the clinical arena (Leonelli and Tempini); and the way personal genetic testing is being uneasily and unevenly 'mainstreamed' into health care practices (Kelly, Wyatt and Harris). The final two chapters, each using a somewhat different lens, focus on one area of medical intervention, reproductive genetics, which has been profoundly changed by developments in genomics. These two chapters provide a complementary perspective focusing on the historical evolution of genetic testing and counseling (Löwy) and an in-depth examination of prenatal genetic testing in contemporary urban China (Zhu and Dong).

Our second section, edited by Claire Marris, addresses 'Genomic Technologies in the Bioeconomy'. As Marris points out in her introduction to the section, what a social science perspective brings to the table here starts with a different take on the notion of bioeconomy. While in the life science community, economics, and policy, the bioeconomy is typically seen as the system of production, exchange and consumption of renewable biomaterials such as fish, wood, or human materials, critical social science work has challenged some of the assumptions and expectations underpinning dominant discourses on the bioeconomy. It has in fact scrutinised the productive (in the literal sense of the word) role of expectations in the bioeconomy as such. Contributions to this section show – from different perspectives – how the bioeconomy is not just about getting value from biological things, but is also about transforming the organisation and conduct of science and innovation (Martin; Chiapetta and Birch). An example of this is the rise of public-private partnerships as a template for innovation due to the alleged 'productivity crisis' in drug development (Nik-Khah), which is accompanied by a de-politicisation of science and research governance (Felt). Because the biological is seen as mutable, interventions into the biological seem a particularly promising form of investment (Pickersgill); value cannot only be extracted from the processing, selling and other uses of bio-materials but also from intervening into the renewal of these materials. Expectations from, and interventions into, the 'renewal' of biomaterials in turn has tangible social dimensions. Not only because these expectations and interventions are shaped by ideas about what purpose materials should serve, but also because in the human domain, biological materials are always also racialised and gendered (Lamoreaux).

Our third section, edited by Stephen Hilgartner, addresses the 'Governance of Medical Genomics'. As the chapters in this section show, governance of genomics is a contested domain in a variety of sites. The first chapter reviews the history and legacy of the Human Genome Project's ethics programmes (ELSI and ELSA), arguing that the institutionalisation of such programmes – a novel mode of governing emerging technologies – is a significant development in contemporary politics (Hilgartner). The next chapter (Parthasarathy) uses comparative analysis to reveal and explain differences in how patent offices in the United States and the European Union managed the politics of biotechnology patents on human genes and life forms. Turning to the governance of clinical research, Cambrosio et al. argue that genomic platforms are sites where new technologies, new conceptualisations of the goals of clinical trials, new organisational routines, and new regulatory landscapes are taking shape and being brought into alignment. The subsequent chapter focuses on how regulatory agencies have addressed the challenges of evaluating genomic diagnostics in Europe and the United States (Hogarth), illustrating the complexity of this domain and outlining areas for future research. Control over personal genomic information in European Union law is the subject of the chapter by Edward Dove, who argues that traditional regulatory approaches, such as 'consent or anonymise', have serious limitations in the new environment in which genomic and other medical information circulates. The final chapter (Hurlbut) addresses the politics of governing technologies for editing the human germ line, such as CRISPR. Taken together, these chapters offer a picture of the dynamic changes in modes of governance now underway, while at the same time pointing out some continuities (e.g., in allocations of epistemic authority) that have contributed a degree of stability to this area of transformational change.

The fourth section edited by Sahra Gibbon and Barbara Prainsack revisits the theme of 'Diversity and Justice', which has both endured and diversified in the time between the first Handbook and the current edition. Chapters in this section examine how questions of equity, ethics and rights interweave and are folded into a range of genomic developments. Two chapters show how the complex inter-relationships between disability, eugenics and enhancement have been and should continue to be of central and ongoing relevance and concern for social scientists examining genomic technologies and medical interventions (Scully; Cavaliere and Camporesi). Other chapters provide perspectives on the framing of genetic discrimination by the insurance industry (van Hoyweghen) and the new form of participation in genomic research created by state and corporate interests (Prainsack). Both these chapters illuminate the inherent politics and power asymmetries that are entailed in foregrounding issues of 'solidarity' and a wider collective inclusion across an evolving terrain of genomics. Unpicking the complex issues and challenges that endure at the interface between race, genomics and health disparities is also the central concern in one of the chapters in this section (Lee). Our final chapter in this section comparatively examines how different histories of medical and population genetic research in Brazil and across the African region shape contemporary engagement with genomic research. This not only informs questions of 'diversity' but how wider concerns with ethics, social justice and inclusion are central to engagement with genomics in emerging and developing economies (Fullwiley and Gibbon).

Our final section, edited by Janelle Lamoreaux, brings together a collection of chapters through the new theme of 'Crossing Boundaries'. Each chapter in this section investigates emergent areas of genomic research – from epigenetics to microbiomes to synthetic biology – and discusses the ways in which such areas (claim to) cross a variety of boundaries – from disciplinary limits to species borders. The first two chapters discuss epigenetic research. The first provides a history of epigenetics and thinks through how Waddington's epigenetic landscape remains relevant in contemporary behavioural genetic research (Lock). The second introduces a critical perspective on epigenetics through the idea of scale, concentrating specifically on studies of suicide risk (Lloyd and Raikhel). The next chapter discusses the global and local characteristics of stem cells, emphasising ethnographic findings from research in Taiwan (Liu). We then turn to a conversation on the increasingly complex considerations of environments in genomic research utilising animal models (Friese), which is followed by a chapter on the natural and social sciences of the microbiome (Benezra). The next chapter provides a discussion of the historical importance of studying how epistemic cultures are both protected and crossed through the example of behaviour genetics (Nelson and Panofsky). The section ends with a critical analysis of a field that once epitomised boundary-crossing, but that authors provocatively now suggest might be best left behind: synthetic biology (Scott, Berry and Calvert). Collectively these chapters suggest that while emergent genomic research often crosses epistemological and ontological boundaries in unexpected ways, some categorical and empirical limits are more obdurate.

Notes

1 This heading is taken from the title for an article in *Nature* by Bustamante et al. (2011).
2 See work being undertaken by Gaudilliere, J.P and Beaudevin, C. et al. as part of the ERC Advanced Grant 'GLOBHEALTH; From international public health to global health', ERC advanced grant 2014–2019, http://globalhealth.vjf.cnrs.fr.

References

Bharadwaj, A. (2013) Subaltern Biology? Local Biologies, Indian Odysseys and the Pursuit of Human Embryonic Stem Cell Therapies, *Medical Anthropology*, Vol. 32, No. 4, pp. 359–373.

Callard, F. and Fitzgerald, D. (2015) *Rethinking Interdisciplinarity Across the Social Science and Neurosciences.* New York: Palgrave Pivot.

Franklin, S. (2005) 'Stem Cells R Us: Emergent Life Forms and the Global Biological', in A. Ong and S. J. Collier, eds., *Global Assemblages: Technology, Politics and Ethics as Anthropological Problems.* New York and London: Blackwell, pp. 59–78.

Fullwiley, D. (2011) *The Enculturated Gene: Sickle Cell Health Politics and Biological Difference in West Africa.* Princeton: Princeton University Press.

Haraway, D. J., (2008). *When Species Meet.* Minneapolis, MN: University of Minnesota Press.

Hajer, M. A. (2009) *Authoritative Governance. Policy Making in the Age of Mediatization.* Oxford: Oxford University Press.

Helmreich, S. (2015) *Sounding the Limits of Life: Essays in the Anthropology of Biology and Beyond.* Princeton: Princeton University Press.

Hilgartner, S. (2017) *Reordering Life: Knowledge and Control in the Genomics Revolution.* Cambridge, MA: MIT Press.

Ingold, T. and Palsson, G. (2013) *Biosocial Becomings: Integrating Social and Biological Anthropology.* Cambridge: Cambridge University Press.

Jasanoff, S. and S-Y. Kim. (2015) *Dreamscapes of Modernity: Sociotechnical Imaginaries and the Fabrications of Power.* Chicago: University of Chicago Press.

N. A. (2017). The World's Most Valuable Resource Is No Longer Oil, But Data. *The Economist* (6 May 2017). Available at: www.economist.com/news/leaders/21721656-data-economy-demands-new-app roach-antitrust-rules-worlds-most-valuable-resource (accessed 29 June 2017).

Nading, A. (2016) Evidentiary Symbiosis: On Paraethnography in Human–Microbe Relations. *Science As Culture* 25(4): 560–581.

Neff, G. (2013). Why Big Data Won't Cure Us. *Big Data* 1(3): 117–123.

Nelkin, D., and Lindee, S. M. (1995) *The DNA Mystique: The Gene as Cultural Icon.* Ann Arbor: University of Michigan Press.

Nettleton, S. (2004) The Emergence of E-scaped Medicine? *Sociology* 38(4): 661–679.

Prainsack, B. (2017). *Personalized Medicine: Empowered Patients in the 21st Century?* New York City: New York University Press.

Reardon, J. (2011). The 'Persons' and 'Genomics' of Personal Genomics. *Personalized Medicine* 8(1): 95–107.

Roberts, E. F. S. (2015). Bio-Ethnography: A Collaborative, Methodological Experiment in Mexico City. *Somatosphere.* Available at: http://somatosphere.net/2015/02/bio-ethnography.html (Accessed 23 June 2017).

Sleeboom-Faulkner, M. (ed.) (2011) *Frameworks of Choice: Predictive and Genetic Testing in Asia.* Amsterdam: Amsterdam University Press.

Sunder Rajan, K. (2017). *Pharmocracy: Value, Politics, and Knowledge in Global Biomedicine.* Durham, NC: Duke University Press.

Thompson, C. (2010) Asian Regeneration? Nationalism and Internationalism in Stem Cell Research in South Korea and Singapore . *In Asian Biotech: Ethics and Communities of Fate*. A. Ong and N. Chen, eds. Pp 95–117. Durham, NC: Duke University Press.

Tsing, A. L. (2005) *Friction: An Ethnography of Global Connection*. Princeton: Princeton University Press.

Wade, P., Beltran, C. L., Restrepo, E. and Ventura Santos, R. (eds) (2014) *Meztiso Genomics: Race Mixture, Nation and Science in Latin America*. Durham, NC: Duke University Press.

Weber, G. M., Mandl, K. D., and Kohane, I. S. (2014). Finding the Missing Link for Big Biomedical Data. *Journal of the American Medical Association*, 331(24): 2479–2480.

Part 1

Genomics and DNA-based technologies in the clinic and beyond

2

Introduction

Sabina Leonelli

The role of genomics in society has become ever more entrenched within the last decade. This is partly due to advances in technologies and particularly sequencing tools, which have transformed the act of obtaining an individual's genetic pedigree from an esoteric, labour-intensive, costly exercise to a largely automated, relatively affordable and mundane practice. It is also a result of the increasing globalisation of DNA-based technologies, which have been picked up by health systems, governments, insurance companies and data analysts all over the world, thus becoming more and more of a platform for international dialogue around understandings of health and disease. Given these developments, it is tempting to think of this historical moment as a "postgenomic" one in the sense that DNA-based tools – and related ideas about human nature and wellbeing – have irrevocably and fully established themselves across cultures and social customs, thus creating a uniform reference point for biomedical practice. And yet, talk of postgenomics can be interpreted very differently. It can indicate the increasing awareness by the public at large (including biomedical researchers and physicians) of how difficult and complex it is to interpret genetic results, and thus to create any generalised understanding of biological process and therapeutic intervention (Richardson and Stevens 2015) or ways to exploit advances in genomic understanding to effectively target the unique characteristics of individuals and groups (Green and Voigt 2016). Postgenomics can also be taken to mark the increasingly hegemonic institutional, corporate and regulatory landscapes in which DNA-based technologies are taking root (Peterson 2014, Sunder Rajan 2017, Murphy 2017), and the myriad questions surrounding their prominent status within and beyond biomedical and clinical environments around the globe (Bliss 2017, Reardon 2017). In short, postgenomics can and arguably should be viewed as marking the tension between the growing entrenchment of genomic practices within medical practices and markets, and the enormous logistical, scientific and moral challenges posed by enacting genomic knowledge and tools across very different medical regimes, skillsets and patients.

It is becoming ever more apparent that while technology may help enormously in making genomic sequencing and related tests increasingly affordable and available, the challenges involved in managing and interpreting the resulting data have in no way lessened. If anything, the hard problems confronted when attempting to decide who should have access to genetic data, and to do what, have acquired a new degree of severity due to the overarching shift in discourse, practices and commercial interests around big and open data (Ebeling 2016, Leonelli 2016). Calls to view, handle and value data – and particularly personal health-related data – as sources of economic and political power are proving increasingly convincing, in the face of systems of data gathering and dissemination that exclude large parts of the population, are subject to very few restrictions concerning potential misuse and misinterpretation of the data at hand, and function as systems of citizen surveillance, protection and service provision all at the same time. Furthermore, some types of data are systematically favoured over others for technical and commercial reasons, again calling for an assessment of what such selective gaze implies for different social groups and activities. Genomic data in particular are often prioritised, thanks to their digital, portable format, their embedding in well-entrenched and highly marketable medical technologies and ways of knowing, and their continuing privileged status as quantitative documents of people's biological inheritance.

What at the turn of the century was mostly a worry for technical experts is now becoming acknowledged in public discourse: creating data may be relatively easy, but preserving them, protecting them and analysing them is difficult and expensive. Even more difficult is deciding who has the expertise and power to take care of such data and use them to extract biomedical insights, ground political and economic decisions, and shape perspectives on the future – and what responsibilities and accountabilities are involved in this process. This is particularly true in a world where internet communication and related technologies are ubiquitous, and yet the capacities to exploit those technologies are very unevenly distributed, and the ways in which they manifest themselves in specific situations, across geographical locations and cultural norms, continues to vary enormously. Contemporary information and computing technologies may make it easier for people to communicate, but who is communicating what to whom around the use of genetics in society, and with which results, interpretations and purposes? And how does such communication unfold in the contemporary political context, where nationalism and populism are on the rise in several high-income countries, and attitudes to national borders, cultural diversity, the threat posed by movements of people and related biological materials, and the legitimacy of scientific expertise are shifting?

These are the questions that motivate the first section of this new handbook, where authors examine the most recent developments in genomic and DNA-based health technologies, and bring fresh perspective on their social and scientific role that takes account of the evolving political landscape. Catherine Bliss starts off the section with a comprehensive review of contemporary instantiations of *biomedicalisation*, which she proposes as the lens that 'helps us ascertain the major shifts in today's social order around the expansion of biomedicine'. Her discussion includes the role and goals of medical care and the pervasiveness of molecular conceptions of biomedicine in everyday life, with significant implications for conceptualisations of the body, personal identity and the pathological. Her chapter situates the development of DNA-based technologies within a rich social, cultural, economic and political context, thus providing a textured landscape for many of the themes that other *Handbook* contributors will address in some detail in what follows.

The chapter by Niccolo Tempini and Sabina Leonelli then zooms into the ways in which the emergence of big data discourse, infrastructures and practices has affected – and, arguably, boosted – the role of genomics in biomedical research and care. They emphasise the ways in which attention to data practices shifts the focus of STS researchers interested in biomedicine beyond the clinic, to embrace the vast variety and multiplicity of social environments (digital or physical) in which the production, dissemination and interpretation of data of medical relevance is happening, and in which data practices are valued in a variety of different ways by different actors. They conclude that whether and how the transformative promise of big data can be delivered for biomedicine and health care depends on the tools and assumptions used when assembling, interlinking and integrating genomic data with other types of biomedical data – a task fraught with technical, ethical and social challenges.

Susan Kelly, Anna Harris and Sally Wyatt also place emphasis on the ways in which genomics is escaping the biomedical and clinical context, by examining in detail the rise and commercialisation of personal genetic testing. Drawing from studies of the dynamics and usage of the internet, as well as research on the enactment of direct-to-consumer testing by online providers such as *23andMe,* this chapter stresses the extent to which bringing genetics out of the clinic is impacting the identity of patients (who are at once consumers and contributors of a service with multiple goals and accountabilities), their relationship with health care professionals and their understanding of the value of data produced through medical interactions. In particular, Kelly, Harris and Wyatt emphasise the disruptive effect of this evolving landscape on the various relations of trust that underpin and facilitate biomedical knowledge and care.

What do these developments mean for clinical work? This question is tackled by Ilana Löwy's chapter, which builds on the history of genetics and its clinical applications to portray how genetic testing and counselling practices have evolved over the last fifty years, the relation between such developments and broader societal trends, and the effects of these shifts on contemporary clinical practice. Löwy focuses on the increasing diversity of tests and targets developed for clinical use, the strong link between test availability and patients' uptake, and the implications for understandings of risk, parenthood and parental responsibility. She also points to the challenges raised by unanticipated secondary findings, as for instance in the case of hereditary conditions that affect whole families, and the resulting climate of "managed fear" in which human reproduction is now planned and enacted.

The last chapter in the section moves away from the Global North and Anglo-American trends, and takes a close look at the historical development and contemporary social implications of the implementation of genomic technologies in China. Jianfeng Zhu, Shiyi Xiong and Dong Dong focus specifically on prenatal genetic testing, a particularly sensitive issue in China given its birth control policies as well as the Confucian approach to responsibility and bonds within the family. The chapter considers recent changes in governmental discourse around the regulation of maternal and infant care, and particularly the role played by genetic knowledge in the current shift from a policy centred on population 'quality control' to an opening towards personal choice around 'reproductive insurance'. As the authors point out, this intersects in complex ways with the expectations and preferences of the families affected by state policies, as well as the training and everyday practices of health care professionals tasked with delivering treatment and assistance. It also exemplifies the surprising speed with which a country can reposition its role as participant and contributor to global health discourse and practices, with China now rapidly moving to establish genomic collection facilities to capture data and samples from its population and make them visible and potentially accessible internationally.

Genomic practices continue to be caught in a web of technological acceleration, societal changes and logistical chaos, with financial resources and market forces driving both the direction and the location of innovation in medical care. As pointed out in all the chapters within this section, this has substantial and uneven effects on popular attitudes and discourse on reproductive technologies and genetic testing. In particular, the increasing alignment of service provision, commercialisation and medical care derives in confusion around who is responsible and trustworthy, for what and in which way – a confusion that affects not only prospective patients, but also health care professionals and regulators. In this moment of transition and change, it is critical for scholarship in the history, philosophy and social studies of genomics and biomedicine to support the development of civil epistemologies around biomedicalisation, and inform emerging regulatory efforts, legal frameworks and commercial strategies.

Acknowledgements

Funding from the European Research Council (grant award 335925) and the UK Economic and Social Research Council (award ES/P011489/1) has supported the research on which this introduction is grounded. I am grateful to the editors, especially Sahra Gibbon, Janelle Lamoreaux and Barbara Prainsack, for their excellent support, prompt assistance and helpful suggestions throughout the editorial process for this section of the *Handbook*; the authors of the chapters in the section for their willingness to work to a tight schedule; and to Aaron Parkhurst for precious editorial assistance.

References

Bliss, C. 2017. *Social by Nature: How Sociogenomics is Redefining What it Means to be Human*. Stanford, CA: Stanford University Press.

Ebeling, M. 2016. *Healthcare and Big Data: Digital Spectres and Phanthom Objects*. London, UK: Palgrave McMillan.

Green, S. and Voigt, H. 2016. Personalising Medicine: Disease Prevention in Silico and in Socio. *Humana. mente Journal of Philosophical Studies* 30: 105–145.

Leonelli, S. 2016. *Data-Centric Biology: A Philosophical Study*. Chicago, IL: Chicago University Press.

Murphy, M. 2017. *The Economization of Life*. Durham and London: Duke University Press.

Peterson, K. 2014. *Speculative Markets*. Durham and London: Duke University Press.

Reardon, J. 2017. *The Postgenomic Condition: Ethics, Justice, Knowledge After the Genome*. Chicago, IL: Chicago University Press.

Richardson, S. and Stevens, H. 2015. *Postgenomics*. Durham and London: Duke University Press.

Sunder Rajan, K. 2017. *Pharmocracy*. Durham and London: Duke University Press.

3

Biomedicalization in the Postgenomic Age

Catherine Bliss

Introduction

In the last quarter century, medicine has acquired a technological and scientific makeover just as it has encroached into ever more aspects of human life. Today, it is often referred to as "biomedicine," connoting the turn toward the burgeoning fields of genomics, bioengineering, biotechnology, and biostatistics. These transformations have generated new ramifications of its growing influence, what researchers now refer to as "biomedicalization."

This chapter introduces the concept of biomedicalization, presenting some major developments in social theory and research. It opens with a look at what biomedicalization is and how it relates to other social processes, such as geneticization and pharmaceuticalization. It then discusses emerging developments in molecular science and the characterization of risk, emerging avenues in social movements and identity politics, and the rise of consumer genomics. The chapter concludes with a look at how biomedicalization entwines with broader forms of capitalism and sociality in the New Millennium.

What is biomedicalization?

"Biomedicalization" has been used to explain a wide gamut of social phenomena, from specific diseases such as HIV and cardiovascular disease to overarching conditions such as mental and sexual health, as well as to life processes such as aging and dying and social characteristics such as race, gender, and sexuality. Researchers examining these phenomena have noticed the ways in which biomedicine has come to define them, casting such phenomena with a bioscientific imprimatur.

The term "medicalization" was first introduced in the early 1970s to characterize the social process by which medicine increases its jurisdiction, entering into formerly nonmedical domains of life (Conrad 1975; Zola 1972). Theorists at the time trained their attention on the dominance of the medical profession, encouragement by social movements and patient advocacy groups, and changing institutional or organizational structures that supported the increasing reach of medicine. Yet by the turn of the Millennium, scholars were finding that medical authority wasn't what it used to be, medicine itself was again restructuring in critical ways, and the forms that health organizing was taking were also changing drastically (Conrad 1992, Starr 2008). Due to government cost control measures and the success of social movements to create patient-based and holistic healthcare, the professional dominance of doctors was rapidly eroding. A corporate brand of managed care was taking the place of self-administered physicians networks. The pharmaceutical industry rushed in to sell drugs directly to managed care organizations and to patients themselves. Third-party payers also grew in influence, assuming a gatekeeping role in healthcare (Prainsack, this volume).

The late 1990s was a critical time for medicine, as the world braced itself for the first mapping of the human genome. The Human Genome Project published a draft map in 2000, and medicine underwent a paradigm shift, becoming rationalized by bioscientific interests and aims (Clarke et al. 2010). A substantial biotechnology industry arose, replete with new, closer academic–industry relations and a growing market of direct-to-consumer goods.

One major aspect of the new bioscientific medicalization, what Clarke et al. (2003: 162) summarize as "the increasingly complex, multisited, multidirectional processes of medicalisation that today are being both extended and reconstituted through the emergent social forms and practices of a highly and increasingly technoscientific biomedicine," is economic restructuring toward the consolidation of a "biomedical technoservice complex." This new system is characterized by multinational corporations and privatized rather than state-funded research and healthcare. Again, managed care systems have replaced physician-dominated systems. Meanwhile, increases in fee-for-service options and the devolution of healthcare management to individual patients, as well as new population-based practices that rely on new concepts of identity, have created a uniquely stratified form of medicalization.

A second aspect of biomedicalization is a concerted focus on health, risk, and surveillance. Health has become a moral imperative, something individuals must work toward. Governments, medical organizations, and individuals themselves constantly monitor risks, and they do so in terms of genetic diagnoses and molecular categorizations. As Clarke et al. (2003: 172) note, "it is impossible not to be 'at risk.'"

Third, biomedicalization is characterized by a "technoscientization" of biomedicine, that is, a rationalization by technology and science at the same time. Processes and systems are computerized and standardized by new technologies and scientific classification systems. Healthcare is driven by evidence-based biomedicine, which is itself characterized by statistical reasoning and molecular science. A host of "Omic" sciences prevail, encouraging biological engineering "from the inside out" (Clarke et al. 2003: 176). Healthcare is delivered by way of electronic record systems, digitized biotechnologies, and bioengineered applications.

Fourth, biomedicalization works by way of new transformations of information and the production and distribution of knowledges. Biomedicine dominates the media, where it dispels alternative notions of health and wellbeing and alternative systems of knowledge. Responsibility for one's health is cast as an individual problem that is to be addressed with consumer applications. A range of cottage industries has cropped up offering consumers DIY (Do-It-Yourself) goods and services, while health gurus and high-profile medical experts have encouraged patients to go online, self-diagnose, and purchase products. At the same time, pharmaceutical commercials and news of clinical trials trump alternative and complementary medicine, now joining with healthcare professionals and patient movements to popularize genetic tests and cures (Löwy, this volume).

Finally, biomedicalization entails a transformation of bodies and identities. Biomedicalization is no longer about controlling a pathological entity, but rather hinges on customizing bodies to be the best they can be. There are two fundamental rhetorics, that of choice and lifestyle optimization. Individuals are pressed to opt for healthy behaviors and habits in their everyday lives, and to use biotechnologies to constantly monitor their improvement from moment to moment, in real time. Norms have multiplied such that there is no longer a singular definition of what it is to be "normal." A plethora of identities have emerged, forged through the interaction with biotechnology and new medical classifications. Increasingly, identity is cast in genetic terms, and it is done so amid a global network that has important international ramifications (see Jingfeng in this volume).

Medicalization theorists have debated whether these transformations warrant a shift in terminology (Conrad 2005; Rose 2007). Still, all agree that where medicalization occurs, it is unique in how it links up with processes like geneticization, molecularization, and pharmaceuticalization, the processes by which reductionist genetic explanations are used to describe differences between individual and group traits and behaviors (Lippman 1991), molecular models advance in science and society (Chadevarian and Kamminga 1998), and the way that conditions are transformed into opportunities for pharmacological interventions (Williams et al. 2011). Biomedicalization increasingly involves the redefinition of characteristics and conditions as inherently found in one's DNA code and thus knowable and treatable by biomedicine, propels molecular models forward in the various fields and subfields of biomedicine as well as in healthcare and society, and bolsters pharmacological hegemony in the basic structures of bioscience and health delivery systems, often leading to an individualization of intervention and a depoliticization of care (Bell and Figert 2015).

From genetics to genomics: new avenues in risk

Genetics has been a leading science for over a century, emerging soon after Darwin published *The Origin of Species* in 1859 and gaining ascendancy in the twentieth century. In the late 1970s, geneticists discovered ways to splice and recombine genomes, and the new science of genomics was born. Since then, scientists have delved into the human genome looking for genes responsible for common chronic illnesses and traits that are relevant to everyone. Still, today genomics has just barely begun to enter the clinic, and has done so more in terms of conversations about potential susceptibilities than actual drugs and diagnostics.

Risk has become a central motif in light of these changes (Tulloch and Lupton 2003). Though genetics also dealt in susceptibilities in its testing for carrier status of single-gene diseases, genomics' focus on multifactorial matters has created a form of biomedicine in which every marker provides some degree of risk status (Rose 2009). Now it is the responsibility of each and every individual to learn about their susceptibilities, to manage them, and optimize their health and wellness with this information.

Some analysts have hailed the ushering in of a Risk Society, a system in which all of our social institutions enforce surveillance and self-surveillance (Adam et al. 2000). However, many note that it is the individual and the individual body that is the main conduit for risk management in the genome age (Lemke 2015). Neoliberal ideologies of individuality and self-government combine with a depoliticized form of public health, devolving responsibility from the state to local experts and lay citizens (Tulloch and Lupton 2003; Prainsack, this volume). The state is still interested in maximizing human potential for state interests, but this is envisioned in terms of a multiplicity of norms by which each and every individual must judge her own capacity and determine the best way to optimize it (Clarke et al. 2010). Individuals are trained to read their own bodies in light of ever refined genomic population affiliations (Rabinow and Rose 2006). Again, the focus is not on clinical branding or cordoning off diseased populations from the healthy as it has been in the past, but rather self-analysis toward self-imposed regiments based on evidence-based biomedical data (Niewöhner 2013). Managing risk requires a "new prudentialism" involving analysis of the complete array of lifestyle choices available to an individual (Nadesan 2010). It requires both buying and buying into a certain way of life that can promote optimal health and wellbeing.

Time is compressed in that markers of the past, the innate traits passed down via one's ancestry become a matter for present-day manipulation in the interests of creating a healthful future (Tutton 2014). Living to one's fullest potential requires gathering all the probabilistic data out there and using it to manage, modify, and mutate, not so much as to stabilize in a new form but rather to adapt the past toward a future good (Rose 2009).

Yet, these new avenues in risk create social inequality. Biopower, power's "capillary form of existence, the point where power reaches into the very grain of individuals, touches their bodies and inserts itself into their actions and attitudes, their discourses, learning processes and everyday lives" (Foucault and Gordon 1980: 39), has become all encompassing. The regulatory dimension of biopower, the biopolitics of the population, looms larger than ever in the form of global genomic sequencing projects, including state-run projects that attempt to collect the DNA of all citizens in a given body politic, and genomic public health initiatives designed to replace social epidemiological ones (Rapp 2013; Shostak 2013). Furthermore, governments use biomedicine to engage in "dividing practices" in which certain populations are classified as different and in need of specific forms of treatment, and biomedicine relies on state taxonomies to conduct research and administer and deliver healthcare (Braun et al. 2007). In liberal western democracies, race is one dominant mode of classification. Engagement and care is also gendered and distributed unevenly according to sexual orientation and socioeconomic status (Happe 2013).

Biopower's disciplinary dimension, the anatomo-politics of the body, is equally imposing as internal control over bodily functions throughout daily life becomes paramount to "living the good life" and "being a good citizen." Eugenics reappears in today's biomedicalized form of citizenship, or "biological citizenship" (Heinemann and Lemke 2015; Petryna 2013). The sharp increase in pharmaceutical consumption, body sculpting, and plastic surgeries, means that social norms are completely internalized and branded into the body (Dumit 2012; Menon 2017). As Mamo and Fosket (2009: 927) argue, "bodies are not born; bodies are made." People view themselves as projects to be eternally worked at. In line with the immanent nature of risk management, bodies are seen as always already incomplete. The focus of self-making is less on disease status than on health and wellness in the moment, thus biomedical discipline reaches into every aspect of life leaving nothing to spare.

Biomedical influence and corollary ills such as genetic determinism, molecular supremacy, and pharmaceutical dominance may not be entirely complete (Franklin and Roberts 2006; Lock 2008). What scholars are calling "postgenomic science," fields like epigenetics (the study of noncoding DNA sequences and their regulation of genes) and gene-environment interactions science, including social genomics (the study of DNA's role in social behavior and outcomes), conceive of genes as being situated in and responsive to their environments (Meloni 2015). They promote "a break from the gene-centrism and genetic reductionism of the genomic age…an emphasis on complexity, indeterminacy, and gene-environment interactions" (Richardson and Stevens 2015). Moreover, studies of conditions undergoing geneticization have found that many conditions do not end up defined in molecular terms or treated by pharmacological means (Gibbon and Novas 2007; Shostak, Conrad, Horwitz 2008). Studies of genetic testing also have shown that people often question the predictivity of tests, and rely on other sources of knowledge to determine paths forward (Hedgecoe 2009). As Clarke et al. (2010: 28) argue, biomedicalization "is punctuated by contradictions and complications of power, knowledge and social action."

But while the new focus on genes in context is altering biomedicine's course, and creating new meaning around genomic causality, "next generation" sequencing projects, such as whole genome sequencing projects and multinational gene-environment sequencing projects, continue to be racialized (M'Charek, Schramm, and Skinner 2014; Shim et al. 2015). The pharmaceutical industry has delivered race-based medicine to populations around the world while continuing its practice of stratifying by gender (Annandale and Hammarström 2015; Inda 2016; Kahn 2013; Pollock 2012). Biotechnology is increasingly focused on behavioral traits in ways that encourage eugenics in all areas of life (Bliss 2017; Reardon 2011).

A recent surge in gene-editing technology has only compounded these trends, making the advent of "designer babies" a fast-approaching reality. In 2015, researchers from around the world descended upon the U.S. National Academy of Science to decide whether to prohibit gene editing from being applied to the human germline. While the scientists present agreed to put a temporary moratorium on germline editing, only four government science bodies were represented, and no official policies were made. Currently, gene editing is unfolding unregulated in the global market, where startups dedicated to applying the technology to everything under the sun abound, thus it remains to be seen how conceptions and practices around risk will change in the new postgenomic climate (Ishii 2015).

Identity and health activism

For centuries, medical knowledge has shaped the social categories of difference that make up the building blocks of who we are with public health drives, censuses, population controls, pronatalist policies, and eugenic strategies (Raman and Tutton 2009). But biomedicalization has brought the relationship between identity and biomedicine into a closer relationship than ever before (Clarke et al. 2010). Technoscience is changing identities in fundamental ways, as medicine is sold to the individual based on privately profiled information, and distributed through informatic networks connected to intricate research and health databases (Tepini and Leonelli, this volume). Pharmaceutical and biotechnology markets not only determine what tools are at hand in the crafting of new identities but also influence how patient advocacy organizations and other health groups produce a politics of health (Biehl 2013; Nguyen 2010).

As identity is reconstituted through the dominant goods and services available, race, gender, sexuality, and class is also biomedicalized (Duster 2007; Schramm et al. 2012). Markets are not only stratified in terms of participation and labor, but goods and services are also targeted at specific niches of the population thereby reifying social differences along identity lines (Epstein 2008). There is an intersectional dimension to this. People of varying class backgrounds are differentially gendered and raced, and vice versa, in the process of buying and selling reproductive services (Almeling 2011; Benjamin 2013; Waldby and Cooper 2008). "Categorical alignment," or the fusion of ascription and identification processes around hegemonic taxonomies has ensured that the dominant category sets used in biomedicine are realized in lay identities (Epstein 2008).

Bioscientists have also been shown to encourage the use of particular characterizations of human difference, even down to the level of taxonomy and label (Panofsky and Bliss 2017). Due to a specific rollout in American Public Health, and the directorial role of the United States in global genomics, all of the world's global genome projects since the turn of the century have classified according to U.S. federal racial standards (Bliss 2012). The field of genomics has not only adhered to these constructions, but the field's leaders have pushed for minority inclusion in genomic research with a heightened attention to racial identity (Reardon 2009; Montoya 2011; Smart et al. 2008). Bioscientists have gone so far as to market blackness or Latino-ness and particular notions of diaspora through the rubric of DNA with tests and drugs aimed at particular groups (Fullwiley 2008; Lee, this volume; Nelson 2016).

Laypeople have equally fought to construct identity along such lines, and to have their constructions recognized and reflected in biomedicine, by creating and joining illness-based social movements, generating lay knowledge about their own medical conditions, and forging new communities based on illness identities (Brown 2013; Brown et al. 2011). They exchange information about their treatment options, participate in national fundraising events, and mobilize in support of relevant policies and initiatives (Klawiter 2008). Such health activism mobilizes research, and oftentimes researchers (Jasanoff 2011; Panofsky 2010). Political groups similarly mobilize grand-scale community health efforts through the proliferation of alternatives to racist and sexist research (Nelson 2016).

Indeed, a completely biomedicalized form of sociality, a "biosociality," has sprung up in which people are organizing around their DNA code (Rabinow 1992: 244). Genetics-based health social movements have proliferated, demanding further biomedicalization (Clarke et al. 2009). Feedback loops form between health activism, research on human variation, and individuals' own struggles to learn how to manage the array of "probabilities, predictions, and preventative interventions" arising from genomic knowledge (Rose 2009: 161).

Dovetailing with this is a global market of direct-to-consumer genetic tests with which people engage in a range of activities, including family planning, parenting, and political bargaining (Kelly et al., this volume). This market creates an apparatus of commodification and standardization around the categorical building blocks of identity that further essentialize traits and behaviors (Bliss 2017). The idea is for all humans to begin tailoring their lives to their genomes from day one, in a biomedicalized form of eugenics.

Conclusion

In sum, biomedicalization is an analytic that helps us ascertain the major shifts in today's social order around the expansion of biomedicine. Medicine is now molecularized, focused largely on genetic essences, and ubiquitous. It is a part of everyday life.

The key shifts this brings are around commodification and economic restructuring, the move from deviance or pathology to personal susceptibility and risk management or enhancement, technoscientization of all aspects of life, the knowledge economy and means of knowing about oneself, and identity formations. As such, the real power of biomedicalization is its ability to impact what people believe is good, logical, and right. Biomedical knowledge influences cultural and political struggles taking place around the world, and it is the basis with which bodies are perceived, interpreted, and understood. It is the zeitgeist of the New Millennium, a "new imaginary" for the globalizing world (Franklin 2003).

References

Adam, Barbara, Ulrich Beck, and Joost Van Loon. *The Risk Society and Beyond: Critical Issues for Social Theory*. Sage, 2000.

Almeling, Rene. *Sex Cells: The Medical Market for Eggs and Sperm*. University of California Press, 2011.

Annandale, Ellen, and Anne Hammarström. "A New Biopolitics of Gender and Health? 'Gender-Specific Medicine' and Pharmaceuticalization in the Twenty-First Century." In *Reimagining (Bio)Medicalization, Pharmaceuticals and Genetics: Old Critiques and New Engagements*, (Eds.) Susan Bell and Anne Figert, 41–55. Routledge, 2015.

Bell, Susan, and Anne Figert. *Reimagining (Bio)Medicalization, Pharmaceuticals and Genetics: Old Critiques and New Engagements*. Routledge, 2015.

Benjamin, Ruha. *People's Science: Bodies and Rights on the Stem Cell Frontier.* Stanford University Press, 2013.

Biehl, João. *Vita: Life in a Zone of Social Abandonment.* University of California Press, 2013.

Bliss, Catherine. *Race Decoded: The Genomic Fight for Social Justice.* Stanford University Press, 2012.

Bliss, Catherine. *Social by Nature: How Sociogenomics Is Redefining What It Means to Be Human.* Stanford University Press, 2017.

Braun, Lundy, Anne Fausto-Sterling, Duana Fullwiley, Evelynn Hammonds, Alondra Nelson, William Quivers, Susan Reverby, and Alexandra Shields. "Racial Categories in Medical Practice: How Useful Are They?" *PLoS Med* 4(9) 2007: e271.

Brown, Phil. *Toxic Exposures: Contested Illnesses and the Environmental Health Movement.* Columbia University Press, 2013.

Brown, Phil, Rachel Morello-Frosch, and Stephen Zavestoski. *Contested Illnesses: Citizens, Science, and Health Social Movements.* University of California Press, 2011.

Chadevarian, Soraya de, and Harmke Kamminga. *Molecularizing Biology and Medicine: New Practices and Alliances, 1920s to 1970s.* Taylor & Francis, 2003.

Clarke, Adele, Laura Mamo, Jennifer Ruth Fosket, Jennifer Fishman, and Janet Shim. *Biomedicalization: Technoscience, Health, and Illness in the U.S.* Duke University Press, 2010.

Clarke, Adele, Janet Shim, Laura Mamo, Jennifer Ruth Fosket, and Jennifer Fishman. "Biomedicalization: Technoscientific Transformations of Health, Illness, and U.S. Biomedicine." *American Sociological Review* 68(2) 2003: 161–194.

Clarke, Adele, Janet Shim, Sara Shostak, and Alondra Nelson. "Biomedicalising Genetic Health, Diseases and Identities." In *Handbook of Genetics and Society: Mapping the New Genomic Era.* (Eds.) Paul Atkinson, Peter Glasner, and Margaret Lock, 21–40. Routledge, 2009.

Conrad, Peter. "Medicalization and Social Control." *Annual Review of Sociology* 18(1) 1992: 209–232.

Conrad, Peter. "The Discovery of Hyperkinesis: Notes on the Medicalization of Deviant Behavior." *Social Problems* 23(1) 1975: 12–21.

Conrad, Peter. "The Shifting Engines of Medicalization." *Journal of Health and Social Behavior* 46(1) 2005: 3–14.

Dumit, Joseph. *Drugs for Life: How Pharmaceutical Companies Define Our Health.* Duke University Press, 2012.

Duster, Troy. "Medicalisation of Race." *The Lancet* 369(9562) 2007: 702–704.

Epstein, Steven. *Inclusion: The Politics of Difference in Medical Research.* University of Chicago Press, 2008.

Foucault, M., and C. Gordon. *Power/Knowledge: Selected Interviews and Other Writings, 1972–1977.* Pantheon, 1980.

Franklin, Sarah. "Life Itself: Global Nature and the Genetic Imaginary." In *Global Nature, Global Culture,* (Eds.) Sarah Franklin, Celia Lury, and Jackie Stacey. Sage, 2003.

Franklin, Sarah, and Celia Roberts. *Born and Made: An Ethnography of Preimplantation Genetic Diagnosis.* Princeton University Press, 2006.

Fullwiley, Duana. "The Biologistical Construction of Race: Admixture Technology and the New Genetic Medicine." *Social Studies of Science* 38(5) 2008: 695–735.

Gibbon, Sahra, and Carlos Novas. *Biosocialities, Genetics and the Social Sciences: Making Biologies and Identities.* Routledge, 2007.

Happe, Kelly. *The Material Gene: Gender, Race, and Heredity after the Human Genome Project.* NYU Press, 2013.

Hedgecoe, Adam. *Geneticization: Debates and Controversies.* Wiley, 2009.

Heinemann, Torsten, and Thomas Lemke. "Biological Citizenship Reconsidered: The Use of DNA Analysis by Immigration Authorities in Germany." *Science, Technology, & Human Values* 39(4) (July 1, 2014): 488–510.

Inda, Jonathan Xavier. *Racial Prescriptions: Pharmaceuticals, Difference, and the Politics of Life.* Routledge, 2016.

Ishii, Tetsuya. "Germline Genome-Editing Research and Its Socioethical Implications." *Trends in Molecular Medicine* 21(8) 2015: 473–481.

Jasanoff, Sheila. *Reframing Rights: Bioconstitutionalism in the Genetic Age.* MIT Press, 2011.

Kahn, Jonathan. *Race in a Bottle: The Story of BiDil and Racialized Medicine in a Post-Genomic Age.* Columbia University Press, 2013.

Klawiter, Maren. *The Biopolitics of Breast Cancer: Changing Cultures of Disease and Activism.* University of Minnesota Press, 2008.

Lemke, Thomas. "Susceptible Individuals and Risky Rights: Dimensions of Genetic Responsibility." *Biomedicine as Culture: Instrumental Practices, Technoscientific Knowledge, and New Modes of Life*. Routledge, 2007.

Lippman, A. "Prenatal Genetic Testing and Screening: Constructing Needs and Reinforcing Inequities." *American Journal of Law and Medicine* 17 (1991): 15–50.

Lock, Margaret. "Biosociality and Susceptibility Genes." In *Biosocialities, Genetics and the Social Sciences: Making Biologies and Identities* (Eds.) Sahra Gibbon and Carlos Novas, 2008, 56–78.

Mamo, Laura, and Jennifer Ruth Fosket. "Scripting the Body: Pharmaceuticals and the (Re)Making of Menstruation." *Signs* 34(4) 2009: 925–949.

M'charek, Amade, Katharina Schramm, and David Skinner. "Topologies of Race: Doing Territory, Population and Identity in Europe." *Science, Technology, & Human Values* 39(4) 2014: 468–487.

Meloni, Maurizio. "Epigenetics for the Social Sciences: Justice, Embodiment, and Inheritance in the Postgenomic Age." *New Genetics and Society* 34(2) 2015: 125–151.

Menon, Alka. "Reconstructing Race and Gender in American Cosmetic Surgery." *Ethnic and Racial Studies* 40(4) (March 16, 2017): 597–616.

Montoya, Michael. *Making the Mexican Diabetic: Race, Science, and the Genetics of Inequality*. University of California Press, 2011.

Nadesan, Majia Holmer. *Governmentality, Biopower, and Everyday Life*. Routledge, 2010.

Nelson, Alondra. *The Social Life of DNA: Race, Reparations, and Reconciliation after the Genome*. Beacon Press, 2016.

Nguyen, Vinh-Kim. *The Republic of Therapy: Triage and Sovereignty in West Africa's Time of AIDS*. Duke University Press, 2010.

Niewöhner, Jörg. "The Material Gene: Gender, Race, and Heredity after the Human Genome Project." *New Genetics and Society* 32(4) 2013: 459–461.

Panofsky, Aaron. "Generating Sociability to Drive Science: Patient Advocacy Organizations and Genetics Research." *Social Studies of Science*, 2010.

Panofsky, Aaron, and Catherine Bliss. "Ambiguity and Scientific Authority: Population Classification in Genomic Science." *American Sociological Review* 82(1) 2017: 59–87.

Petryna, Adriana. *Life Exposed: Biological Citizens after Chernobyl*. Princeton University Press, 2013.

Pollock, Anne. *Medicating Race: Heart Disease and Durable Preoccupations with Difference*. Duke University Press, 2012.

Rabinow, Paul. *Artificiality and Enlightenment: From Sociobiology to Biosociality*, Princeton, NJ: Princeton University Press, 1992.

Rabinow, Paul, and Nikolas Rose. "Biopower Today." *BioSocieties* 1(2) 2006: 195–217.

Raman, Suthra and Richard Tutton. "Life, Science and Biopower." *Science, Technology and Human Values* 17(2009): 1–24.

Rapp, Rayna. "Commentary: Thinking through Public Health Genomics." *Medical Anthropology Quarterly* 27(4)2013: 573–576.

Reardon, Jenny. *Race to the Finish: Identity and Governance in an Age of Genomics*. Princeton University Press, 2009.

Reardon, Jenny. "The 'persons' and 'genomics' of Personal Genomics." *Personalized Medicine* 8(1)2011: 95–107.

Richardson, Sarah, and Hallam Stevens. *Postgenomics: Perspectives on Biology after the Genome*. Duke University Press, 2015.

Rose, Nikolas. "Beyond Medicalisation." *The Lancet* 369(9562) 2007: 700–702.

Rose, Nikolas. *The Politics of Life Itself: Biomedicine, Power, and Subjectivity in the Twenty-First Century*. Princeton University Press, 2009.

Schramm, Katharina, David Skinner, and Richard Rottenburg. *Identity Politics and the New Genetics: Re/Creating Categories of Difference and Belonging*. Vol. 6. Berghahn Books, 2012.

Shim, Janet, Katherine Darling, Sara Ackerman, Sandra Soo-Jin Lee, and Robert Hiatt. "Reimagining Race and Ancestry." In *Reimagining (Bio)Medicalization, Pharmaceuticals and Genetics: Old Critiques and New Engagements*, edited by Susan Bell and Anne Figert. New York: Routledge, 2015.

Shostak, Sara. *Exposed Science: Genes, the Environment, and the Politics of Population Health*. University of California Press, 2013.

Shostak, Sara, Peter Conrad, and Allan V. Horwitz. "Sequencing and Its Consequences: Path Dependence and the Relationships between Genetics and Medicalization." *American Journal of Sociology* 114(S1) 2008: S287–S316.

Smart, Andrew, Richard Tutton, Paul Martin, George TH Ellison, and Richard Ashcroft. "The Standardization of Race and Ethnicity in Biomedical Science Editorials and UK Biobanks." *Social Studies of Science* 38(3) 2008: 407–423.

SmartDNA. "DNA Testing Kit – Parents and Children," 2017. http://dnatests-me.elasticbeanstalk.com/parents-children.php.

Starr, Paul. *The Social Transformation of American Medicine: The Rise Of A Sovereign Profession And The Making Of A Vast Industry*. Basic Books, 2008.

Tulloch, John, and Deborah Lupton. *Risk and Everyday Life*. Sage, 2003.

Tutton, Richard. *Genomics and the Reimagining of Personalized Medicine*. Ashgate, 2014.

Waldby, Catherine, and Melinda Cooper. "The Biopolitics of Reproduction: Post-Fordist Biotechnology and Women's Clinical Labour." *Australian Feminist Studies* 23(55) 2008: 57–73.

Williams, Simon, Paul Martin, and Jonathan Gabe. "The Pharmaceuticalisation of Society? A Framework for Analysis." *Sociology of Health & Illness* 33(5) 2011: 710–725.

Zola, Irving Kenneth. "Medicine as an Institution of Social Control." *The Sociological Review* 20(4) 1972: 487–504.

4

Genomics and big data in biomedicine

Niccolò Tempini and Sabina Leonelli

Introduction

The ease with which genomic data can be generated and disseminated, and particularly the novel opportunities offered by the fast and cheap genotyping of individual patients, is resulting in the building of large-scale online databases. These include infrastructures supporting the exchange, integration and interpretation of health-related data as well as tools designed to facilitate the analysis of clinical situations (Merelli et al., 2014; Staes et al., 2009; Wang & Krishnan, 2014). These databases collate and provide access to data produced at high *velocity*, thanks to increasing levels of automation and standardisation in the sequencing pipeline; in great *volume*, since a single genome off the sequencer can require up to 180 gigabytes of disk space; and coming from a vast *variety* of sources, both in the institutional and geographical sense of encompassing many different locations of data production and the epistemic sense of including many types of high-throughput biomedical data beyond the genome, such as transcriptomics, proteomics, metabolomics and so on. Thus exemplifying the 3Vs definition often used to characterise Big Data (Kitchin 2013), genomics is widely regarded as a prominent example of how Big Data collection and analysis can result in novel approaches to healthcare, such as personalised and precision medicine (Hey et al. 2009, Collins 2010, Hood & Friend 2011).

The capacity to generate, organise and interpret vast datasets is rooted in the history of molecular biology, and has long been associated with opportunities for innovation in medical knowledge and interventions. Medical genetics before genomics had already uncovered the aetiology of major genetic diseases and hereditary cancer syndromes, as well as stimulating the development of advanced technology to characterise DNA and study genes. The ensuing wave of innovation in genomics, linked to technical and methodological developments such as Next Generation Sequencing, has facilitated novel approaches to the genetic study of populations, with increasing numbers of Genome-Wide Association Studies (GWAS) aiming to understand genotype–phenotype associations. The data resulting from these studies have been used to develop and routinise diagnostic tests, which in some well-publicised cases have considerably shortened the process required to produce a reliable diagnosis (Eisenstein, 2014; Khoury et al., 2010). The future of genome-based diagnostics and interventions looks particularly promising within cancer research, with hundreds of identified oncogenes and some preliminary success in developing targeted therapies. Prima facie, it thus seems reasonable to point to the large-scale use of genomic data as a source of real and positive transformation for both biomedical research and resulting diagnostic processes and treatments.

This chapter considers the challenges encountered when conceptualising genomics and related technologies as a model for data-driven research. We start with a brief introduction to the role of Big Data in biomedical research and healthcare, including a brief overview of the types of data involved and the extent to which biomedical databases manage to encompass them and make them usable. We then discuss how genomic data fit the category of 'Big Data', taking into account the problematic nature of this label and the increasing awareness among biomedical researchers of the limits of an approach to health solely or even primarily focused on genomics. In closing, we point to the search for mechanisms to integrate different types of genomic data, as well as genomic data with other relevant data sources, as a key open challenge for attempts to develop and implement precision medicine.

Big data in biomedical research and healthcare

Big health data include a vast variety of data types and sources, ranging from data collected through lab studies of non-human models (including animals as well as plants and microorganisms) to data provided by patients through social media or generated through direct-to-consumer genetic services, data obtained through clinical and longitudinal studies on various human populations, and data obtained by sites of diagnosis and therapy such as hospitals and individual GP practices (Lucivero & Prainsack 2015; Green & Vogt 2016; Prainsack & Vayena 2013; Kallinikos & Tempini 2014; Harris et al. 2016). Not only do such data vary dramatically in their format, target and provenance, but also in the methods, goals and assumptions used by the communities that generate them, which can often have non-overlapping or even conflicting commitments with regards to what constitutes good, reliable and significant data, under which circumstances, and for what purposes (Ossorio 2011, Leonelli 2012, Demir and Murtagh 2013, Tempini 2015, Helgesson & Krafve 2015, Hogle 2016).

Such diversity is reflected in the requirements that new technological solutions and data-intensive processes of data storage, dissemination, visualisation and analysis need to satisfy. To characterise big health data today is thus also to capture the flexibility with which sophisticated computational tools and infrastructure architectures are being applied to very different situations and problems of knowing. In genomics and the biomedical fields in which genomic data have been most extensively used (such as oncology), high levels of standardisation and automation of data-related processes have enabled the construction of complex data pipelines, which transfer, modify, analyse, and share in turn the data across numerous different contexts and through several networked data infrastructures. In cancer genomics, for instance, databases that collect and share identified genetic mutations such as The Cancer Genome Atlas (TCGA) feed data into databases such as the Catalogue of Somatic Mutations in Cancer (COSMIC), which in turn enrich such information with functional interpretations sourced from published literature and panel studies – a task which itself required consultation and data mining from yet another set of databases and data sources. Enriched databases then provide data to clinical research systems, labs, and interpretations services that offer first or second opinion and consultancy to customers taking genome sequencing tests (Forbes et al. 2011; Keating et al. 2016; An et al. 2014; Ramos et al. 2015; Greshake et al. 2014; Cerami et al. 2012).

While such chains of data processing and enrichment are crucial to the use of genomic data for medical purposes, these complex networks of systematic relationships between data infra-structures have not yet materialised in areas of biomedicine that are less visible and funded than oncology. Many projects focusing on rare or under-researched diseases are struggling with how to source, aggregate, organise and share knowledge that has been produced and shared unsystematically by a variety of disconnected sources. One of the reasons for this struggle is that the intermediate stages of developing such networks are all but automated. Crucial and labour-intensive epistemic valuations are involved each time that data are moved across sites, as the data need to be refashioned for the kind of work and goals that any given situation is set up to support. The curation of genomic data for biomedical use bears strong theoretical implications and operational path-dependencies (Bowker 2000, Tempini 2015, Leonelli 2016). For instance, curators in cancer genomics need to distinguish 'driver' from 'passenger' mutations. This decision depends on researchers' awareness and interpretation of knowledge available on normal human variation, the availability of cancer genome data and analytical tools. However, it is hard to reach this decision algorithmically without appeal to human judgement. A cancer genome often presents mutations in the thousands. After comparison with reference (non-cancerous) genomes, only a handful of them are singled out as cause for the cancerous processes (the drivers). The remaining mutations are the effects of the mutagenesis (the passengers) and are considered 'noise'. Singling out the drivers enables researchers and clinical services to study how to impact the disease through targeted therapies. Yet this is necessarily contingent on the availability of data on cancerous and non-cancerous genomes, and the ways in which the two sets are compared. The processes by which these distinctions are made are often traceable, but embedded in work flows where fast-paced analysis is a priority.

Furthermore, end users of the data, such as clinicians or drug development labs, usually have a variety of different requirements, which often leads to users pulling the development of data infrastructures in opposite directions. The emergence of networks of interdependent database projects puts pressure on the organisational requirements and the types of skills needed to facilitate data movements among them, while making strategic decisions about long-term database development in such a complex and fast-evolving environment is increasingly difficult. Additionally, the problem of sourcing the appropriate amount of skills in genomics extends far beyond the development of data infrastructures. Research on contemporary oncological practices shows that the changing and fragmented knowledge base required to make decisions for personalised treatment regimes necessitates the mediation of expert bodies such as Molecular Tumor Boards, where different specialists convene to piece together the clinical situation (Cambrosio et al. 2017). The unique complexity acquired by each individual case in a time of unparalleled data affluence is posing enormous organisational and methodological challenges. This is addressed by so-called 'interpretation services', a further group of genomic data infrastructures focused on interpreting lab results and adapting existing data resources to individual situations.

Nested data infrastructures and the commodification of health data: reframing genomics as big data

This backdrop points to at least three ways in which research data infrastructures are nested and interdependent: (1) institutionally, through the organisational, commercial and collaborative relationships between the various bodies that foster, support and maintain them; (2) in terms of compatibility or alignment of goals, and through the chain of processes, interdependent practices and related data operations required to make data useful in clinical and care settings; and (3) in terms of expertise and skills, through the ever-transforming relations between the dozens – sometimes hundreds – of individuals involved in sourcing, organising and managing the data in question, as well as the infrastructures themselves (Ribes 2017).

The complex, tortuous 'pipes' that distribute data are not exclusive to genomic health, but are rather a global phenomenon common to many data types and driven by information technology that promises extreme gains in performance with respect to important aspects of organised processes in all social domains. Because of this common infrastructure, data can be mixed, leaked and disseminated in sometimes irrelevant, controversial and inappropriate ways, which is particularly concerning in the case of genomic data featuring in ethically controversial projects (e.g. Regalado 2015). While there are stringent ethical checks in place for data produced by publicly funded genomic research, health data produced in other contexts are released to commercial organisations and marketers in ways that are not transparent nor subject to clear legislation (Schroeder 2014, Aicardi et al. 2016, Mittelstadt and Floridi 2015, Richards et al. 2015). Such data live a life of their own, and can be duplicated, aggregated, and used numberless times for constructing and structuring commercial opportunities (Ebeling 2016). Sectors of the corporate world that have pioneered commercial applications of big data technologies are interested in extending their reach to genomic data, an inclusion that associates genomics ever more closely with the financial market and venture capital (Fortun 2008, Sunder Rajan 2006). With the spread of direct-to-consumer tests and the oft-associated proprietary data capture (e.g. *23andMe*), genomic data are increasingly used in domains of application that are far from its origins in the lab and the clinic.

The travel of genomic data into big data analytics is the most powerful and direct indication of the ways in which genomics is finding a new social and epistemic identity in the big data world – where big data cannot be characterised purely in terms of the 3Vs discussed above, but rather as an expression of market forces and commodification of information (Kitchin and McArdle 2016). However, the evolution of new forms of value creation from genomic data has been patchy and innovation of the field still relies on cycles of public funding and promissory social imaginaries. The aftermath of the Human Genome Project (and its alter-ego Celera Genomics) exemplifies how projecting the results of genomic research into the future, and relating them to elusive goals such as health and prosperity, is crucial to their continued relevance and support among funders, policy-makers and civil society (Hilgartner 2013, Sunder Rajan 2006, Reardon 2009). Equally clear is the extent to which the evolution of value-creating uses of genomic data remains dependent on the discovery and understanding of robust associations between genetic make-up, physiological processes and health outcomes (Hedgecoe, 2004; O'Malley & Stotz, 2011). The integration of genomics with other research domains is widely perceived as necessary to achieve a broader and deeper understanding of biological process including organismic development and evolution (Stevens & Richardson, 2015). The difficulty and slowness in developing innovative therapies, including pharmacogenomic drugs, have alerted the scientific community to the need for a deeper understanding of other organisational structures and processes than those strictly described by the genomic vocabulary (Keating & Cambrosio, 2013; Laksman & Detsky, 2011; Longo, 2012; Samani et al., 2010).

Given these developments, the idea that genomic data are big data that can be purposed for the exploration of an entire landscape of undiscovered connections, risk factors and therapies looks like the answer to a badly formulated question. While genomic data might be certainly of hitherto unforeseen volume, velocity and variety, their characteristics may make them too hard to use for the goals of interest, especially since accomplishing these goals depends on so many factors beyond genomic research and data production themselves.

Towards precision medicine? The challenge of data integration

The vast and rapid expansion of the territory of biomedical phenomena that can be investigated through big data assemblage and integration is a double-edged sword in the quest for truly personalised medicine. On the one hand, it constitutes a starting point for finding new therapeutic solutions, and thus make good on the promise of translational returns (in terms of new diagnostic and therapeutic tools) made in relation to genomic data. On the other hand, the new molecular understanding of most diseases and conditions sees them associated to a staggering diversity of factors. This 'long tail' of unique medical cases will remain difficult to address through high-end pharmacogenetic solutions.

The paramount question has become how to extend the use of sequencing services beyond what physicians currently know how to do, and how to reduce the gap between the variation that is observed, and that which can be interpreted (Ashley et al. 2010). The future of genomic data is tied to their effective integration with other data sources, and their circulation beyond genomic medicine and into all realms of biomedical R&D. However, at least two factors that are crucial to such integration are not yet in place. First, a better understanding of the relationship between genomic and other biological phenomena is required in order to make better sense of observed patterns in sequences, their expression and their relation to other data sources (Dupré 2010; Samani et al., 2010; Wolkenhauer & Green, 2013). Second, data infrastructures are required that can effectively collect, visualise and make searchable datasets of completely different origin, and can also continue to maintain, update and expand those resources over time. The financial effort involved in developing such structures is enormous, not to speak of the challenges of producing standard terminology and tools that fit the ever-changing knowledge and technology landscape around human health.

Better ways to circulate data are also significant for optimising biomedical decision-making. Improvements in the interpretation of available data, and related forms of consulting, depend on the existence of relevant physiological and therapeutical knowledge, but also on the extent to which data can be effectively mobilised and on the availability of expert advice on how to interpret such data. It has been vastly shown that genomic information can have significant health effects, beyond its direct clinical applications, in the behaviour of patients (Harris et al. 2013, Prainsack and Vayena 2013), and yet genomic information is difficult to interpret and act upon for patients and their relatives (Green et al. 2013; Laksman & Detsky 2011; Lyon 2012). In addition to this, people make sense of genomic information in a range of different ways and will arguably do so every time they have access to new information. To help explore the multi-layered meanings and implications of genetic information, it is all the more important that patients have access to appropriate experts, as for instance in Dutch genetic counselling clinics (Taussig 2009).

Differences are developing in this respect between haves and have-nots, with business models of service providers becoming less dependent on a strict, monopolistic style of control of the data as patients are increasingly associated to other revenue streams. The proliferation of direct-to-consumer testing, marketing on promises of unlocking the secrets to one's ancestry and revealing a panel of risk factors, has created a market in the generation and access to one's own genetic information. This meets consumer interest and in many cases enables providers to bypass or anticipate the inclusion of tests in healthcare, raising concerns on the business model, the reliability of the results, and the lack of in-depth consulting for interpretation and sense-making of the results. These services put pressure on genomic sequencing towards structuring the relationship between provider and user as market transaction (Harris et al. 2014, Green & Vogt 2016). Ethics issues are not only related to genomics at the time of testing, but also at later stages of data use including concerns around re-contacting patients (Carrieri et al. 2017) and/or relatives following incidental findings and the discovery of new functional interpretations. Also, the increasing availability of testing has given birth to new kinds of ethically complex social interactions, such as individuals requiring testing of their partners to evaluate chances of inheritable disease in future offspring. The ways in which people with different means will deal with the diverse set of uncertainties that are accompanying the application of genomic services in and outside the clinic risk to further compound inequality. Finally, a lot of the data to be integrated with genomic data do not enjoy a clear legal and institutional status (Schroeder 2014, Mittelstadt and Floridi 2015), as relevant frameworks and policies are struggling to move beyond obsolete interpretations of informed consent (Richards et al. 2015, Aicardi et al. 2016). These uncertainties contribute to a fabric of frictions that make the circulation of different kinds of data particularly difficult.

Acknowledgements

Authors gratefully acknowledge funding by the European Research Council, grant number 335925 (DATA_SCIENCE), and the helpful comments of editors to an earlier draft.

References

Aicardi, Christine, Lorenzo Del Savio, Edward S. Dove, Federica Lucivero, Niccolò Tempini, and Barbara Prainsack. 2016. Emerging Ethical Issues Regarding Digital Health Data. On the World Medical Association Draft Declaration on Ethical Considerations Regarding Health Databases and Biobanks. *Croatian Medical Journal* 57(2): 207–213.

An, O., Pendino, V., D'Antonio, M., Ratti, E., Gentilini, M., Ciccareli, F. D. 2014. NCG 4.0: The Network of Cancer Genes in the Era of Massive Mutational Screenings of Cancer Genomes. *Database* 2014(0): bau0152013;bau015.

Ashley, E. A., Butte, A. J., Wheeler, M. T., Chen, R., Klein, T. E., Dewey, F. E., … Altman, R. B. (2010). Clinical Assessment Incorporating a Personal Genome. *The Lancet* 375(9725), 1525–1535.

Bowker, G. C., 2000. Biodiversity Datadiversity. *Social Studies of Science* 30, 643–683.

Cambrosio, A., Bourret, P., Keating, P., Nelson, N. C., 2017. Opening the Regulatory Black Box of Clinical Cancer Research: Transnational Expertise Networks and Disruptive Technologies. *Minerva* 55(2), 161–185.

Carrieri, D., Dheensa, S., Doheny, S., Clarke, A. J., Turnpenny, P. D., Lucassen, A. M., Kelly, S. E., 2017. Recontacting in Clinical Practice: An Investigation of the Views of Healthcare Professionals and Clinical Scientists in the United Kingdom. *Eur J Hum Gene* 25, 520–521.

Cerami, E., Gao, J., Dogrusoz, U., Gross, B. E., Sumer, S. O., Aksoy, B. A., Jacobsen, A., Byrne, C. J., Heuer, M. L., Larsson, E., Antipin, Y., Reva, B., Goldberg, A. P., Sander, C., Schultz, N., 2012. The cBio Cancer Genomics Portal: An Open Platform for Exploring Multidimensional Cancer Genomics Data. *Cancer Discovery* 2, 401–404.

Collins, F. S. (2010). *The Language of Life: DNA and the Revolution in Personalized Medicine* (1 edition, p. 371). HarperCollins e-books.

Demir, Ipek, and Madeleine J. Murtagh. 2013. "Data Sharing across Biobanks: Epistemic Values, Data Mutability and Data Incommensurability." *New Genetics and Society* 32(4): 350–365.

Dupré, J. (2010). Emerging Sciences and New Conceptions of Disease; or, Beyond the Monogenomic Differentiated Cell Lineage. *European Journal for Philosophy of Science*, 1(1), 119–131.

Ebeling, M. F. E., 2016. *Healthcare and Big Data*. Palgrave Macmillan US, New York.

Eisenstein, M. (2014). Personalized Medicine: Special Treatment. *Nature*, 513(7517), S8–S9.

Forbes, S. A., Bindal, N., Bamford, S., Cole, C., Kok, C. Y., Beare, D., Jia, M., Shepherd, R., Leung, K., Menzies, A., Teague, J. W., Campbell, P. J., Stratton, M. R., Futreal, P. A., 2011. COSMIC: Mining Complete Cancer Genomes in the Catalogue of Somatic Mutations in Cancer. *Nucl. Acids Res. 39*, D945–D950.

Fortun, M., 2008. *Promising Genomics*. University of California Press, London.

Green, R. C., Berg, J. S., Grody, W. W., Kalia, S. S., Korf, B. R., Martin, C. L., … Biesecker, L. G. (2013). ACMG Recommendations for Reporting of Incidental Findings in Clinical Exome and Genome Sequencing. *Genetics in Medicine*, 15(7), 565–574.

Green, S., & Vogt, H. (2016). Personalizing Medicine in Silico and in Socio. Retrieved from http://philsci-archive.pitt.edu/12274.

Greshake, B., Bayer, P. E., Rausch, H., Reda, J. 2014. "openSNP–A Crowdsourced Web Resource for Personal Genomics." *PLoS ONE* 9(3): e89204.

Harris, A., Kelly, S., Wyatt, S., 2016. *CyberGenetics*. Routledge, London.

Harris, A., Kelly, S. E., Wyatt, S., 2014. Autobiologies on YouTube: Narratives of Direct-to-Consumer Genetic Testing. *New Genetics and Society* 33, 60–78.

Harris, A., Kelly, S. E., Wyatt, S., 2013. Counseling Customers: Emerging Roles for Genetic Counselors in the Direct-to-Consumer Genet Test Market. *Journal of Genetic Counseling* 22, 277–288.

Hedgecoe, A. (2004). *The Politics of Personalised Medicine: Pharmacogenetics in the Clinic* (p. 217). Cambridge University Press.

Helgesson, C. -F., Krafve, L. J., 2015. Data Transfer, Values, and the Holding Together of Clinical Registry Networks, in: Dussauge, I., Helgesson, C. -F., Lee, F. (Eds.), *Value Practices in the Life Sciences and Medicine*. Oxford University Press, Oxford, United Kingdom, pp. 224–246.

Heyet al. 2009. *The Fourth Paradigm*. Microsoft Publishing.

Hilgartner, S., 2013. Constituting Large-Scale Biology: Building a Regime of Governance in the Early Years of the Human Genome Project. *BioSocieties* 8, 397–416.

Hogle, L. F., 2016. Data-intensive Resourcing in Healthcare. *BioSocieties* 11, 372–393.

Hood, L., Friend, S. H., 2011. Predictive, Personalized, Preventive, Participatory (P4) Cancer Medicine. *Nat Rev Clin Oncol* 8, 184–187.

Kallinikos, J., Tempini, N., 2014. Patient Data as Medical Facts: Social Media Practices as a Foundation for Medical Knowledge Creation. *Information Systems Research* 25, 817–833.

Keating, P., & Cambrosio, A. (2013). 21st-century Oncology: A Tangled Web. *The Lancet*, 382(9909), e45–e46.

Keating, P., Cambrosio, A. and Nelson, N. (2016). 'Triple Negative Breast Cancer': Translational Research and the (Re)Assembling of Diseases in Post-Genomic Medicine. *Studies in History and Philosophy of Biological and Biomedical Sciences*, 59, 20–34.

Kitchin, R. (2013). *The Data Revolution*. Sage.

Kitchin, R., McArdle, G. (2016). What makes Big Data, Big Data? Exploring the Ontological Characteristics of 26 Datasets. *Big Data & Society* 3, 2053951716631130. doi:10.1177/2053951716631130.

Khoury, M. J., Evans, J., & Burke, W. (2010). A Reality Check For Personalized Medicine. *Nature*, 464 (7289), 680–680.

Laksman, Z., & Detsky, A. S. (2011). Personalized Medicine: Understanding Probabilities and Managing Expectations. *Journal of General Internal Medicine*, 26(2), 204–206.

Leonelli, S. (2016). *Data-Centric Biology: A Philosophical Study*. Chicago, IL: Chicago University Press.

Leonelli, S. (2012). When Humans Are the Exception: Cross-Species Databases at the Interface of Clinical and Biological Research. *Social Studies of Science* 42(2): 214–236.

Longo, D. L. (2012). Tumor Heterogeneity and Personalized Medicine. *New England Journal of Medicine*, 366(10), 956–957.

Lucivero, F., & Prainsack, B. (2015). The Lifestylisation of Healthcare? "Consumer Genomics" and Mobile Health as Technologies For Healthy Lifestyle. *Applied & Translational Genomics* 4, 44–49.

Lyon, G. J. (2012). Personalized Medicine: Bring Clinical Standards to Human-Genetics Research. *Nature*, 482(7385), 300–301.

Merelli, I., Perez-Sanchez, H., Gesing, S., & D'Agostino, D. (2014). Managing, Analysing, and Integrating Big Data in Medical Bioinformatics: Open Problems and Future Perspectives. *Biomed Research International*, 134023, 13 pages. URL http://dx.doi.org/10.1155/2014/134023 (accessed 20 November 2017).

Mittelstadt, B.D., Floridi, L., 2015. The Ethics of Big Data: Current and Foreseeable Issues in Biomedical Contexts. *Sci Eng Ethics* 1–39.

O'Malley, M. A. and Stotz, K. (2011). Intervention, Integration and Translation in Obesity Research: Genetic, Developmental and Metaorganismal Approaches. *Philosophy, Ethics, and Humanities in Medicine*, 6(2). URL www.peh-med.com/content/6/1/2 (accessed 20 November 2017).

Ossorio, Pilar. 2011. "Bodies of Data: Genomic Data and Bioscience Data Sharing." *Social Research* 78(3): 907–932.

Prainsack, B., Vayena, E., 2013. Beyond the Clinic: "Direct-to-Consumer" Genomic Profiling Services and Pharmacogenomics. *Pharmacogenomics* 14, 403–412.

Ramos, A. H., Lichtenstein, L., Gupta, M., Lawrence, M. S., Pugh, T. J., Saksena, G., Meyerson, M., Getz, G., 2015. Oncotator: Cancer Variant Annotation Tool. *Human Mutation* 36, E2423–E2429.

Reardon, J., 2009. *Race to the Finish: Identity and Governance in an Age of Genomics*. Princeton University Press, Princeton.

Regalado, A., 2015. J. Craig Venter's Human Longevity to Offer Health Services [WWW Document]. *MIT Technology Review*. URL www.technologyreview.com/news/541516/j-craig-venter-to-sell-dna-data-to-consumers/ (accessed 10 April 2015).

Ribes, D., 2017. Notes on the Concept of Data Interoperability: Cases from an Ecology of AIDS Research Infrastructures. Presented at the Computer-Supported Collaborative Work, Portland, OR.

Richards, M., Anderson, R., Hinde, S., Kaye, J., Lucassen, A., Matthews, P., Parker, M., Shotter, M., Watts, G., Wallace, S., Wise, J., 2015. *The Collection, Linking and Use of Data in Biomedical Research and Health Care: Ethical Issues*. Nuffield Council on Bioethics, London.

Samani, N. J., Tomaszewski, M., & Schunkert, H. (2010). The Personal Genome – The Future of Personalised Medicine? *The Lancet*, 375(9725), 1497–1498.

Schroeder, R., 2014. Big Data and the Brave New World of Social Media Research. *Big Data & Society* 1.

Staes, C. J., Xu, W., LeFevre, S. D., Price, R. C., Narus, S. P., Gundlapalli, A., … Facelli, J. C. (2009). A Case for Using Grid Architecture For State Public Health Informatics: The Utah Perspective. *BMC Medical Informatics and Decision Making*, 9, 32.

Stevens, H., Richardson, S. S., 2015. Beyond the Genome, in: *Postgenomics*. Duke University Press, London, pp. 1–8.

Sunder Rajan, K., 2006. *Biocapital*. Duke University Press, London.

Taussig, K.S. 2009. *Ordinary Genomes*. Duke University Press.

Tempini, N., 2015. Governing PatientsLikeMe: Information Production and Research Through an Open, Distributed and Data-Based Social Media Network. *The Information Society* 31, 193–211.

Wang, W., & Krishnan, E. (2014). Big Data and Clinicians: A Review on The State of The Science. *JMIR Medical Informatics*, 2(1), e1.

Wolkenhauer, O., & Green, S. (2013). The Search For Organizing Principles as a Cure Against Reductionism in Systems Medicine. *FEBS Journal*, 280(23), 5938–5948.

5

Mainstreaming genomics and personal genetic testing

Susan E. Kelly, Sally Wyatt and Anna Harris

An important development in personalised medicine is the development of the direct-to-consumer (DTC) genetic testing industry. This chapter considers what it means for genetics to 'go online', specifically what happens when individuals are able to purchase personal genome tests directly from companies supplying both sequence data and interpretation of health risks.

This chapter explores a growing body of research about the market for DTC genetic testing from a social science perspective. Focusing on genetic testing for disease, the chapter examines the new social arrangements which emerge when genetic testing, a traditionally clinical practice, is taken into different spaces. We explore the intersections of new genetics and new media by drawing from three different fields of study. Internet studies provide ways of understanding online infrastructures, users' practices, and the materiality of virtual exchanges. Science and technology studies offers insights into the variety of human and non-human actors involved, how to study controversy, and how to critique deterministic understandings of health and genetics. The sociology of health and illness enriches the analysis with attention to healthcare professionals and practices, and the social life of genetic information.

This chapter focuses on two themes which have arisen from our own research and that of others working on DTC genetic testing (e.g. Lee, 2013; McGowan et al., 2010; Nelson and Robinson, 2014; Pálsson, 2009; Prainsack, 2011; Sharon, 2014; Tutton, 2014). The first concerns the new possibilities for spatial-temporal arrangements between patients, consumers, healthcare professionals, and private companies. The second concerns the relations of trust which are continually performed and negotiated in online genetics. We first introduce the DTC genetic testing industry, and the changing role of the internet in social life, particularly related to health.

Brief history of DTC genetic testing

The Human Genome Project was completed in 2003, a moment heralded as 'the genomic era' in an editorial in a leading medical journal (Guttmacher and Collins, 2003). This opened up the possibility of DTC genetic testing, so that people could order genetic tests directly, without necessarily going via a healthcare professional. DTC genetic testing began with the launch of *Knome* by Harvard Professor of Genetics George Church (Singer, 2007). Another early entry into the DTC genetic testing industry was *deCODE*, an Icelandic company, capitalising (controversially) on the genetic and medical information of the largely homogeneous Icelandic population (Fortun, 2008). While *deCODE* focused on genetic research and logged up years of scientific success, it never turned a profit nor discovered a blockbuster therapy. In 2007, the company introduced *deCODEme*, offering a personal genetic test for US$985 that was part medicine and part recreation. Several other companies entered the personal genome market in 2007, offering genetic testing over the internet, directly to consumers, including *23andMe* and *Navigenics*. These companies hoped 'to usher in a new era of personalised genomic medicine by empowering individuals to access and understand their own genetic information' (*23andMe*, 2007, cited in McGuire et al., 2009: 3). The saliva-based personal genome kit sold by *23andMe* was named 'Invention of the Year' in 2008 by *Time Magazine*. The industry has played with a number of ambiguities. For example, the products invite both playful and more serious health-related motivations from consumers. Another ambiguity concerns the tension between individual genetic information and scientific/biobank aspects of the companies' activities, further capitalising upon both genome sequencing and internet technologies.

Playing with these ambiguities, the DTC genetic testing industry has been problematic from both regulatory and market perspectives. In June 2008, California health regulators sent cease-and-desist letters to thirteen genetic testing firms including *23andMe*. The regulators asked the companies to prove both that a physician was involved in the ordering of each test, and that state clinical laboratory licensing requirements were being fulfilled. In late 2013, the US Food and Drug Administration sent a warning letter to *23andMe*. A class action lawsuit was filed against the company, complaining that the information marketed as health-related was neither valid nor useful, and that information from customers is stored in databases that were marketed to the scientific community (Munro, 2013). The company began marketing similar DTC genetic testing products in both Canada and the UK, and in late 2015 the FDA ban on selling health-related tests in the US was lifted. *deCODEme* stopped selling genetic tests via its website in 2013, having declared earlier that it had not made a profit.

Health online

By the mid-1990s, the internet had moved beyond its original academic environment, and became widely available to both commercial and individual users. Within a few years it was widely used to find and exchange health-related information. Not only did it provide people with access to medical literature and officially sponsored health websites, people could set up and use a variety of internet-based fora, such as list-servs, bulletin boards, and personal home pages, to share information and personal experiences. These were often celebrated not only by policy makers but also by scholars for the ways in which they enabled people to become producers of knowledge about health, and to engage in more equal relationships with healthcare professionals (Hardey, 1999; Burrows et al., 2000).

The rise of social media in the early years of the twenty-first century made it even easier for people to share information, and made the role of the internet in healthcare practice increasingly visible (Wathen et al., 2008; Adams, 2010). Many popular accounts of the participatory potential of social media are enthusiastic. *Twitter*, blogs, *YouTube*, *Facebook*, and *Wikipedia* have all been lauded for their capacity to harness people's creativity and knowledge, and for their potential to challenge traditional hierarchies in politics, science, and the media. Questions have been raised by scholars working in more critical traditions, just as they did about the internet in the 1990s, about whether lay inclusion helps to 'democratise' knowledge formation or if existing hierarchies are re-enacted online (Lupton, 2014; Prainsack, 2011; Sharon, 2014; Parthasarathy, 2014).

Social media do have the potential to transform relationships between healthcare professionals, patients, consumers, funding agencies, healthcare systems and industry (Dedding et al., 2011), but whether and how such potential is realised depends on many factors. Notions of 'the clinic' have expanded so that consultations between healthcare professionals and patients are conducted via the internet, using remote monitoring devices and webcams (Christensen and Hickie, 2010; Mort et al., 2003; Oudshoorn, 2012). Online patient groups have used internet platforms to share experiences and resources, discuss research developments, and generate new knowledge (Kaplan et al., 2011). Patient-experience websites such as *HealthTalkOnline* and *PatientsLikeMe* demonstrate other ways in which patients, carers and others, can engage with each other, and potentially conduct their own research (Allison, 2009).

DTC genetic tests are not offered on the 'flat' sites, common in the late-1990s, which were often little more than online brochures and leaflets. Many of the DTC genetic testing companies described earlier use a variety of social media platforms in order to engage and interact with consumers, including blogs, *Twitter* and *YouTube*. User-generated content such as feedback forms and blog comments feed into company research design. Users' web activity is collected through log files and cookies, and such data are used by companies to monitor uses of their websites, to improve their services, and to tailor and customise content for customers, sometimes bringing them into contact with one another.

The internet continues to be celebrated as a tool of empowerment, particularly in scientific and medical fields. But the various forms of internet-mediated healthcare raise issues concerning privacy, expertise, access, exclusion and hypochondria/anxiety. Others have shown that web architecture and engagement with web technologies are more complex than posited, involving the replication of dominant hierarchies, and possibly the introduction of new forms of inequality and exploitation (Goldberg, 2011; König, 2013; Niederer and van Dijck, 2010; Terranova, 2000). Engagement with and through social media is sometimes contradictory, and the infrastructure of the internet both enables and constrains engagement with scientific research.

New social relations

As soon as the DTC genetic testing industry caught the attention of philosophers and ethicists (e.g. Borry and Howard, 2008; Sterckx et al., 2013), and of social scientists (see above), questions have been raised about how healthcare professionals are involved. Genetic information has long been identified as potentially harmful, not least because of its murky history with eugenics (Merz, 2016; Subramaniam, 2014). Genetic data are also difficult to interpret, based on complex science often not immediately meaningful to individuals concerned about their own health risks, and DTC genetic testing challenges the traditional expert (largely medical) control of information.

Clinical genetics, in some countries supported by the profession of genetic counselling, has developed as a mediator of genetic science to the public (although genetics in public health has a much longer history). From the last century, the right to make decisions about one's own genetic risks and a perhaps paternalistic control of the process of interpretation of genetic information by the medical profession have characterised social relations around genetic information. A further set of social relations includes those between the producers and users of genomic knowledge. However, one of the selling points of the DTC genetic testing industry has been the empowerment accompanying direct access to one's "own" genome, related to a broader ethos of 'citizen science', which would allow individuals to go beyond autonomy within the traditional patient/physician relationship and, so empowered, 'take charge' of their own health and futures.

In marketing statements, the DTC genetic testing industry allies itself to emerging notions of rights, citizenship, health outcomes data, and lifestyle choices. This shift is in line with the 'new public health' in which responsibility for health (such as making healthy life choices, but extending to knowledge of one's health-related data) falls upon individual citizens rather than the state (Peterson and Lupton, 1996). The DTC genetic testing industry has taken note of criticisms concerning unmediated access to genetic information, and has responded by re-configuring the relationships on offer between consumers and healthcare professionals. Companies provide a range of relationships with medical professionals, including staffing physicians to order tests, offering genetic counselling either directly or as an add-on service, affiliating with online commercial genetic counselling services, and/or claiming their products do not offer medical diagnostic information but are really only for educational or recreational purposes.

There does not seem to be much evidence that patients are bringing their personal genomic information to their encounters with healthcare professionals in any major way, nonetheless the DTC genetic testing industry has the potential, if not to reconfigure, at least to open new spaces for social relationships around genomic, and more broadly medical, expertise, and for expert/lay relations. For example, in moving online, genetic counselling has altered some of the core tenets of the discipline, such as considering the family unit with regard to genetic information rather than only the individual (Harris et al., 2016).

Further, as Kate O'Riordan correctly notes, DTC genetic tests are more than merely biomedical in the relationships they entangle, as they operate 'at the intersection of digital biosociality, consumption and knowledge production' (2013: 518). As such, they open spaces for social relationships around identity, sociality, economics, and the politics of knowledge. Such spaces support the emergence of playful forms of engagement with biomedical science and the potential futures they portend via personal genomics, a playfulness that is itself enrolled in the marketing and product displays of companies including *23andMe*, which market personal genomics as not only self-knowledge, but material for playful self-creation, extension and representation (Harris et al., 2014).

Changing relations of trust

DTC genetic testing is a powerful example of how digital technologies reconfigure trust relationships concerning health and illness, between people, their bodies, and experiential knowledge. When genetics goes online, trust becomes an important aspect of how relations between users, healthcare providers, industry and governing bodies are assembled and reconfigured. Four main kinds of trust are important when considering the DTC genetic testing industry: in genetics, in the internet, in companies, and in research and research data.

Health, especially the lack of it, raises many questions about trust: in one's body and its responses; in individual healthcare professionals and informal caregivers; in medical knowledge; and in healthcare systems. Digital technologies affect all of these. Trust is brought into relief wherever there are perceived to be exchanges of information, or the possibility of its appropriation. The most obvious exchange in DTC genetic testing concerns individuals sending samples of their saliva to an internet-based company in order to receive genetic information about themselves. In such exchanges customers also provide financial details, and more through their potential ongoing engagement on the site through commenting on fora or blogs. Information is returned to the customer in the form of genetic test results, analysis and interpretation of genetic data, and material on the website, blogs and fora about genetics, genetic testing and other company activities. Information is also shared between users, as they can invite others to view their genetic results, and through the user-generated content on various platforms.

23andMe asks its customers to provide further information about their health, personal traits and behaviour via simple online surveys. This is then linked to the customer's genetic data and contributes towards building a research repository of genetic and health information to be used by researchers and pharmaceutical companies.

In buying a genetic test, the customer is implicitly demonstrating trust in the genetic basis of health conditions and human traits. Companies foster the geneticisation of health and behaviour on their websites, although acknowledging in relatively small print the 'influence' of environmental and other factors. Anthropologists (e.g. Lock et al., 2006) have argued that rather than strengthening beliefs about the genetic basis of life processes, genetic testing is interpreted in the context of existing notions of family inheritance. As with other forms of health information, genetic information that is most consistent with pre-existing expectations about heritable characteristics and diseases is generally perceived as more trustworthy.

Trust in digital technologies, particularly the internet, is also embedded in the online genetic testing process. Commercial sites put significant effort into communicating trust, and emphasis is given to the security of their websites as places to share sensitive information, including genetic, health and financial details. In their privacy statements, the companies provide extensive technical details about secure online payment systems, data encryption, firewalls between genetic and account information, and how they monitor their employees' use of databases. The online genetic testing market presents the internet as a relatively risk-free place in which to explore your genetic information, information that is not necessarily tied to your medical record. Sharing genetic information is encouraged amongst users, in order to find relatives or others with similar genetic profiles. This also includes sharing further information about health and other behaviours for research purposes. The internet is viewed as a way in which to democratise genetic knowledge, and to empower users to do what they want with this information. It is promised to be a platform by which consumers or participants can be freed from institutional power and actively engaged in finding out more about their genetic selves.

The DTC genetic testing market is premised on trust in both genetics and the internet, and each company also needs to foster trust in its own product. Links to trustworthy institutions are made through the promotion of recent funding by national funding agencies, profiles of scientific advisory board members include details of university affiliations, and hyperlinks are incorporated to independent genetic counselling services. Public trustworthiness in the DTC genetic testing enterprise is also achieved through promotion of the companies in traditional media outlets, often with celebrity endorsement.

The databases are an important part of the business model for several DTC genetic testing companies, containing not only the genetic testing results of users/customers but also data generated through online surveys. These databases are used and promoted as important tools in conducting cutting-edge scientific research on the genetic basis of health conditions. The company needs to build trust into its research paradigm in order for customers to share further personal information through the answering of the surveys. This is done through public engagement and the early feedback of research results to participants and would-be participants using various platforms. The stabilisation of controversy, whether this be controversial research techniques or disputes over the genetic basis of disease for example, is also another way in which to build trust (Harris et al., 2016). Trust is fostered by allowing participants and customers to communicate with the company via social media, enabling the company to react quickly to areas of distrust. Companies such as *23andMe* need to establish trust relations with their customers in order to conduct research, as well as trusting their customers, for the research differs from more traditional medical research in a number of ways, particularly in the use of self-reported data (Wyatt et al., 2013).

In order to have the infrastructural arrangements to conduct this research, companies rely on venture capital and a trusting relationship with investors. Funders want to see returns on their investment, and it increasingly appears that profits will not be obtained from the sale of genetic tests, but rather from the potential of the research database to generate revenue from pharmaceutical companies, other biotechnology firms, and through the development of patents. In order to secure these profits however, companies potentially jeopardise trust relations with consumers. The online genetic testing market is one in which the main players must carefully balance profits and participation. As internet users become increasingly aware of the business practices behind sites, maintaining trust amid growing consumer scepticism (e.g. through information strategies that demonstrate transparency in practice) will remain important.

Conclusion

In this chapter, we have examined new possibilities for spatial-temporal arrangements emerging with online genetics, and relations of trust which are continually performed and negotiated in this field. These themes resonate with aspects of personalised medicine and other digital genomic futures such as big data science, participatory science, and user-generated data. The study of the DTC genetic testing industry provides insight into a range of consumer, research and data practices that involve science and the internet; analysing 'cybergenetics' has broader implications for the shape and import of genomic science in contemporary society.

Note: This chapter draws on our book, *Cybergenetics: Health genetics and new media* (Harris, Kelly and Wyatt), published by Routledge in 2016.

References

Adams, S. (2010) 'Sourcing the crowd for health experiences: Letting the people speak or obliging voice through choice?', in Harris, R., Wathen, N. and Wyatt, S. (eds.), *Configuring health consumers: Health work and the imperative of personal responsibility*, Basingstoke: Palgrave Macmillan.

Allison, M. (2009) 'Can web 2.0 reboot clinical trials?', *Nature Biotechnology*, vol. 27, no. 10, pp. 895–902.

Borry, P. and Howard, H. (2008) 'DTC genetic services: A look across the pond', *American Journal of Bioethics*, vol. 8, no. 6, pp. 14–16.

Burrows, R., Nettleton, S., Pleace, N., Loader, B. and Muncer, B. (2000) 'Virtual community care: Social policy and the emergence of computer mediated social support', *Information, Communication and Society*, vol. 3, no. 1, pp. 95–121.

Christensen, H. and Hickie, I. (2010) 'Using e-health applications to deliver new mental health services', *Medical Journal of Australia*, vol. 192, no. 11, pp. S53–S56.

Dedding, C., van Doorn, R., Winkler, L. and Reis, R. (2011) 'How will e-health affect patient participation in the clinic? A review of e-health studies and the current evidence for changes in the relationship between medical professionals and patients', *Social Science and Medicine*, vol. 72, no. 1, pp. 49–53.

Fortun, M. (2008) *Promising genomics: Iceland and DeCODE genetics in a world of speculation*, Berkeley, CA: University of California Press.

Goldberg, G. (2011) 'Rethinking the public/virtual sphere: The problem with participation', *New Media and Society*, vol. 13, no. 5, pp. 739–754.

Guttmacher, A. and Collins, F. (2003) 'Welcome to the genomic era', *New England Journal of Medicine*, vol. 349, no. 10, pp. 996–998.

Hardey, M. (1999) 'Doctor in the house: The Internet as a source of lay health knowledge and the challenge to expertise', *Sociology of Health and Illness*, vol. 21, no. 6, pp. 820–835.

Harris, A., Kelly, S. and Wyatt, S. (2014) 'Autobiologies on YouTube: Narratives of direct-to-consumer genetic testing', *New Genetics and Society*, vol. 33, no. 1, pp. 60–78.

Harris, A., Kelly, S. and Wyatt, S. (2016) *Cybergenetics: Health genetics and new media*, London: Routledge.

Kaplan, K., Salzer, M., Solomon, P., Brusilovskiy, E. and Cousounis, P. (2011) 'Internet peer support for individuals with psychiatric disabilities: A randomized controlled trial', *Social Science and Medicine*, vol. 72, no. 1, pp. 54–62.

König, R. (2013) 'Wikipedia: Between lay participation and elite knowledge representation', *Information, Communication and Society*, vol. 16, no. 2, pp. 160–177.

Lee, S. (2013) 'American DNA: The politics of potentiality in a genomic age', *Current Anthropology*, vol. 54, supplement 7, pp. S77–S86.

Lock, M., Freeman, J., Sharples, R. and Lloyd, S. (2006) 'When it runs in the family: Putting susceptibility genes in perspective', *Public Understanding of Science*, vol. 15, no. 3, pp. 277–300.

Lupton, D. (2014) 'Critical perspectives on digital health technologies', *Sociology Compass*, vol. 8, no. 12, pp. 1344–1359.

McGowan, M., Fishman, J. and Lambrix, M. (2010) 'Personal genomics and individual identities: Motivations and moral imperatives of early users', *New Genetics and Society*, vol. 29, no. 3, pp. 261–290.

McGuire, A., Diaz, C., Wang, T. and Hilsenbeck, S. (2009) 'Social networkers' attitudes toward direct-to-consumer personal genome testing', *American Journal of Bioethics*, vol. 9, nos. 6–7, pp. 3–10.

Merz, S. (2016) '"Health and ancestry start here": Race and presumption in direct-to-consumer genetic testing services', *Ephemera*, vol. 16, no. 3, pp. 119–140.

Mort, M., May, C. and Williams, T. (2003) 'Remote doctors and absent patients: Acting at a distance in telemedicine?', *Science, Technology and Human Values*, vol. 28, no. 2, pp. 274–295.

Munro, D. (2013) 'Class action lawsuit filed against 23andMe', *Forbes Business*, 3 December.

Nelson, A. and Robinson, J. (2014) 'The social life of DTC genetics: The case of 23andMe', in Kleinman, D. and Moore, K. (eds.) *Routledge Handbook of Science, Technology, and Society*, New York: Routledge.

Niederer, S. and van Dijck, J. (2010) 'Wisdom of the crowd or technicity of content? Wikipedia as a sociotechnical system', *New Media and Society*, vol. 12, no. 8, pp. 1368–1387.

O'Riordan, K. (2013) 'Biodigital publics: Personal genomes as digital media artifacts', *Science as Culture*, vol. 22, no. 4, pp. 516–539.

Oudshoorn, N. (2012) 'How places matter: Telecare technologies and the changing spatial dimensions of healthcare', *Social Studies of Science*, vol. 42, no. 1, pp. 121–142.

Pálsson, G. (2009) 'Biosocial relations of production', *Comparative Studies in Society and History*, vol. 51, no. 2, pp. 288–313.

Parthasarathy, S. (2014) 'Producing the consumer of genetic testing: The double-edged sword of empowerment', in Kleinman, D. and Moore, K. (eds.) *Routledge Handbook of Science, Technology, and Society*, New York: Routledge.

Peterson, A. and Lupton, D. (1996) *The new public health: Discourse, knowledges and strategies*, London: Sage.

Prainsack, B. (2011) 'Voting with their mice: Personal genome testing and the "participatory turn" in disease research', *Accountability in Research*, vol. 18, no. 3, pp. 132–147.

Sharon, T. (2014) 'Healthy citizenship beyond autonomy and discipline: Tactical engagements with genetic testing', *Biosocieties*, vol. 10, no. 3, pp. 295–316.

Singer, E. (2007) 'Your personal genome: George Church wants to sequence your genome', *MIT Technology Review*, 6 December. www.technologyreview.com/s/409153/your-personal-genome.

Sterckx, S., Cockbain, J., Howard, H., Huys, I. and Borry, P. (2013) '"Trust is not something you can reclaim easily": Patenting in the field of direct-to-consumer genetic testing', *Genetics in Medicine*, vol. 15, no. 5, pp. 382–387.

Subramaniam, B. (2014) *Ghost stories for Darwin: The science of variation and the politics of diversity*, Urbana: University of Illinois Press.

Terranova, T. (2000) 'Free labor: Producing culture for the digital economy', *Social Text*, vol. 18, no. 2, pp. 33–58.

Tutton, R. (2014) *Genomics and the reimagining of personalized medicine*, London: Ashgate.

Wathen, N., Wyatt, S. and Harris, R. (eds.) (2008) *Mediating health information: The go-betweens in a changing socio-technical landscape*, Houndmills: Palgrave Macmillan.

Wyatt, S., Harris, A., Adams, S. and Kelly, S. (2013) 'Illness online: Self-reported data and questions of trust in medical and social research', *Theory, Culture and Society*, vol. 30, no. 4, pp. 128–147.

6

Bringing genetics into the clinic: the evolution of genetic testing and counselling

Ilana Löwy

Introduction

In the second half of the twentieth century, new biomedical technologies made it possible to detect harmful mutations, including before birth. People with unexplained symptoms learned the cause of their problems, and some received efficient treatment. Other people became aware that they or their offspring were at risk of developing a genetic disease, and were able to act on this knowledge. Moreover, prenatal diagnosis of genetic diseases/impairments, coupled with a possibility of a selective abortion of affected foetuses, and preimplantatory genetic testing, enabled couples at a high risk of transmitting a hereditary disease to have healthy children. Other consequences of clinical genetic testing were more problematic. With the rapid expansion of such testing and a shift from diagnosis of an already existing condition to a diagnosis of risk, people who believed that they were healthy discovered the presence of a previously unsuspected genetic anomaly and faced difficult choices.

In 1950, discussing the eugenicist's aspirations to eliminate "harmful genes" from populations, the geneticist Herman Muller affirmed that, "none of us can cast stones, for we are all fellow mutants together" (quoted in Paul, 1987). Since practically everyone carries some "defective" genes, eugenic measures, such as a forced sterilisation of the "unfit," were pointless (Stern, 2006). Muller believed nevertheless that in the future technical advances would allow the identification of carriers of especially harmful mutations, then persuade them to refrain from having children (Paul, 1987). This chapter tells the story of the fulfilment of Muller's prophecy and its multiple consequences, some foreseen and others not. Innovations such as genetic testing and counselling came into being in a "full world": they are shaped by their past – the trajectory of the introduction of genetics into the clinics from the mid-twentieth century on – and by their context – other developments in present time biomedicine. The first part of this chapter follows the history of genetic diagnosis in Western Europe and North America, regions where this diagnostic approach became a quasi-routine biomedical technology, while the second part examines present-time uses of genetic tests, especially in the domain of reproductive choices.

Genetic counselling before genetic tests

The observation that some diseases are transmitted in families is very old. From the eighteenth century on, physicians attempted to distinguish inborn impairments which were the result of accidents of pregnancy and childbirth from those produced by "morbid heredity" ("bad blood") (Lopez-Beltran, 2007). In the early twentieth century geneticists studied hereditary diseases and differentiated dominant autosomal, recessive autosomal and sex-linked transmission of these diseases. At that time, the main consequence of identification of such disease in the family was the possibility to tell mutation carriers what their chances of having an affected child were. Such information, geneticists explained, allowed prospective parents to make informed choices. In practice, many geneticists strongly recommended to some people to refrain from procreation (Bosk, 1992; Stern, 2012: Paul, 2017).

In the 1960s and 1970s, the view that some women/couples should be advised not to have children reflected a large consensus about "responsible reproduction." Experts and lay people agreed that reproductive choices have social consequences and are a legitimate matter of social concern. Moreover, individuals who sought the advice of a genetic counsellor were often middle class and educated. It was assumed that the choice to undergo genetic counselling indicated that they were responsible people who took their duties to their future offspring and society seriously, and will make the "right" decision when given accurate facts. Problems had arisen when it was difficult to provide such facts: when is was not sure whether the disease present in the family and/or in one of the parents was hereditary; it was not certain what the pattern of its inheritance was; or it was difficult to assess how severe the impairment of the child who inherited a mutation present in the family would be (Bosk, 1992). When the experts were certain that the condition was hereditary, severe, and all the people who inherit it are affected, they felt confident they could provide a directive guidance to the users of their services. Until the late 1960s, such advice was grounded in probabilistic calculus. Prospective parents were told what their chances of having an impaired child were – typically one out of four for a recessive condition, one out of two for a dominant condition, and one out of two boys for a sex-linked one. Such advice was expected to be given prior to a pregnancy. Not infrequently, however, a woman consulted a genetic counsellor when she was already pregnant. Before the legalisation of abortion this was theoretically "too late," although it is reasonable to assume that some among the women who learned they had a high probability to give birth to a severely impaired child terminated the pregnancy. In the late 1960s and early 1970s, when abortion was legalised in the majority of Western countries, these women could freely elect to have an abortion. However, at that time, abortion for a risk of a hereditary disease became increasingly unnecessary. Physicians were able to determine directly whether a fetus was affected or not. Detection of a risk of hereditary pathology was transformed into a diagnosis of an already existing genetic anomaly.

Chromosomes, genes and diseases: from risk to diagnosis to screening

In 1959, scientists were able, for the first time, to link inborn anomalies to specific mutations. This demonstration was first made in conditions – Down syndrome (trisomy 21), Klinefelter syndrome (47XXY) and Turner syndrome (45X0), which usually are not transmitted in families, but are accidents of the production of sperm and egg cells (Harper, 2006). Nevertheless, the evidence that it is possible to link specific inborn impairments with well defined changes in genetic material of the cell was seen as a watershed. As the geneticist Clarke Fraser put it, from an abstract notion, genes became something that people could see (Fraser, 2008). At first, the possibility to display the presence of an abnormal number of chromosomes was a useful diagnostic device but did not play a role in genetic counselling. Its status had changed dramatically in 1968, with the development of amniocentesis (the sampling of amniotic fluid) and culture of foetal fibroblasts suspended in this fluid (Nadler, 1968). The new method made possible pre-natal detection of an abnormal number of chromosomes, but also of metabolic diseases which modify the chemistry of the cell, such as Tay Sachs Disease (Davidson and Rattazi, 1972).

In the 1970s, prenatal testing was directed to two populations of pregnant women: those aware of the presence of hereditary disease in the family, and those of "advanced maternal age," at higher risk of having a child with Down syndrome, one of the most frequent genetic anomalies. The second group was much larger than the first. The generalisation of prenatal genetic testing was closely intertwined with attempts to identify Down syndrome foetuses. In the early 1970s, the majority of the experts, but also the majority of users of genetic services, implicitly accepted the principle that the prevention of birth of a child with a severe impairment such as Down syndrome would benefit families and the society. In the late 1960s and early 1970s, amniocentesis was a risky procedure, because of a danger of induction of a spontaneous abortion. The use of ultrasound to guide the sampling of amniotic fluid reduced its risks and favoured a widespread diffusion of this approach. In many countries amniocentesis was recom-mended to women over 35, since experts argued that starting at that age, women's risk to have a child with Down syndrome was higher than her risk to lose a healthy foetus (Resta, 2002). Age alone was, however, an imperfect measure of the "risk of Down," since many children with this condition were born to young women. In the 1980s and 1990s, scientists developed blood tests which indicated that a pregnant woman had an increased probability to have a Down syndrome child. These tests were combined with ultrasound "signs of Down." In the early twenty-first century many Western European countries introduced screening for Down, proposed to all the pregnant women. Women were invited to undergo such a screening between 11 and 13 weeks of pregnancy, and those found to be at an increased risk of carrying a trisomic foetus were offered the possibility of undergoing amniocentesis (Wald and Hackshaw, 1997). The balance of prenatal genetic testing shifted again from a diagnosis of an already existing condition to the evaluation of a risk (Löwy, 2014a; Löwy, 2014b).

One of the unanticipated consequences of a widespread diffusion of screening for Down was an increased frequency of diagnoses of other chromosomal anomalies. In some cases counting the number of foetal chromosomes led to detection of a very severe genetic condition, often incompatible with life, such as trisomy 13 (syndrome of Patau) or trisomy 18 (syndrome de Edwards). In other cases it revealed the presence of an abnormal number of sex chromosomes such as Turner syndrome (45X0) and Klinefelter syndrome (47XXY), both linked with infertility and mild to moderate health and cognitive problems. Women who wanted to know whether their future child would have Down syndrome learned instead that their future child would have an inborn condition they had often never heard about, and whose precise consequences were difficult to predict. In the twenty-first century, another typical scenario which leads to unanticipated finding of a genetic anomaly is the detection of a suspected foetal anomaly during a routine ultrasound examination. When this happens, the pregnant woman is often invited to undergo amniocentesis, and genetic testing of foetal cells: not only the counting of the number of chromosomes, but a "whole genome testing" – search for genetic anomalies of the foetus, often through comparative genomic hybridisation (CGH). CGH compares the tested DNA with a reference sample of "normal" DNA, and visualises differences between the two samples. CGH (also called chromosomal microarrays) is especially efficient for the detection of deletions or duplications of chromosome segments (Hogan, 2016). Pregnant women who learn that something can be wrong with their future child rarely refuse an invitation for genetic testing. Sometimes such tests find a genetic anomaly of the foetus, sometimes no such anomaly is found, and sometimes the test uncovers the presence of "variants of unknown/uncertain clinical significance" (VUS), which may either reflect a normal variability of the human genome, or indicate the presence of yet unknown pathology (Timmermans et al. 2017). In other cases still, genetic testing of the foetus diagnoses the presence of an anomaly, which, upon future investigations, reveals the – until then unsuspected – presence of a hereditary condition in the family, a "secondary" or "incidental" result of genetic testing (Robyr et al. 2006).

Genetic testing in the clinics and "secondary findings"

Unintended and unanticipated "secondary" findings are one of the most challenging consequences of the massive diffusion of issues of genetic tests, especially those conducted in a reproductive context. Besides prenatal diagnosis, secondary findings may be the result of attempts to find out the reason for the presence of a pathology or unexplained developmental delay in a child, infertility, and study of health complaints in adults. When the genetic condition is not hereditary – for example when an infertile man discovers that he has Klinefelter syndrome – the consequences of its diagnosis are akin to those of a diagnosis of other chronic impairments. However when a condition is transmitted in the family, the consequences of its detection go beyond the diagnosed individual, especially when family members learn about significant reproductive risks or a presence of serious health problem which may appear late in life. Since many hereditary conditions cannot be cured, their diagnosis may be a source of important distress, especially for people who previously saw themselves as healthy (Robyr et al. 2006).

Another source of secondary finding is curiosity-driven genetic tests, like those performed by the company *23andMe*. Such tests can reveal unsuspected mutations, like mutations in BRCA 1 and 2 genes, which greatly increase a woman's risk of developing breast and ovarian cancer. Until recently, the main indications for undergoing BRCA testing were a family history of BRCA associated malignancies and a diagnosis of breast cancer, especially in a younger woman, or of histological type frequently associated with BRCA mutations. Some researchers also proposed a systematic screening for BRCA mutations of all the individuals belonging to a population with identified "founder" (population specific) BRCA genes, such as the Ashkenazi Jews (King et al., 2014). From 2008 on, *23andMe* included testing for BRCA mutations in their genetic disorder risk assessment. Researchers linked to the *23andMe* company claimed that the great majority of the people who accidentally learned about the increased risk of cancer for themselves or members of their family were very pleased to learn that it is possible to "do something" – mainly a preventive ablation of ovaries or breasts – to reduce the danger of the development of a malignancy (Francke et al., 2013). In November 2013 the FDA banned the provision of medically relevant information by *23andMe* in the US (although this service remained available in other countries). FDA experts justified their decision with the claim that *23andMe* fails to provide adequate counselling for people who discovered that they are mutation carriers. Some leading US geneticists protested against this ban. They argued that consumer-directed genetic testing democratises health care, empowers users of health services and encourages health-promoting behaviour (Green and Farahany, 2014). Their argument faithfully follows the logic which led to the development of direct to consumer (DTC) genetic testing (Kelly et al., this volume).

Dilemmas produced by "secondary findings" of genetic tests, those performed in the clinics and those marketed directly to consumers, are predicted to intensify following the diffusion of whole exome/whole genome testing, and of non-invasive prenatal testing (NIPT). The cost of whole exome testing diminished greatly from mid-2000s on, thanks to the diffusion of next generation sequencing (NGS). New genomic technologies are increasingly employed in the search of suspected genetic origins of human pathologies. Thus, in 2005 the American College of Medical Genetics, recommends a systematic inclusion of genetic and genomic studies in each clinical evaluation of intellectual delays in a child, with the underlying assumption that in many cases such delays originate in a genetic anomaly (Shaffer et al., 2005).

From the 1970s on, physicians, aware of the presence of foetal cells in the pregnant woman's blood, tried to isolate these cells in order to study their hereditary material without the risks to the foetus linked with amniocentesis (Hertzenberg et al., 1979). These efforts were unsuccessful, as were the early attempts to isolate sufficient purified foetal DNA from the maternal blood to allow reliable diagnosis of genetic defects. In the twenty-first century advances in molecular biology made possible the successful separation of the signal (foetal DNA) from the noise (maternal DNA), an approach named non-invasive prenatal testing (NIPT) (Go et al., 2011). In 2014, the main use of NIPT was screening for the presence of an abnormal number of chromosomes, above all Down syndrome, Patau syndrome and Edwards syndrome (Bianchi et al., 2014a). Upon its introduction in 2012 NIPT was presented above all as a safer approach to the already existing screening for Down syndrome. The great advantage of NIPT, the companies which manufactured this test argued, is to greatly reduce the number of amniocenteses for Down risk. However, in spring 2014 several commercial producers of non–invasive tests had introduced testing for several genetic syndromes that involve deletions (the loss of a fragment of a chromosome); such testing is expected to be extended to other chromosomal anomalies as well. Syndromes links with some of these anomalies, such as DiGeorge syndrome (22q11.2del), have variable expression. Genetic data do not predict the severity of the expected impairment, making decisions about the continuation of pregnancy challenging (Bianchi et al., 2014b).

In 2015 European and US societies for Human Genetics expressed doubts about the widespread use of NIPT to diagnose rare or moderately severe genetic conditions and advised – at least temporarily – to limit the use of this technology to screening for trisomies 21, 18 and 13 (Dondorp et al., 2015). Nevertheless, in countries in which the uptake of NIPT is mainly regulated by the market, such as the United States or Latin American countries, many women undergo a "complete" NIPT testing, which includes testing for the presence of an abnormal number of sex chromosomes and deletions (Zeng et al. 2017). Experts predict that in the not too normal far future it will be possible to offer pregnant women a risk-free "reading" of the entire foetal DNA early in the pregnancy (Yurkiewicz et al., 2013). Pregnant women and their partners may receive a flood of confusing and potentially destabilising information about the possible physical and intellectual problems of their future child, and face difficult choices (Greely, 2011; Ravitsky, 2015). For women who live in industrialised countries, and important segments of those who live in intermediary ones, decision of whether to become pregnant, how to become pregnant, and what the fate of an already existing pregnancy will be, became increasingly entangled with new genetic and genomic technologies. At the same time, an increasing number of people learn that their health problems stem from changes in their genetic material, and this knowledge may change their perception of self and modify their family dynamics.

On being "mutants together": biosociality in the era of managed fear

Muller's 1950 prediction that advances in genetic knowledge will facilitate the identification of carriers of deleterious mutations was accurate, but not his prediction that carriers of "harmful" mutations could then be persuaded to refrain from having children a development which will reduce the number of children born with severe hereditary diseases. Such a reduction did take place, at least in Western societies, but it resulted, at least partly, from possibilities open to mutation carriers, to elect biomedical interventions which will allow them to have healthy biological children: preconception, preimplantation and prenatal testing (Cowan, 2006). This statement needs to be qualified. Not all the people who carry potentially pathological mutations wish to employ these techniques. Some are influenced by the "expressionist argument," that the prevention of birth of a child with a given genetic impairment is a powerful statement that life of people with this impairment – often other children of the couple or close relatives – is not worth living. Some women/ couples refuse therefore prenatal or preimplantatory diagnosis and affirm that they are willing to accept any child they will have (Asch, 1999; Boardman, 2014). Others decide not to have biological children, adopting an attitude theoretically made obsolete by new genetic and reproductive technologies (Kelly, 2008). It is also important to remember that these new genetic and reproductive technologies are accessible only to a small fraction of the world's population. In developing and intermediary countries, these technologies are accessible only to the upper middle class. Moreover abortion is illegal in many countries, severely limiting womens' reproductive choices. In many societies only the affluent mutation carriers can decide whether they wish to halt the transmission of their mutation to the next generation.

Even in societies in which the majority of health care users have access to new medical technologies, advances in clinical genetic testing came with a price. Knowledge of one's biological risks is not always a blessing (Rosenberg, 2009). We may be all "fellow mutants together," but, especially in reproductive health area, each "fellow mutant," may face complicated and stressful choices. With the predicted increase of the number of prenatally diagnosed mutations, such choices may become even more challenging. Technological progress, the Dutch media expert José Van Dijk proposes, does not necessarily make the decision process more transparent or rational, nor do they allow the users of these technology more freedom to choose. Sometimes just the opposite may be true: more advanced diagnostic technologies, such as genetic testing, may make choices more complex and may limit patients' autonomy (Van Dijck, 2005, p. 116).

Geneticists and genetic counsellors are aware of the presence of problems and dilemmas generated by technological change. They are also aware of the professionals' tendency to support the rapid extension of scope of genetic tests. The official explanation for such extension is that patients/ users of health services push for more testing. The reality is more complex. Professionals promote genetic tests because they believe that they are useful, but also for other reasons: because commercial laboratories propose them, because such tests expand the domain of their professional jurisdiction, and because some fear legal consequences of failing to offer a test. Many patients assume that since specialists are offering a test, it must be a good thing (Resta, 2017), and yet it is not necessarily so. Genetic tests help many patients, especially those who receive a diagnosis of a curable/treatable condition but also those who can better understand the source of their or their children's health problems (Latimer, 2013). On the other hand, as the genetic counsellor Susan Markens explains, with the continuous focus on maximisation of health and minimisation of risk via "responsible" individual choices, and the specialists' enthusiastic adoption of ever-expanding scope of genetic testing, "it just becomes much more complicated" (Markens, 2013).

References

Asch, A. (1999). Prenatal diagnosis and selective abortion: a challenge to practice and policy. *American Journal of Public Health*, 89(11), pp. 1649–1657.

Bianchi, D., Oepkes D. and Ghidini, A. (2014a). Current controversies in prenatal diagnosis 1: Should noninvasive DNA testing be the standard screening test for Down syndrome in all pregnant women? *Prenatal Diagnosis*, 34(1), pp. 6–11.

Bianchi, D., Van Mieghem, T., Shaffer, L. G., *et al.* (2014b). In case you missed it: The Prenatal Diagnosis section editors bring you the most significant advances of 2013. *Prenatal Diagnosis*, 34(1), pp. 1–5.

Boardman, F. K. (2014). The expressivist objection to prenatal testing: The experiences of families living with genetic disease. *Social Science and Medicine*, 197, pp. 18–25.

Bosk, C. (1992). *All God's Mistakes: Genetic Counseling in a Pediatric Hospital*, Chicago University Press.

Cowan, R. (2006). *Heredity and Hope: The Case for Genetic Screening*, Harvard University Press.

Davidson, R. and Rattazi, M. C. (1972). Review: Prenatal diagnosis of genetic disorders. *Clinical Chemistry*, 18, pp. 179–187.

Dondorp, W., de Wert, G., Bombard, Y. *et al.* on behalf of the European Society of Human Genetics (ESHG) and the American Society of Human Genetics (ASHG) (2015). Non-invasive prenatal testing for aneuploidy and beyond: challenges of responsible innovation in prenatal screening. *European Journal of Human Genetics*, 23, pp. 1438–1450.

Francke, U., Dijamco, C., Kiefer, A. K. *et al.* (2013). Dealing with the unexpected: consumer responses to direct-access BRCA mutation testing. Feb. 12; 1:e8. doi: doi:10.7717/peerj.8.

Fraser, F.C. (2008). On mice and children: Reminiscences of a teratogenicist. *American Journal of Medical Genetics*, part A, 146A (2008), pp. 2179–2202.

Go, A. T., van Vogt, J. M. and Oudejans, C. B. (2011). Non-invasive aneuploidy detection using free fetal DNA and RNA in maternal plasma: Recent progress and future possibilities. *Human Reproduction Update*, 17, pp. 372–382.

Greely, H. (2011). Get ready for the flood of fetal gene screening. *Nature*, 469, pp. 289–290.

Green, R. and Farahany, N. (2014). Regulation: The FDA is overcautious on consumer genomics. *Nature* 505(7483), pp. 286–287.

Harper, P. (2006). *First Years of Human Chromosomes: The Beginnings of Human Cytogenetics*, Oxford: Scion.

Hertzenberg, L., Bianchi, D., Schroder, J. *et al.* (1979). Fetal cells in the blood of pregnant women: detection and enrichment by fluorescence-activated cell sorting. *Proceedings of the National Academy of Science*, 76, pp. 1453–1459.

Hogan, A. (2016). *Life Histories of Genetic Diseases: Patterns and Prevention in Postwar Medical Genetics*, Baltimore: Johns Hopkins University Press.

Kelly, S. (2008). Choosing not to choose: reproductive responses of parents of children with genetic conditions or impairments. *Sociology of Health and Illness*, 31(1), pp. 81–97.

King, M. C., Levy-Lahad, E. and Lahad, A. (2014). Population-Based Screening for BRCA1 and BRCA2. *JAMA*, 312(11), pp. 1091–1092.

Latimer, J. (2013). *The Gene, the Clinics and the Family*, London: Routledge.

Lopez-Beltran, C. (2007). The medical origins of heredity. In Muller – Wille, S. and Rheinberger, H.J. (eds), *Heredity Produced: At the Crossroad of Biology, Politics and Culture, 1500–1870*. Cambridge Mass.: MIT Press, pp. 105–132.

Löwy, I. (2014a). How genetics came to the unborn: 1960–2000. *Studies in History and Philosophy of Biological and Biomedical Sciences*, 47, pp. 154–162.

Löwy, I. (2014b). Prenatal diagnosis: The irresistible rise of the 'visible foetus'. *Studies in History and Philosophy of Biological and Biomedical Sciences*, 47, pp. 290–299.

Markens, S. (2013). "It just becomes much more complicated": Genetic counselors' views on genetics and prenatal testing. *New Genetics and Society*, 32, pp. 302–321.

Nadler, H. (1968). Antenatal detection of hereditary disorders. *Pediatrics*, 42, pp. 912–918.

Paul, D. (2017). Norm change in genetic services: How the discourse of choice replaced the discourse of prevention. *Varia Historia, Belo Horizonte*, 33(2017), pp. 21–47.

Paul, D. (1987). Our "load of mutation" revisited, *Journal of the History of Biology*, 20(3), pp. 321–335, quotation p. 331.

Ravitsky, V. (2015). Non invasive prenatal testing (NIPT): identifying key clinical, ethical, social, legal and policy issues. Background paper, Nuffield Council on Bioethics, November 2015. http://nuffield bioethics.org/wp-content/uploads/NIPT-background-paper-8-Nov-2015-FINAL.pdf. (accessed March 31, 2017).

Resta, R. (2002). Historical aspects of genetic counseling: Why was maternal age 35 chosen as the cut-off for offering amniocentesis? *Medicina nei Secoli*, 14, pp. 783–811.

Resta, R. (2017). Are we ready for this? The DNA exchange, February 6, 2017, https://thednaexchange.com/2017/02/06/are-we-ready-for-this/ (accessed March 31, 2017).

Robyr, R., Bernard, J. P., Roume, J. and Ville, Y. (2006). Familial diseases revealed by a fetal anomaly. *Prenatal Diagnosis*, 26, pp. 1224–1234.

Rosenberg, C. (2009). Managed fear. *The Lancet*, 373(9666), pp. 802–803.

Shaffer, L. G. on behalf of the American College of Medical Genetics (ACMG) Professional Practice and Guidelines Committee (2005). American College of Medical Genetics guideline on the cytogenetic evaluation of the individual with developmental delay or mental retardation. *Genetics in Medicine*, 7, pp. 650–654.

Stern, A. (2006). *Eugenic Nation: Faults and Frontiers of Better Breeding in America*, Berkeley: California University Press.

Stern, A. M. (2012). *Telling Genes: The Story of Genetic Counseling in America*. Baltimore: Johns Hopkins University Press.

Timmermans, S., Tietbohl, C., and Skaperdas, E. (2017). Narrating uncertainty: Variants of uncertain significance (VUS) in clinical exome sequencing. *BioSocieties*, 12(3), pp. 439–458.

Van Dijck, J. (2005). *The Transparent Body: A Cultural Analysis of Medical Imaging*, Seattle: The University of Washington Press.

Wald, N.J. and Hackshaw, A. K. (1997). Combining ultrasound and biochemistry in first semester screening for Down's syndrome. *Prenatal Diagnosis*, 17, pp. 821–829.

Yurkiewicz, Ł.,Korf, B. and Soleymani Lehmann, L. (2013). Prenatal whole-genome sequencing – is the quest to know a fetus's future ethical? *New England Journal of Medicine*, 370(3), pp. 196–197.

Zeng, X., Zannoni, L., Löwy, L. and Camporesi, S. (2017). Localizing NIPT: Practices and meanings of non-invasive prenatal testing in China, Italy, Brazil and the UK. *Ethics, Medicine and Public Health*, http://dx.doi.org/10.1016/j.jemep.2016.06.004.

7

From quality control to informed choice: understanding "good births" and prenatal genetic testing in contemporary urban China

Jianfeng Zhu and Dong Dong

Genetic testing and genetic screening as two selective reproduction technologies (Gammeltoft and Wahlberg 2014) are frequently used in prenatal health care around the world, and many scholars have examined prenatal genetic testing in different local contexts (e.g., Rapp 1999; Rothman 1993; Sleeboom-Faulkner 2010). In this chapter, we will trace the historical changes in prenatal genetic testing practices in China since the 1990s. We focus particularly on China as an exemplar of the significance that selective reproduction technologies have in social, cultural and regulatory terms, and the role that they play in shaping governance practice, particularly within traditions and communities exposed to non-Western medical cultures. Scholars writing about Chinese genetic testing often take family as an important element of traditional Confucianism culture (Sui and Sleeboom-Faulkner 2010). While agreeing with this approach, we argue that health professionals often play an important role in families' decision-making. Hence, our overview includes not only the viewpoint of patients and families affected by prenatal genetic testing (and resulting interventions), but also the voices of bioethicists and frontline health workers from local hospitals. These are important actors who not only contribute to producing discourses regarding population quality and "good births," but also actively practice them in their daily work.

National context: population control

One of the most important national agendas in post-Mao China is to use all possible technologies to speed up the process of modernisation, the goal of which cannot be achieved without having the quality of population improved. The government states that "the issue of population is essentially a problem of development and can only be solved ultimately through economic, social and cultural development" (The State Council Information Office 2000). As a result, "control population growth and improve population quality" constitutes a major discourse underlining China's population policy and is an integral part of the state's modernisation program (Greenhalgh 2005). Officially initiated in 1980, the one-child policy created and projected intense hope and anxiety onto pregnancy. When it was fully abandoned at the end of 2015 and replaced by the two-child policy, the state's control over the "quality" (*suzhi* 素质) of its population started to place more emphasis on selective reproduction technologies, in order to raise the quantity of births without sacrificing the quality of the whole population. Such a national dream of strengthening a China Nation through rationally planning population quality and quantity appears to be extremely appealing not only to the state but also to families and individuals; for the individual family considers their fate is closely tied to the future of the Chinese nation. In China, the desire for a perfectly healthy baby is widespread among urban middle-class families. This desire is grounded in their imagination of the future, hierarchical deployment of resources and power at the global level. Beneath this desire, there also lies urban Chinese couples' deeper fear of "losing the competition at the starting line." The post-Mao Chinese government has put a great deal of effort into speeding up modernisation, laying out the ladder of progress at the global level, with the capitalist Euro-American world on top and the former socialist world in the middle. The couples we have met so far have taken this official narrative of linear development to heart and genuinely believe that China has already been left behind, due to the country's semi-colonialist history and Mao's chaotic communist past. The political narrative of "catching up" with the West and the discourse of population control have both become predominant in ordinary people's everyday lives, which shapes how bioethicists, medical professionals and couples understand their stances, practices and decisions. This is a very similar case to those documented by Prainsack (2006) in her studies of the prevalent political narratives of "demographic threat" in Israel, related to embryonic stem cell research and human cloning. Concerns about population quantity and quality are frequently used to justify reproduction decisions regarding wanting one "perfect" baby, because having a perfectly healthy baby means granting the newborn certain economic, social and cultural prestige that can lead to a good start in the future global competition (Zhu 2013).

International criticism and national response

In 1995, China adopted the Maternal and Infant Health Care Law (the Law). According to the Chinese state, the Law aims to ensure the health of mothers and infants and improve the quality of the newborn population. This law immediately evoked a considerable amount of criticism in Western media, due to the mandatory pre-marital medical check-ups it established. According to the Law, it was mandatory for to-be-married couples to go through medical check-ups before registration; for those who "have been diagnosed with certain genetic diseases of a serious nature, which is considered to be inappropriate for child-bearing from a medical point of view, the two may be married only if both sides agree to take long-term contraceptive measures or to take ligation operations for sterility" (Article 10). This particular article was and still is interpreted as a type of Chinese eugenics, solidified in law. Scholars who examine China's *yousheng* policy either define it as "new eugenics" (Handwerker 2002) or footnote in their work the debate over the English translation of *yousheng* as eugenics (Wahlberg 2016). It is worth providing more detailed discussions of the debate because, perhaps for the first time, Chinese bioethicists are currently directly engaged in Chinese state policymaking and are interacting with the international community in this field. Furthermore, we hope to use our ethnographic encounters with doctors in clinics to demonstrate how deeply nationalist ideas of Chinese characteristics are embedded in the medical field and, once again, to stress that medical knowledge and practices cannot be separated from their cultural contexts.

In responding to the Western media's criticism, Renzong Qiu, the Director of the Bioethics Program at the Chinese Academy of Social Sciences, claimed that translating the Law as eugenics was misunderstanding the Chinese word *yousheng* (Qiu 1999). According to Qiu, *yousheng* has two meanings: healthy births and eugenics programs. The former is associated with child-rearing and the latter is linked to Nazi policies. He believes that the English translation of the Law as eugenics reflects this second meaning. He argued that, for a policy to entail eugenics, it "must first reject individual consent and, second, be based on racism" (Qiu 1999, 30). Since Qiu believes that China's Maternal and Infant Health Care Law lacks these two conditions, it therefore cannot involve eugenics. Qiu also emphasised that the Law was not motivated by racism but by "a desire to reduce birth defects" (Qiu 1999). Qiu's opinions were widely broadcast in China and his arguments were not only supported by the Chinese state but were also embraced by most OB/GYNs and clinical geneticists in China.

Discussions with frontline medical professionals about *yousheng* and their strong objections to seeing *yousheng* as eugenics, as well as our own observations of various *yousheng* practices in contemporary China, indicate that eugenics is a rather contested concept in different cultural contexts. Undoubtedly, *yousheng* can be translated as eugenics when looking at texts of official documents, laws and rules. It can also be argued that *yousheng* is a new form of governmentality that normalises and disciplines people's lives. In addition to this, we are well aware that, under *yousheng*'s influence, there will always exist the potential for discrimination based on the idea of "normal/abnormal" or "low/high quality bodies." Such potentiality is always ready to be actualised. However, when attempting to replace *yousheng* in Chinese with the English term "eugenics," we were contested by the doctors that we met in our fieldwork; this became a significant ethnographic moment, during which we were forced to check our taken-for-granted academic assumptions. Is there anything lost in translation? Is it simply a question of translation?

Obviously, the doctors' resistance speaks directly to the Euro-American media's accusation that China solidifies its eugenics practices in law. The doctors' struggles with the English translations shows their strong intentions of refusing to be subjected to the same discourses rooted in a foreign language. By refusing to speak the same language, both literally and metaphorically, they are indeed protesting against being enrolled in a game in which only Westerners have the right to set up the rules. Nevertheless, their resistance is perhaps nothing but a product of the national subject-making project of the Chinese state. Frank Dikötter (1998) examined Chinese biology and reproduction science texts from Late Imperial China to the People's Republic and shows how the understanding of eugenics in China reflects nationalist thoughts. Indeed, we can still detect nationalist elements when encountering informants in our field. We can further argue that doctors' insistence on calling their practices and the state's population control policy *yousheng* – in contrast to the Western discourse of eugenics – comes from their subjection to the Chinese state's post-Mao nationalist concerns. From the very beginning of China's economic and social reforms in the early 1980s, in order to legitimise the communist party's political control over China, the Chinese government maintained that its market-oriented reforms were not identical to capitalist reforms; instead, it aimed to establish a socialist society with "Chinese characteristics." Labelling something national as having "Chinese characteristics," in order to distinguish the self from the other, reflects Chinese nationalism, a relational ideology based on imagining China with respect to others (Dai 2001). Within such a context, we argue that the moment at which medical professionals juxtapose *yousheng* in Chinese with eugenics in English reifies their national subject position, as the state intends. Such subjects are willing to morally justify the practice of *yousheng* for the sake of an imagined national body as the Chinese population at an aggregated level. From this perspective, the Chinese body needs to be strengthened by all possible means. This means particularly the adoption of Western technologies to ensure that every individual is born perfectly healthy.

In October 2003, the government abolished the compulsory premarital health check-up that couples were required to have since 1986 with the intent of improving the quality of births. This change is regarded as a great improvement and is welcomed by humanitarians both within and outside China. However, medical professionals and *yousheng* experts are worried about the increasing defected birth rate and they constantly voice their concerns in the mass media. In order to ensure the "quality" of newborns, the Chinese state has introduced and promoted various prenatal testing technologies through its well-established perinatal care system. To have a baby in China requires the state's permission. A birth permit (*zhun sheng zheng*, 准生证) is indispensable and must be obtained before delivering the baby in a public hospital. The same permit is also required for a birth certificate application and the child's *hukou* registration (Li et al. 2010). In addition to this, a perinatal-care card (PCC, *weichan baojian ka* 围产保健卡) has also become mandatory, since 1996. Acquiring the birth permit and the PCC often takes weeks or even months. In order to get the PCC, couples have to go through various procedures, including attending maternal and infant health care education classes and perinatal medical check-ups. The cumbersome bureaucratic procedures promote the message that maternal health is not an individual business at all, but the state's affair. Through these bureaucratic documents and procedures, the Chinese government exerts its regulatory power over the whole population, attempting to recruit every possible newborn into its system of quality control.

When couples are recruited into the prenatal health care system, it starts what Susan Green-halgh called "bureaucratic actualization" (2003, 199), whereby disciplinary power takes over, as opposed to the state's top-down mandatory rules. The pink PCC booklet guides expectant parents through maternal health education classes, as well as prenatal medical check-ups, offering a platform for the disciplinary power of knowledge to work. In the education classes, expectant couples are informed of possible issues that might occur if they do not undertake the proper recommended health care. They are taught to think of health, the body, the self and the foetus in terms of probabilities, risks and economy, which subjects them to the medical gaze, a field of reproduction knowledge and a field of visibility. Women, particularly those from the urban middle class, learn to know themselves, their bodies and their babies, and they believe it is their responsibility to take care of the self and the baby within. In this field of power, only knowledge, expert, modernity and economy matter. Knowledge is replacing the state's mandatory laws. As women desire more knowledge, time, space and emotion, their desire, along with everything else, is under the gaze of the state. In short, Foucault's panoptic schema works in this context to make every individual woman internalise power relationships and finally become "the principle of [her] own subjection" (Foucault 1979, 203). The Chinese government's regulatory power over population quality as a whole combines with medical experts' knowledge and power in regard to urban couples and their families, who up until 2016 were only allowed to have one "perfect" child. The co-existence of, first, the state's top-down surveillance over expectant couples' pregnancies and, second, medical professionals' permeable medical advice is the most salient feature of population quality control during the first decade of the twenty-first century in urban China.

The two-child policy and healthy China

In 2016, the Chinese government relaxed the family planning policy and allowed all couples to have a second child. The two-child policy facilitates the spread of the discourse of choice in regard to reproduction. The government justifies this policy change by presenting statistical evidence of China as an increasingly aging society and the consequent urgent need to increase its future labor force. From August 19 to August 20, 2016, in a national meeting on health in

Beijing, the Chinese president, Xi Jinping, called for the full protection of the people's health. He stated that an "all-round moderately prosperous society could not be achieved without people's all round health" (Xinhua News 2016). He emphasised the need to promote healthy lifestyles, strengthen health services, improve health protection, build a healthy environment and develop health-related industries. He urged the inclusion of health in government policies, to ensure that all people enjoy the health benefits of these policies. He stressed the ways in which the government should perform its duties in basic medical services and how the market should be revitalised in the fields of non-basic services. The concept of a "healthy China" emerged as one crucial aspect of the Chinese dream tied to each individual's lifestyle.

Although a family can now choose to have one or two children, the notion of "population quality" has not loosened. On September 12, 2016, the 12th "Birth Defects Prevention Day," data reported by the Ministry of Health were widely circulated; approximately 5.6 percent of newborns in China had birth defects, rising from 4.0 percent in the 1990s and affecting approximately 900,000 infants each year. Experts believe that this increase in birth defects may be linked to the withdrawal of compulsory premarital health checks in 2003, as well as an increase in the number of women who have their children at an older age. What is new here is that environmental pollution and individual lifestyle choices, such as addictions to tobacco, alcohol, drugs and online games, are laid out as the major causes of birth defects. In other words, lifestyle choices are emphasised as important contributions to birth defects. To connect this angle of individual responsibility to Jinping's speech on the concept of the "healthy China" described above, the stress placed on population health as a precondition of the Chinese dream of resurging as a strong nation obliges every individual to take a prenatal check-up and abandon an unhealthy lifestyle, to assure the birth of a healthy baby. "Only a healthy baby can lead to a happy family and then to a harmonious society," says a propaganda film made and aired in Henan province (the Health and Family Planning Commission of Henan Province and the Office of Talents Work of Henan Province 2017). Under the framework of a "healthy China," one's personal lifestyle choices are tied to the resurgence of a national Chinese future and family; it becomes a medium through which the individual must carry out their daily activities and through which the government exerts its power.

From controlling quality to informed choice

Accordingly, the Chinese state is attempting to gradually change its discursive strategies in the eugenics form of "quality control" to the notion of "reproductive insurance." Personal choice is emerging from this insurance-oriented regime. This discursive strategy is facilitated by a series of policies in the fields of reproduction and medicine, including the two-child policy, the concept of a healthy China, governmental initiatives associated with precision medicine and the government's support of the genetic sequencing industry. In parallel with the state's *yousheng* propaganda, which attributes personal lifestyle as one of the causes of an increase in birth defects, is the emergence of a range of "choice" discourses, which are not only promoted by the state but are also expedited by the involvement of the market, especially within the emergent health industry, for example, with genetic companies that further provide various price-tagged "choices" for families.

In clinical practice, geneticists now employ a non-directive style of communication and provide "options" for couples and their family members. Doctors who used to work for "good birth clinics" (*yousheng menzhen* 优生门诊) are now taking "re-educational" training programs on prenatal genetic screening and diagnosis. The departments they work at are now called "prenatal diagnosis clinics" (产前诊断门诊) or "genetic counselling clinics" (遗传咨询门诊).

These types of names emphasise the role of doctors as medical professionals, rather than as implementers of the state's policies, even though, in reality, both roles often co-exist. The communication style of doctors has also changed from, for example, discussing a baby with birth defects as an "economic burden" to society, to a discourse filled with the notion of "informed choices." Doctors often maintain that they need to be neutral and make no judgments, offering professional information and leaving decisions to be made by patients.

Choice aims to cultivate a self-managing subject (see Bliss, Chapter 3). Governing is supposed to cost less and be far more efficient. However, the self-managing subject, in reality, does not always follow the state's plan of producing a "high quality" baby. In reality, choices made by couples and their families vary, as shown in many other scholars' work (e.g., Ahmed et al. 2012 in the UK; García et al. 2008 in the Netherlands; and Santalahti et al. 1998 in Finland). As Rayna Rapp states, authoritative knowledge becomes "a way of organising power relations" and makes the choices in question "literally unthinkable in any other way" (Rapp 1992, quoted from Jordan 1997, 55). Therefore, we would like to point out that, without addressing the structural challenges faced by the family, simply proposing to enhance the communication skills of doctors by means of "informed choices" will only scratch the surface of the deeply embedded social problems in the field of prenatal genetic testing in China.

Conclusion

To summarise, by tracing the historical trajectory of prenatal testing in China from the 1990s to the twenty-first century, we show how the family's desire for a perfectly healthy baby and the national desire for a "high quality" population are closely linked to legitimising *yousheng* practices in clinics. It appears that there is a very clear trend involving changing regimes over the last 20 years in the field of prenatal genetic testing, leading towards a neoliberal ideology. Nevertheless, following Rofel's (2007) suggestion that one should not assume any uniformity of experiences in regard to globalisation or neoliberalism in any countries or regions, we attend to local contexts and practices and show the complexity of notions of eugenics, "good births" and informed choices. As we have demonstrated throughout this chapter, individual families' goals of having a healthy baby and the state's goal of raising the "quality" of the population seemingly fit together perfectly. However, in reality, family members negotiate their understandings of a "high quality" baby with the state's understanding of a "high quality" population through the mediation of the doctors they meet in clinics. As important mediators, clinical doctors act as links between family and state. Our ethnographic engagement clearly reveals that doctors' opinions and practices regarding *yousheng* have been changing in correspondence with policy and official discourses.

During the 1990s, through to the early twenty-first century, doctors have considered themselves to be "quality inspectors" (Zhu 2013). By the time we have written this chapter, they will have continued to emphasise "informed choice" in one's daily routine. Before, they did not give a second thought to aborting an "abnormal" foetus, due to China's strict one-child policy; now, they frequently use "life" to justify their decisions regarding what kind of information they want to convey to the family. These rapid and significant changes witnessed in our field compel us to rethink China's role in the current postgenomic era. Obviously, with further decentralisation of global capitalism and the changing political landscapes everywhere in the world, the Chinese state has changed from a passive reactor, as exemplified above, in the crisis regarding eugenics laws, to an active leader in regard to embracing the genetics industry and advocating a national gene bank. China is more willing than ever to participate in this new game at any cost, despite ordinary people's confusion and the resulting chaos on the ground.

We hope this chapter can open up more discussions in regard to the topics of genetic testing, which could move beyond the prior framework of globalisation and localisation, Western technology and local culture.

References

Ahmed, Shenaz, Louise D. Bryant, Zahra Tizro, and Darren Shickle. 2012. "Interpretations of Informed Choice in Antenatal Screening: A Cross-Cultural, Q-Methodology Study." *Social Science & Medicine* 74(7): 997–1004.

Dai, Jinhua. 2001. "Behind Global Spectacle and National Image Making: The Tides of Nationalism." *Positions: East Asia Cultures Critique* 9(1): 161–186.

Dikötter, Frank. 1998. *Imperfect Conceptions: Medical Knowledge, Birth Defects, and Eugenics in China*. New York: Columbia University Press.

Foucault, Michel. 1979. *Discipline and Punish: The Birth of the Prison*. Translated by Alan Sheridan. New York: Vintage.

Greenhalgh, Susan. 1993. "The Peasantization of the One-Child Policy in Shaanxi." In *Chinese Families in the Post-Mao Era*, edited by Deborah Davis and Stevan Harrell, 219–250. Berkeley: University of California Press.

Greenhalgh, Susan. 2003. "Planned Births, Unplanned Persons: 'Population' in the Making of Chinese Modernity." *American Ethnologist* 30(2): 196–215.

Greenhalgh, Susan. 2005. "Missile Science, Population Science: The Origins of China's One-Child Policy." *The China Quarterly* 182: 253–276.

Gammeltoft, Tine M., and Ayo Wahlberg. 2014. "Selective Reproductive Technologies." *Annual Review of Anthropology* 43: 201–216.

García, Elisa, Danielle R. M. Timmermans, and Evert van Leeuwen. 2008. "The Impact of Ethical Beliefs on Decisions about Prenatal Screening Tests: Searching for Justification." *Social Science & Medicine* 66(3): 753–764.

Handwerker, Lisa. 2002. "The Politics of Making Modern Babies in China: Reproductive Technologies and the 'New' Eugenics." In *Infertility Around the Globe: New Thinking on Childlessness, Gender, and Reproductive Technologies*, edited by C. M. Inhorn and F. van Balen, 298–314. Berkeley: University of California Press.

Jordan, Brigitte. 1997. "Authoritative Knowledge and its Construction." In *Childbirth and Authoritative Knowledge: Cross-Cultural Perspectives*, edited by R. Davis-Floyd and C. F. Sargent, 55–79. Berkeley: University of California Press.

Li, Shuzhuo, Yexia Zhang, and Marcus W. Feldman. 2010. "Birth Registration in China: Practices, Problems and Policies." *Population Research and Policy Review* 29(3): 297–317.

Prainsack, Barbara. 2006. "'Negotiating Life': The Regulation of Human Cloning and Embryonic Stem Cell Res in Israel." *Social Studies of Science* 36(2): 173–205.

Qiu, Renzong. 1999. "A Concern for Collective Good." *The UNESCO Courier* 52(9): 30.

Rapp, Rayna. 1992. "Commentary on 'Birth in Twelve Cultures: Papers in Honor of Brigitte Jordan.'" Paper presented at the annual meeting of the American Anthropological Association, San Francisco, December 1992.

Rapp, Rayna. 1999. *Testing Women, Testing the Fetus: The Social Impact of Amniocentesis in America*. New York: Routledge.

Rofel, Lisa. 2007. *Desiring China: Experiments in Neoliberalism, Sexuality, and Public Culture*. Durham: Duke University Press.

Rothman, Barbara Katz. 1993. *The Tentative Pregnancy: How Amniocentesis Changes the Experience of Motherhood*. New York: Norton.

Santalahti, Päivi, Elina Hemminki, Anne-Maria Latikka, and Markku Ryynänen. 1998. "Women's Decision-Making in Prenatal Screening." *Social Science & Medicine* 46(8): 1067–1076.

Sleeboom-Faulkner, Margaret, ed. 2010. *Frameworks of Choice: Predictive and Genetic Testing in Asia*. Amsterdam: Amsterdam University Press.

Sui, Suli, and Margaret Sleeboom-Faulkner. 2010. "Genetic Testing for Duchenne Muscular Dystrophy in China: Vulnerabilities Among Chinese Families." In *Frameworks of choice: Predictive and Genetic Testing in Asia*, edited by Margaret Sleeboom-Faulkner, 167–182. Amsterdam: Amsterdam University Press.

The Health and Family Planning Commission of Henan Province and the Office of Talents Work of Henan Province. 2017. "Hold up Tomorrow's Sun with Love (用爱托起明天的太阳)." Available March 16 2017. https://v.qq.com/x/page/d0336s9esp4.html.

The State Council Information Office, PRC. 2000. "White Paper on China's Population and Development in the 21st Century (中国二十一世纪人口与发展白皮书)." Accessed May 17, 2017. http://english1.english.gov.cn/official/2005-07/27/content_17640.htm.

Wahlberg, Ayo. 2016. "The Birth and Routinization of IVF in China." *Reproductive Biomedicine & Society Online* 2: 97–107.

Xinhua News. 2016. "President Xi Calls for Full Protection of People's Health." Accessed March 1 2017. http://news.xinhuanet.com/english/2016-08/20/c_135618492.htm.

Zhu, Jianfeng. 2013. "Projecting Potentiality." *Current Anthropology* 54(S7): S36–S44.

Part 2
Genomic technologies in the bioeconomy

8

Introduction

Claire Marris

This section of the *Handbook* brings together a set of chapters that provide an overview of social science scholarship about "the bioeconomy," especially as it pertains to health and human genetics. As pointed out by Paul Martin in his chapter, "[t]he notion of the bioeconomy only emerged in the 2000s and was largely absent from public and policy discourse before then, but has since become one of the most important justifications for public investment in genomics" (Martin, p.79, this volume). In 2012 both the United States (US) and the European Commission (EC) published key documents setting out their strategy for promoting the bioeconomy (European Commission, 2012; The White House, 2012). The definition of the bioeconomy in the US *National Bioeconomy Blueprint* is representative of how it is portrayed in policy documents produced by governmental institutions:

> the bioeconomy [is] economic activity powered by research and innovation in the biosciences [...]. The bioeconomy emerged as an Administration priority because of its tremendous potential for growth and job creation as well as the many other societal benefits it offers. A more robust bioeconomy can enable Americans to live longer and healthier lives, develop new sources of bioenergy, address key environmental challenges, transform manufacturing processes, and increase the productivity and scope of the agricultural sector while generating new industries and occupational opportunities.
>
> *(The White House, 2012)*

Scientific institutions involved in biosciences, notably those involved in genetics, have also embraced the concept. As just one example, the Chairs of the UK's three bioscience Leadership Councils produced this definition:

> All economic activity derived from bio-based products and processes which contributes to sustainable and resource-efficient solutions to the challenges we face in food, chemicals, materials, energy production, health and environmental protection.
>
> *(BBSRC, n.d.)*

Thus, governmental and bioscientific institutions have defined the bioeconomy in terms of its *materiality* as economic activity derived from biological research and innovation, and the products and processes that are expected to arise from that research/innovation. They have also expressed immense optimism about the prospects for the bioeconomy for solving a wide spectrum of environmental and health problems; although, ultimately, for governmental institutions, "it is about growth and jobs" (European Commission, 2012). Portrayals of the bioeconomy by its advocates are very positive, in several senses of the word. They are clearly "hopeful and confident, and think of the good aspects of a situation rather than the bad ones" (Anon, n.d.). In addition, the bioeconomy is portrayed as positive in the sense that it is "real" and this reality is built upon tangible "actual or specific qualities" (Anon., n.d.), namely the value that can be extracted from biological materials. In contrast, social science scholarship does not take the positive nature of the bioeconomy for granted. It is more critical about the potential outcomes, both about whether the optimistic hopes will be realized in the foreseeable future, and whether these hoped-for futures should be seen as positive and unreservedly welcomed, especially for/by the least powerful, such as patients and women. Social science scholars also stress the intangible attributes of the bioeconomy. They do not focus only on its biological basis. Instead, they show the ways in which the bioeconomy depends on, and generates, new or particular political, economic and social realities. In 2006, in one of the first policy documents about the bioeconomy, the Organisation for Economic Cooperation and Development (OECD) did seem to recognize that the bioeconomy was not just about particular kinds of economic activities, but also about the transformation of our world: "the bioeconomy can be thought of as a world where biotechnology contributes to a significant share of economic output" (OECD, 2006, p. 22). The chapters in this section review how social scientists have described and analyzed these transformations. Although the authors express scepticism about both the reality and the universally positive impact of the bioeconomy, they all seek to move beyond sweeping celebrations or castigations of the hoped-for futures portrayed by the advocates of this new world in order to analyze the underlying transformations co-produced alongside the political economy of the bioeconomy.

As the chapters in this section demonstrate, the bioeconomy is not simply "powered by" biological research and innovation: the bioeconomy *transforms* the organization and conducts research and innovation that it purportedly depends on. In particular, it involves an increased permeability between public and private spheres. Thus, for Chiappetta and Birch "the bioeconomy is more that an analytical or descriptive category; it is also a political-economic project embedded in particular capitalist logics" (Chiappetta and Birch, this volume). They stress that "these policies and visions represent more than the conceptualisation of the potential benefits of biology itself; they are also implicated in policy attempts to shape the direction of biological R&I as a capitalist enterprise." Their chapter, like others in this section, builds upon important prior work by social scientists on "biocapital," that has stressed the intimate links between biotechnologies and the market, notably by Marxist-inspired scholars Sunder Rajan and Catherine Walby (Sunder Rajan, 2006; Waldby & Mitchell, 2006); and feminist scholars Sarah Franklin and Margaret Lock (Franklin & Lock, 2003). In 2008 Stefan Helmreich published a comprehensive review of what he termed "species of biocapital," sketching out a "variety of terms [that] have been forwarded to name how 'life' in the age of genomics, stem cell research, and reproductive technology has become enmeshed in market dynamics" (Helmreich, 2008, p. 463). A decade on, the chapters in this section of the *Handbook* illustrate how science studies scholars have continued to "generate their own accounting of the bioeconomy" (Helmreich, 2008, p. 463).

The chapters in this section also build upon prior scholarship on the sociology of expectations by Nik Brown, Paul Martin and others, focusing on the role of (generally unfulfilled) hope (Borup et al., 2006; Brown & Michael, 2003; Martin et al., 2008). Paul Martin's chapter starts from the observation that although the field of genomics has attracted massive public and private investment over the last 20 years, progress in translating this into widely used clinical applications has been slow. Martin therefore asks: "how has it been possible to sustain high levels of public and private support and investment for genomics over a long period in the face of limited tangible benefits for healthcare?" and "how has support for the idea of the bioeconomy gathered momentum in the context of this slow progress in realizing the benefits of genomics?" These questions also permeate through the other chapters and endure as a key interest in social science analyses of the bioeconomy. Martin's chapter shows how this high level of hope has been maintained around a perpetually reconstructed *future potential* of genomics. It analyzes the performative function of this "genomic hope" in winning support for a particular imagined future and shows how genomic hope, the application of technology in healthcare and the creation of the bioeconomy have been and continue to be *co-produced*.

Martyn Pickersgill's chapter describes an "imagined biological" based on a range of assumptions and understandings of articulations of genes, brains, and bodies. His chapter analyzes how assumptions about the plasticity of this imagined biological influence policy rhetoric and policy interventions – especially for "the early years" (the first three years of a child's life); and on the creation of capital associated with the plasticity of this imagined biological. He shows how bodies are understood to shift and change, and how understandings of the imagined biological also shift over time. When the biological is understood as mutable, interventions on the biological are promoted as a source of societal and economic value: producing citizens who are healthier and better adapted to what is expected of them in a capitalist society, and who are not a burden on "the tax payer." Thus, like "genomic hope," discourse around the "era of the brain" and epigenetics presents the (imagined) biological as promissory matter.

Edward Nik-Khah's chapter focuses on the way in which economic ideas are used to justify particular forms of organization of science and medicine. He argues that the most distinctive – but often overlooked – feature of neoliberalism concerns *knowledge*. Unlike classical liberals, neoliberals do not argue that markets should be praised for their presumed superiority in allocating goods or enhancing productivity, but for their capacity to generate knowledge: their "*epistemic* virtues." He demonstrates how a phalanx of academic departments and think tanks around the Chicago School of Economics has set about promoting their particular neoliberal view of pharmaceutical science and regulation. A key feature of this vision is that science must endorse the epistemic superiority of the marketplace. This leads them, for example, to argue that post-market surveillance should replace large-scale Randomized Clinical Trials for new pharmaceuticals: real-world data and the Internet of Things is promoted as a far more effective means of producing evidence than Randomized Clinical Trials and publications in academic journals. These neoliberals envision partial "incomplete" approvals of new drugs, that would allow the medical marketplace to generate knowledge. They recognize that this may conflict with the pursuit of patient health but rather provocatively argue that "the problem is not that we have too many thalidomides, but too few" (Nik-Khah, p.96, this volume).

Lamoreaux identifies "gendered bioeconomies" as an important "species of capital." Her chapter reviews the literature from feminist science studies scholars, who have stressed the importance of analyzing the interlinkages between gender (and ethnicity) and biotechnology. Thus, "[n]ot only are biotechnologies intertwined with markets, but they are also entangled with gendered and racialized bodies and biological substances" and "[r]eproductive biovalue is incalculable without consideration of gender" (Lamoreaux, p.73, this volume). Feminist ethnographic researchers have revealed systematic asymmetries between women and men, as well as their associated germ cells. For example, Lamoreaux draws on Almeling's (2011) work that shows the differential values attributed to egg and sperm, where men who "donate" sperm are understood as employees performing work, whereas women who donate eggs are viewed as altruistic gift givers. Lamoreaux points out that the commodification of germ cells and reproduction does not just produce babies, but also gendered understandings of the body, of reproductive expectations, and new gendered pharmaceutical, medical, scientific and labor markets. Lamoreaux also emphasizes the intersectional dimension of feminist analyses of the bioeconomy. While many social studies scholars (e.g. Chiappetta and Birch, p.63, this volume) emphasize the importance of focusing on "the political-economic" rather than only the biological or material, analyses by feminist scholars tend to reject this dichotomy altogether. They analyze the circulation and valuation of material and immaterial phenomenon in tandem, and thus stretch understandings of both reproduction and the economy.

The concept of the bioeconomy, as promoted by its advocates, is based on a notion of "latent value" that can be extracted from biological material and processes. The chapters in this section illustrate how different conceptions and implications of this notion of latent (bio)value can be identified in the social science literature on the bioeconomy. Chiappetta and Birch argue that for some science studies scholars, notably Waldby, latent value (in the form of biotechnological products) is *inherent* in living matter (e.g. tissues, cell lines) "as surplus or waste." Here the notion of "latent value" is quite close to that used by advocates of the bioeconomy from governmental and scientific institutions whereby latent value is "incumbent in biological products and processes" (OECD, 2006, p. 1). Chiappetta and Birch stress that, from their perspective, latent value is not inherent in biological material and processes; rather, it is conferred by the networks working on these objects. For Martin, and other work that focuses on the promissory or speculative nature of the bioeconomy, value is latent in the sense that it is perpetually positioned in a hoped-for future. As discussed by Chiappetta and Birch in their chapter, Sunder Rajan is interested in what he terms an "inversion" of production and value apparent in the bioeconomy: "he is interested in how promise, hope, and other future claims come to constitute present value as a result of the expectation that modern biotechnology will lead to commercial products" (Chiappetta and Birch, this volume). This can be seen as another form of "latency." In Pickersgill's chapter, another form of "latent value" emerges, where the creation of capital emerges from the plasticity of the "imagined biological."

The last chapter in the section, by Ulrike Felt, is about the concept of "Responsible Research and Innovation" (RRI). RRI has emerged in the last five years or so as the latest in a series of attempts by scientific institutions to address what they perceive as problematic links between "science" and "society." Previous attempts, under the banner of Ethical, Legal and Issues (ELSI or ELSA), were criticized by some science studies scholars because they only allowed a narrow set of questions to be raised, often pre-formatted by the physical/natural scientist, and ultimately appeared more focused on smoothing out social concerns rather than addressing them. The ELSI/ELSA model was also associated with a division of moral labor: "scientific" researchers are seen to be responsible for the creation of knowledge and technology, while ELSA/ELSI researchers, who come for the social sciences and humanities, care for and reflect upon the social and ethical implications of the knowledge and technology created. When RRI first emerged around 2012, it was presented by some authors as a response to these critiques and there was some hope among some science studies scholars (myself included) that RRI might provide the means to move away from the "science and society" ELSI/ELSA model into a more open, more participative and more reflexive model that would recognize the intricate interlinkages of "science in society." Just a few year later, these hopes appear to be dashed. Arie Rip (2016), for example, has suggested that RRI can be seen as the emperor's new clothes. Felt shows how RRI could perhaps have become a means to help disentangle – and challenge – the links between sociotechnology and capitalism generated by/in the bioeconomy that the other chapters in this section explore. However, this would have meant enabling RRI programs to ask uncomfortable questions, for example around the taken-for-granted optimism about "scientific progress" and its necessarily positive contribution to healthy and sustainable livelihoods. Many RRI programs have attempted to overcome ELSI/ELSA's "division of moral labor" by enforcing collaboration between researchers and other scholars from the social sciences, humanities, and the arts, but this has usually still (like ELSI/ELSA programs that came before) been as part of co-funded programs led by the natural/physical scientists. Together with colleagues involved in RRI programs in the field of synthetic biology in the UK, I have found that such collaborations seriously limited the kinds of roles that can be played by science studies scholars (Balmer et al., 2015, 2016).

Felt also explains how RRI emerged at a time when innovation came to be seen as the key driving force for societal development. As Felt says, during this period "[t]he importance of a steady flow of innovations to assure international competitiveness has become the core mantra in European discourse." The chapters by Martin and Nik-Khah also reveal how the discourse around the bioeconomy is associated with a shift in language from "research and development" to "innovation," which was in part a response to the unfulfilled hopes about the delivery of biotechnology products onto the market. In the pharmaceutical sector, the so-called "productivity crisis" (a decreasing number of novel drugs at the same time as rising R&D costs) has been used as justification for a new innovation model based on public–private partnerships (Nik-Khah, this volume). The slow progress in the rate of adoption of genomics in the clinic is used to justify major organizational change and investment "not only on the grounds of improved patient care, but also in promoting innovation and entrepreneurship within the bioeconomy" (Martin, this volume). In this context, it is difficult for RRI initiatives to ask questions that challenge the very need for this continuous flow of innovations, and the overall organizational changes associated with this push for scientific entrepreneurship. As I have argued elsewhere, social scientists working within RRI programs may wish to bring to the fore "uncomfortable knowledge" that their collaborators from the natural sciences actually need to ignore in order to function (Marris et al., 2014). As a result, RRI programs, like ELSI/ELSA programs before them, tend to address only narrowly defined concerns about the "responsible" nature of specific innovations or fields of technoscience (essentially "do no harm"). As Felt suggests, it seems that, rather than opening up discussions about the (capitalist) politics entangled in the bioeconomy, RRI is destined to serve neoliberal ideas through depoliticizing debates. Therefore, despite so much social science scholarship on the ways in which the bioeconomy is necessarily entangled and co-produced with social, political and economic realities, the notion of the bioeconomy promoted by policy actors remains positive in yet another sense: it is portrayed as ostensibly "free of ethical, political, or value judgments" (Anon., n.d.).

References

Almeling, R., 2011. *Sex Cells: The Medical Market for Eggs and Sperm*. University of California Press, Berkeley.

Anon. (n.d.). Positive *Collins Online English Dictionary* [Online] www.collinsdictionary.com/dictionary/english/positive [Accessed 21 June 2017]).

Balmer, A., Calvert, J., Marris, C., Molyneux-Hodgson, S., Frow, E., Kearnes, M., ... Martin, P. (2015). Taking Roles in Interdisciplinary Collaborations: Reflections on Working in Post-ELSI Spaces in the UK Synthetic Biology Community. *Science and Technology Studies*, 28(3), 3–25.

Balmer, A., Calvert, J., Marris, C., Molyneux-Hodgson, S., Frow, E., Kearnes, M., ... Martin, P. (2016). Five Rules of Thumb For Post-ELSI Interdisciplinary Collaborations. *Journal of Responsible Innovation*, 3(1), 73–80. doi:10.1080/23299460.2016.1177867

BBSRC. (n.d.). *The Bioeconomy*. Retrieved from www.bbsrc.ac.uk/research/briefings/bioeconomy.

Borup, M., Brown, N., Konrad, K., & Van Lente, H. (2006). The Sociology of Expectations in Science and Technology. *Technology Analysis & Strategic Management*, 18(3), 285–298. doi: 10.1080/09537320600777002

Brown, N., & Michael, M. (2003). A Sociology of Expectations: Retrospecting Prospects and Prospecting Retrospects. *Technology Analysis & Strategic Management, 15*(1), 3–18. doi:10.1080/0953732032000046024.

European Commission. (2012). *Innovating for Sustainable Growth: a Bioeconomy for Europe*. Retrieved from Brussels: https://publications.europa.eu/en/publication-detail/-/publication/1f0d8515-8dc0-4435-ba53-9570e47dbd51.

Franklin, S., & Lock, M. (Eds.). (2003). *Remaking Life and Death: Toward an Anthropology of the Biosciences*. Santa Fe: SAR Press.

Helmreich, S. (2008). Species of Biocapital. *Science as Culture, 17*(4), 463–478. doi:10.1080/09505430802519256.

Marris, C., Jefferson, C., & Lentzos, F. (2014). Negotiating the Dynamics of Uncomfortable Knowledge: The Case of Dual Use and Synthetic Biology. *BioSocieties, 9*(4), 393–420. doi:10.1057/biosoc.2014.32.

Martin, P., Brown, N., & Kraft, A. (2008). From Bedside to Bench? Communities of Promise, Translational Research and the Making of Blood Stem Cells. *Science as Culture, 17*(1), 29–41. doi:10.1080/09505430701872921.

OECD. (2006). *The Bioeconomy to 2030: Designing a Policy Agenda*. Retrieved from Paris: www.oecd.org/sti/biotech/34823102.pdf.

Rip, A. (2016). The Clothes of the Emperor. An Essay on RRI In and Around Brussels. *Journal of Responsible Innovation, 3*(3), 290–304. doi:10.1080/23299460.2016.1255701.

Sunder Rajan, K. (2006). *Biocapital: The Constitution of Post-Genomic Life*: Duke University Press.

The White House. (2012). *National Bioeconomy Blueprint*. Retrieved from Washington, DC: https://obamawhitehouse.archives.gov/sites/default/files/microsites/ostp/national_bioeconomy_blueprint_april_2012.pdf.

Waldby, C., & Mitchell, R. (2006). *Tissue Economies: Blood, Organs and Cell Lines in Late Capitalism*. Durham: Duke University Press.

9

Limits to biocapital

Margaret Chiappetta and Kean Birch

Introduction

The entanglement of technoscience and capitalism has been discussed at some length in the social sciences (e.g. Dasgupta and David, 1994; Sunder Rajan, 2006; Mirowski, 2011; Tyfield, 2012b and 2012c; Stephan, 2012; Tyfield, 2012a, 2012b, 2012c; Birch and Tyfield, 2013; Birch 2017). This literature illustrates the change in the political economy of research and innovation (R&I), specifically the increasing commodification, commercialization, privatization, and marketization of scientific research (Lave et al., 2010).

This entanglement is particularly evident in relation to the biological and biotechnological sciences following the emergence of the biotech sector in the late 1970s. The most recent manifestation reflects the expansion of the *-omics* fields (e.g. genomics, proteomics, metabolomics) since 2000. This latest phase has involved the reorientation of public funding in universities toward public/private partnerships, amongst other initiatives, thereby (re)configuring research agendas toward both the growth of intellectual property (IP) and open science initiatives that increasingly impact the production and dissemination of scientific knowledge (Lave et al., 2010).

This chapter focuses on this entanglement of technoscience and capitalism, specifically as it relates to the concept of *biocapital*. The rapid growth of the *-omics* sciences necessitates a reexamination of how we understand the "bio-economy." We frame the bioeconomy as not primarily based on biophysical materialities or "liveliness," but rather as underpinned by an array of social practices and networks. We begin with an analysis of key concepts such as *biocapital* and *biovalue*, before outlining the concept of *mediating devices* as an alternative analytical tool for understanding genomics and techno-economic networks.

Theorizing biocapital

Numerous bio-concepts have emerged in the literature concerning the political economy of biological sciences and innovation, notably as a way to conceptualize the so-called *bioeconomy* (see Birch, forthcoming). From "biovalue" through "biocapital" and beyond, these concepts provide a useful contribution to debates regarding the relationship between the biological sciences and capitalism (e.g. Waldby, 2000, 2002; Rose, 2001, 2007; Sunder Rajan, 2006, 2012; Waldby and Mitchell, 2006; Cooper, 2008; Brown et al., 2011; Brown, 2012; Vermeulen et al., 2012; cf. Birch and Tyfield, 2013; Birch, 2017). In this section, we outline these concepts and their theoretical underpinnings as they relate to their conceptual discussion of capital, value, and valuation in the life sciences or biotech sector.

The emergence of the biotech sector in the late 1970s drove the search to harness the potential of scientific knowledges in the development of new (biological) products and processes as part of a broader concern with declining profit rates and stagnant economic growth in the Global North (Cooper, 2008). A number of countries, especially the USA, sought to foster and develop a domestic biotech sector. Many countries followed suit. Later, in the mid-2000s, the OECD endeavored to build an international policy agenda based on the idea that modern biotechnology represents a potential tool to extract the "*latent value* incumbent in biological products and processes" (OECD, 2006: 1; emphasis added). Policy-makers in the European Union and USA subsequently developed their own "bioeconomy" agendas (e.g. EC, 2012; White House, 2012). As Hilgartner (2007) argues, however, these policy agendas and visions represent more than the conceptualization of the potential benefits of biology itself; they are also implicated in policy attempts to shape the direction of biological R&I as a capitalist enterprise. As such, the bioeconomy is more than an analytical or descriptive category; it is also a political-economic project embedded in particular capitalist logics (Goven and Pavone, 2015). The bioeconomy therefore represents a fertile ground on which to think through the relationship between technoscience and capitalism. In this section, we focus primarily on two of the most influential bio-concepts; Catherine Waldby's (2000, 2002) notion of *biovalue* and Kaushik Sunder Rajan's (2006, 2012) conception of *biocapital*.

First, Waldby (2002) argues that contemporary biotechnology has led to the "reorganization of the boundaries and elements of the human body" (p. 308) such that "biotechnology produces a margin of biovalue, a surplus of fragmentary vitality" (p. 310). Waste tissues derived from marginal subjects are especially prone to commodification as said tissues are extracted and re-engineered, and can be "leveraged biotechnically so that they can *become more* prolific or useful" (Waldby and Mitchell, 2006: 32). Biovalue, then, "refers not to the stable and known properties of tissues but to the capacity of tissues to lead to new and unexpected forms of value" (ibid: 108). In Waldby's formulation, it is the character of the tissue itself, suitably enhanced through biotechnologies, which is the source of biovalue. This perspective reiterates the notion of latent value inherent in living matter with Waldby implying that value is embedded in the products of biological matter and processes. Hence, it is understandable why she characterizes it as biovalue, since value is identified as the revenues or income derived from the sale of commodities that modern biotechnology has enabled us to develop from biological matter. However, this perspective comes close to that of "biotech boosters" highlighted by Helmreich (2007: 293), in that it promotes the idea that biotechnological products are "latent" in biological processes and that any such process necessarily "*constitutes a form of surplus value production.*" Helmreich's analysis problematizes Waldby's conception of value as being inherent or latent within particular materialities or material objects (e.g. tissues, cell lines) as surplus or waste, rather than value being conferred by the networks working on these objects, in addition to various other legal and financial instruments (ibid.).

Second, biocapital is another important analytical contribution, originally formulated by Franklin and Lock (2003), but more often associated with Sunder Rajan's (2006) version of the concept. In their version, Franklin and Lock (2003: 8) theorize biocapital as the wealth dependent on a "form of extraction that involves isolating and mobilizing the primary reproductive agency of specific body parts" (such as cells). Here, we focus on Sunder Rajan, especially his emphasis on the "market potential of bioproducts" – or the future promises that are constitutive of value in the present (Helmreich, 2009: 127). Sunder Rajan discusses what he terms an "inversion" of production and value; that is, how "the future [is] always being called in to account for the present" (Sunder Rajan, 2006: 116). In this sense, he is interested in how promise, hope, and other future claims come to constitute present value as a result of the expectation that modern biotechnology will lead to commercial products (Sunder Rajan, 2006). Fortun (2008) also discusses the notion of value as being *promissory* or *speculative* in nature. In the case of genomic science, for example, Fortun argues that genomic information may elucidate the probability of developing a disease, and thus offers "promises" of commercial application, thereby mobilizing actors outside of the laboratory. Like Waldby's concept of biovalue, Sunder Rajan's formulation of biocapital concerns the identification of value in the bioeconomy. However, whereas Waldby focuses on the materiality of biological processes leading to biological products as the site of value, Sunder Rajan is more interested in value derived from biotechnological information and knowledge. These intangible assets can be commodified as products whose value is constituted by their future promises.

While Waldby and Sunder Rajan may have different starting points and different takes on where value is in the bioeconomy, they can both be characterized as *production-* or *commodity-* based approaches, since they are conceptually grounded in Marxist conceptions of value. So, despite their differences, their respective theoretical frameworks concentrate on the production of commodities, whether or not these commodities are material or immaterial, are actual or promissory, and so on. Moreover, their perspectives reflect a conception of value in which value is identified with biological material – e.g. tissues, cells, bodies, etc.

Our intention now is to tease apart some of the ambiguities in these two concepts. These can be characterized as: (1) the confusion of political-economic terms; (2) the conceptual gap in theorizing (tangible or intangible) assets; and (3) the evaluation process of these assets.

First, bio-concepts such as biovalue and biocapital frequently conflate political-economic terms, especially in discussions of capital, capitalization, etc. The use of these terms can give rise to confusion, as "capital" can refer to a number of different things; these include finance capital deployed by venture capitalists and other investors, business investment in physical assets like buildings, machinery and laboratory space, and tangible or intangible assets (e.g. IP) themselves, all of which are used to develop and create products. When Waldby and Mitchell (2006) refer to capital they are primarily concerned with using the term to refer to venture capital (i.e. finance) while also focusing on the supply side of capital investment (i.e. financial *and* business investment). In the case of Sunder Rajan (2006), the term capital is almost erased from discussion as a result of his development of the new concept of *bio*capital, and where he does allude to capital in relation to the circulation of capital flows, it is not clear what this necessarily refers to.

The confusion around capital is compounded by the use of the term capitalization by these and other scholars analyzing the bio-economy; it is a concept that is infrequently defined yet frequently invoked. For example, Sunder Rajan (2012) refers to the capitalization of "life," though what this capitalization actually means is unclear. The closest definition might be something like the process of transforming things (e.g. life, tissue) into capital (without clarifying what kind of capital).

Second, it is increasingly evident that value in the bioeconomy is not (and cannot be) derived primarily from commodities, material or immaterial, since there are so few products on the market (Nightingale and Martin, 2004; Pisano, 2006). Rather, a number of scholars have started to theorize value in relation to tangible and intangible assets that are both resources in production *and* alienable private property in their own right (e.g. platform technologies, scientific expertise, intellectual property, etc.) (see Birch and Tyfield, 2013; Cooper and Waldby, 2014;

Lezaun and Montgomery, 2015; Birch, 2017). In particular, there is a need to theorize the value and valuation of intangible assets (Birch, 2012; Birch and Tyfield, 2013). Intangible assets include things like software and data, intellectual property rights, brand value, etc. As we will discuss in the following section, these intangible assets also form the basis of *mediating devices* that are constitutive of social relations in scientific collaboration, knowledge production, and dissemination in the biological sciences.

Finally, it is evident that many scholars have not yet addressed the ways that value is accounted for or calculated by political-economic actors like venture capitalists, other financiers, investors, etc. who are integrally involved in techno-economic networks. As mentioned, there is an assumption that value is latent within particular material objects (e.g. Waldby, 2000, Waldby and Mitchell, 2006). However, there are few attempts to think through how the value of new biotechnologies is calculated and capitalized, especially in relation to (intangible) assets like intellectual property (see Martin, 2015). It is important to examine the social practices of accounting, calculating, and valuation that come to constitute techno-economic networks. In taking these practices seriously, it is necessary to think about how value is understood and managed, and by whom, and how these social practices are carried out. Crucially, Birch (2017) argues it is not primarily scientists who undertake these accounting, calculation, and valuation practices; rather, it is financial investors, business managers, technology transfer officers, etc. Attempts to conceptualize economic value in the bioeconomy need to incorporate the practices of venture capitalists, stockbrokers, institutional investors, and others making decisions about how to manage the value of new biotechnologies and new biotechnology firms.

Beyond biocapital

In this section, we aim to examine the complexity of modern biotechnology and its relationship to capitalism through a discussion of the role of different *mediating devices* at play in the organization of (biological) R&I. We begin with a brief overview of techno-economic networks and then outline what we mean by "mediating devices," extending the existing definition in the literature beyond financial instruments to include a discussion of open and proprietary mechanisms in the context of -*omics* sciences.

Callon (1991: 133) outlines the concept of *techno-economic networks* as "a coordinated set of heterogeneous actors … [who] interact more or less successfully to develop, produce, distribute and diffuse methods for generating goods and services." His definition highlights the complex relationship between science, technology, and capitalism. Following the passage of the *Bayh-Dole Act* (1980) in the USA, public research institutions were able to attract investment from private industry by patenting research results (Mirowski, 2011), creating a new marketplace for scientific knowledge and its byproducts. This marketplace was marked by increased heterogeneity in terms of the organizations involved in the process of R&I as well as a diversification of the practitioners, funders, and vendors of science. As a result, complex techno-economic networks extended well beyond a single university or firm to include organizations ranging from technology transfer offices, funding agencies, policy-makers, venture capitalists, and so forth (Popp-Berman, 2012).

We use the term "mediating device" to represent the techno-economic *intermediaries* that enable these actors to interact and collaborate, and which then come to constitute the value relations in these networks. As Miller and O'Leary (2007: 710) note, "these [intermediaries] give material content to the links uniting the actors. They may be written documents, technical artefacts, human beings, or money." Cognate concepts like "mediating instruments" have been discussed at length in the context of the broad range of practices of calculation, valuation, budgeting, and computation (Miller and O'Leary, 2007; Miller and O'Leary, 1987; Morrison and Morgan, 1999). These instruments serve to mediate the interactions between science and capitalism, shaping the ways in which science is governed and regulated, as well as the means by which the products of R&I may be exchanged (Miller, 2007; see also Power, 1994). They represent "the material and discursive assemblages that intervene in the construction of markets," suggesting these objects encompass "analytical techniques to pricing models, . . . purchase settings to merchandising tools, . . . trading protocols to aggregate indicators" (Muniesa et al., 2007, p. 2; see also Muniesa, 2014).

Meditating devices include various forms of open (e.g. open science licensing, open access databases) and proprietary (e.g. material transfer agreements, patent) rights and mechanisms that enable public and private research groups, government regulators, policy-makers, and investors to interact, collaborate, and innovate. These *mediating devices* constitute value in the articulation of biotechnology and capitalism. Essentially, they are the linchpins in contemporary techno-economic networks that enable collaboration, commercialization, and knowledge transfer, as well as the management of value in the development and commercialization of assets and products in the life sciences.

These devices are particularly salient in the context of genomics. As we have highlighted elsewhere, the growth of the -*omics* sciences has dramatically altered the organization of biological R&I (Birch et al., 2016). With the shift in focus from objects (e.g. the gene) to systems (e.g. the genome), R&I has become data-accelerated, globally collaborative, and interdisciplinary. Research in fields such as genomics or proteomics is reliant on "communally available biobank databases to interrogate nucleotide sequences of interest, compare protein sequences, and search for sequence data in particular disease contexts" (ibid.: p. 38), and extends beyond the confines of a single laboratory to include multi-organizational collaboration across organizations with very different value regimes. However, the data-accelerated and interdisciplinary nature of R&I in fields such as genomics is limited by the complex entanglement of biotechnology and capitalism, and the commercialization of science itself. On the one hand, duplicating research findings is prohibitively costly, creating an imperative for ("pre-competitive") collaboration; on the other hand, enclosing research results behind IP rights provides incentives for private investment in R&I. The concept of mediating devices is of particular importance here. These devices, situated within complex techno-economic networks, play a crucial role in configuring the organization, governance, and valuation of biological R&I. They constitute the ways in which knowledge and information are circulated and in which disparate groups collaborate. They determine at what point in the process of R&I actors within these networks interact, attach value to these interactions (e.g. by stipulating potential royalties), and regulate how value may be appropriated from the products of these interactions (Birch et al., 2017).

There is some debate as to whether proprietary mediating devices (e.g. patents, copyrights, MTAs) impede innovation in the biological sciences by limiting avenues of inquiry, excluding certain groups from the R&I process, and creating bottlenecks through patent thickets (e.g. Heller and Eisenberg, 1998; Mirowski, 2011; Caulfield et al., 2012). MTAs, for instance, are contracts governing the transfer of research materials between groups, whereby one group owns the rights to said material and the other intends to use it for research purposes (Mirowski, 2011).

Conversely, open mediating devices offer a potential solution to the limits imposed by their proprietary counterparts. Structural genomics in particular has witnessed a push for increased use of open science initiatives in recent years, in part due to the laborious nature of sequencing (Sá and Tamtik, 2011). To mitigate the apparent drawbacks of proprietary devices, open devices in biological R&I seek to reduce entry costs (e.g. fees required to access data or publications), expand ownership (of knowledge, research products, etc.), increase sharing, and ensure that more fundamental/less commercially oriented research is not neglected (Feldman and Nelson, 2008; Hope, 2008). Open access databases (e.g. drug screening platforms or sequence databases) stipulate that reusable scientific data is released into the public domain, such that international actors at diverse institutes may contribute their own research findings, and ensure that researchers are not limited by confidentiality clauses or publication restrictions (ibid.). Perhaps most importantly, these devices ensure that commercial barriers to entry are low, thus avoiding the prohibitively costly repetition of studies or sequencing efforts. For example, X-ray crystallography, NMR spectroscopy, and the numerous other high-throughput methods required to systematically decipher gene and protein structures are exceptionally costly, and funding for R&I is scarce (Sá and Tamtik, 2011).

It is important to note that open mediating devices do not necessarily offer a panacea to the "ills" of their proprietary counterparts (Birch et al., 2017). As Hayden (2010: 98–99) argues, "the public domain extends only so far as property regimes do" and the notion of "making things public does not mean setting them into circulation outside of monopolistic deployments of intellectual property." This inevitably has implications for R&I conducted in the Global South (e.g. clinical trials, generic drug formulations, etc.) (ibid.). Hilgartner (2009) also points out that open and proprietary devices play a significant role in determining the allocation of control over emerging technologies: these devices "can serve as a vehicle for asserting managerial dominion . . . over the invention itself" as well as over the broader social and technical orders in which new technologies are entwined (p. 213).

Conclusion

Current conceptions of the relationship between capitalism and technoscience, broadly speaking, are often commodity-based approaches, especially in discussions of the "bioeconomy." In this chapter, we have highlighted the range of social practices and networks that underpin R&I in the biological sciences. The expansion of the *-omics* sciences has necessitated a new evaluation of the mediating devices that are constitutive of social and value relations in technoscientific collaboration, production, and dissemination.

Further research is needed to theorize the value and valuation of intangible assets in these techno-economic networks (Birch, 2012; Birch and Tyfield, 2013), and to understand the mediating devices discussed above. As noted, there is significant room in these debates to analyze how the value of new biotechnologies is calculated and capitalized, particularly in relation to (intangible) assets like intellectual property (Birch, 2017).

References

Birch, K. (2012). Knowledge, place and power: Geographies of value in the bioeconomy. *New Genetics and Society* 31(2): 183–201.

Birch, K. (2017). Rethinking value in the bio-economy: Finance, assetization and the management of value. *Science, Technology and Human Values* 42(3): 460–490.

Birch, K. (forthcoming). The Problem of Bio-concepts and the Political Economy of Nothing. *Cultural Studies of Science Education.*

Birch, K. and Tyfield, D. (2013). Theorizing the bioeconomy: Biovalue, biocapital, bioeconomics or … what? *Science, Technology and Human Values* 38(3): 299–327.

Birch, K., Tyfield, D. and Chiappetta, M. (forthcoming, 2017). From neoliberalizing research to researching neoliberalism: STS, rentiership and the emergence of commons 2.0. In D. Cahill, M. Konings and M. Cooper (eds), *The SAGE Handbook of Neoliberalism.* London: SAGE.

Birch, K., Dove, E., Chiappetta, M. and Gursoy, U. (2016). Biobanks in oral health: Promises and implications of 'post-neoliberal' patterns of science and innovation. *OMICS: A Journal of Integrative Biology* 20(1): 36–41.

Brown, N. (2012). Contradictions of value: Between use and exchange in cord blood bioeconomy. *Sociology of Health and Illness* 35(1): 97–112.

Brown, N., Machin, L. and McLeod, D. (2011). Immunitary bioeconomy: The economisation of life in the international cord blood market. *Social Science and Medicine* 72: 1115–1122.

Callon, M. (1991). Techno-economic networks and irreversibility. In J. Law (Ed.), *A Sociology of Monsters: Essays on Power, Technology and Domination*. London: Routledge.

Caulfield, T., Harmon, S., and Joly, Y. (2012). Open science versus commercialization: A modern research conflict? *Genome Medicine* 4(17): 1–11.

Cooper, M. (2008). *Life as Surplus*. Seattle: University of Washington Press.

Cooper, M. and Waldby, C. (2014). *Clinical Labor*. Durham: Duke University Press.

Dasgupta, P., and David, P. (1994). Towards a new economics of science. *Research Policy* 23(5): 487–521.

EC (2012). *Innovating for Sustainable Growth: A Bioeconomy for Europe*. Brussels: European Commission.

Feldman, R. and Nelson, K. (2008). Open source, open access, and open transfer: Market approaches to research bottlenecks. *Northwestern Journal of Technology and Intellectual Property* 7(1): 14–32.

Fortun, M. (2008). *Promising Genomics*. Berkeley: University of California Press.

Franklin, S. and Lock, M. (eds) (2003). *Remaking Life and Death: Toward an Anthropology of the Biosciences*. Santa Fe: SAR Press.

Goven, J. and Pavone, V. (2015). The bioeconomy as political project: A polanyian analysis. *Science, Technology and Human Values* 40(3): 302–337.

Hayden, Cori (2010). The Proper Copy: The insides and outsides of domains made public. *Journal of Cultural Economy* 3(1): 85–102.

Helmreich, S. (2007). Blue-green capital, biotechnological circulation and an oceanic imaginary: A critique of biopolitical economy. *BioSocieties* 2(3): 287–302.

Helmreich, S. (2008). Species of biocapital. *Science as Culture* 17(4): 463–478.

Helmreich, S. (2009). *Alien Ocean: Anthropological Voyages in Microbial Seas*. Berkeley: University of California Press.

Heller, M. A., and Eisenberg, R. S. (1998). Can patents deter innovation? The anticommons in biomedical research. *Science* 280(5364): 698–701.

Hilgartner, S. (2007). Making the bioeconomy measurable: Politics of an emerging anticipatory machinery. *BioSocieties* 2(3): 382–386.

Hilgartner, S. (2009). Intellectual property and the politics of emerging technology: Inventors, citizens, and powers to shape the future. *Chicago-Kent Law Review* 84(1): 197–224.

Hope, J. (2008). *Biobazaar: The Open Source Revolution and Biotechnology*. Cambridge, MA: Harvard University Press.

Hopkins, M., Martin, P., Nightingale, P., Kraft, A. and Mahdi, S. (2007). The myth of the biotech revolution: an assessment of technological, clinical and organisational change. *Research Policy* 36(4): 566–589.

Lave, R., Mirowski, P., and Randalls, S. (2010). Introduction: STS and Neoliberal Science. *Social Studies of Science* 40(5): 659–675.

Lezaun, J. and Montgomery, C. (2015). The pharmaceutical commons: Sharing and exclusion in global health drug development. *Science, Technology and Human Values* 40(1): 3–29.

Martin, P. (2015). Commercialising neurofutures: Promissory economics, value creation and the making of a new industry. *BioSocieties* 10(4): 422–443.

Miller, P. and O'Leary, T. (1987). Accounting and the construction of the governable person. *Accounting, Organizations and Society* 12(3): 235–265.

Miller, P. and O'Leary, T. (2007). Mediating instruments and making markets: Capital budgeting, science and the economy. *Accounting, Organizations and Society* 32: 701–734.

Mirowski, P. (2011). *ScienceMart*. Cambridge: Harvard University Press.

Morrison, M., and Morgan, M. S. (1999). Models as mediating instruments. In M. S. Morgan and M. Morrison (Eds.), *Models as Mediators: Perspectives on Natural and Social Science*. Cambridge: Cambridge University Press.

Muniesa, F. (2014). *The Provoked Economy*. London: Routledge.

Muniesa, F., Yuval, M., and Callon, M. (2007). An introduction to market devices. *The Sociological Review* 55: 1–12.

Nightingale, P. and Martin, P. (2004). The myth of the biotech revolution. *Trends in Biotechnology* 22(11): 564–569.

OECD (2006). *The Bioeconomy to 2030: Designing a Policy Agenda.* Paris: Organisation for Economic Co-Operation and Development.

Pisano, G. (2006). *Science Business.* Cambridge: Harvard University Press.

Popp-Berman, E. (2012). *Creating the Market University: How Academic Science Became an Economic Engine.* Princeton: Princeton University Press.

Power, M. (1994). From the science of accounts to the financial accountability of science. *Science in Context* 7(3): 355–387.

Rose, N. (2001). The politics of life itself. *Theory, Culture and Society* 18(6): 1–30.

Rose, N. (2007). Molecular biopolitics, somatic ethics and the spirit of biocapital. *Social Theory and Health* 5(1): 3–29.

Sá, C., and Tamtik, M. (2011). Structural genomics and the organisation of open science. *Genomics, Society and Policy* 7: 20–34.

Stephan, P. (2012). *How Economics Shapes Science.* Cambridge: Harvard University Press.

Sunder Rajan, K. (2006). *Biocapital.* Durham: Duke University Press.

Sunder Rajan, K. (ed.) (2012). *Lively Capital.* Durham: Duke University Press.

Tyfield, D. (2012a). A cultural political economy of research and innovation in an age of crisis. *Minerva* 50(2): 149–167.

Tyfield, D. (2012b). *The Economics of Science: A Critical Realist Overview (Vol.1) – Illustrations and Philosophical Preliminaries.* London: Routledge.

Tyfield, D. (2012c). *The Economics of Science: A Critical Realist Overview (Vol.2) – Towards a Synthesis of Political Economy and Science & Technology Studies.* London: Routledge.

Vermeulen, N., Tamminen, S. and Webster, A. (eds) (2012). *Bio-objects: Life in the 21st Century.* Farnham: Ashgate.

Waldby, C. (2000). *The Visible Human Project: Informatic Bodies and Posthuman Medicine.* London: Routledge.

Waldby, C. (2002), Stem cells, tissue cultures and the production of biovalue. *Health: An Interdisciplinary Journal* 6(3): 305–323.

Waldby, C. and Mitchell, R. (2006). *Tissue Economies: Blood, Organs and Cell Lines in Late Capitalism.* Durham: Duke University Press.

White House, The (2012). *National Bioeconomy Blueprint.* Washington, DC: The White House.

10
Gendered bioeconomies

Janelle Lamoreaux

In their 2003 edited volume *Remaking Life and Death*, Sarah Franklin and Margaret Lock used the term *biocapital* to describe the increasing entanglement of biotechnology with the economy. Much subsequent research, including Kaushik Sunder Rajan's *Biocapital: The Constitution of Postgenomic Life*, has explored how emergent biotechnologies intersect with processes of commercialization and market frameworks (Sunder Rajan 2006, p. 33; see also Chiapetta and Birch, this volume). In a 2008 review of biocapital research, Stefan Helmreich wrote, "The store of science studies work theorizing the conjuncture of economic action and contemporary biotechnology is now well stocked" (Helmreich, 2008, p. 463). Nearly a decade later, scholars continue to stock what have become multiple stores of bioeconomic research, further contributing to what Helmreich described as "articulations of biocapital and its kin" (Helmreich, 2008, p. 463).

Within this still growing area of research, Lock and Franklin's 2003 articulation of biocapital continues to stand out. Their edited collection emphasized the importance of biocapital analyses as new instantiations of longstanding feminist efforts to shift the focus of economics toward considerations of reproduction, gender, and kinship. As Lock and Franklin state, "Biocapital is not just dependent on reproduction, it is constituted by it… Reproduction – like gender, nature and kinship, often feminized – has been wrongly marginalized in accounts of economic change and development" (Franklin and Lock, 2003, pp. 10–11). For Franklin, Lock, and feminist science studies scholars to follow, it is crucial to understand not only the role of production in (bio)capital accumulation, but also the role of reproduction. More than a "natural" act, in today's biocapital landscape reproduction itself is "harnessed" as a mode of accumulating capital (Franklin and Lock, 2003, p. 10).

With this emphasis on reproduction in mind, in this chapter I explore *gendered bioeconomies* as an important "species of biocapital" (Helmreich, 2008) for at least two reasons, related to two different understandings of gender. First, a consideration of gendered bioeconomies is important because, as argued by Franklin and Lock above, reproductive objects of analysis often remain feminized and are therefore granted less importance in economic research than their "productive" counterparts. Second, contemporary bioeconomic activities related to reproduction such as fertility preservation, surrogacy or genetic testing continue to work through gendered bodies and bodily substances, and therefore have stratified gender impacts. Furthermore, gendered bioeconomic entanglements have been accelerated by the recent proliferation of reproductive technologies, both in the classic sense of *assisted reproductive technologies* (ARTs), and in the broader sense of biotechnologies that reproduce and regenerate life in many forms.

With this dual meaning of gender in mind, I start this chapter with the "substantial beginnings" of research on reproduction and biocapital, focusing on the social science of eggs and sperm in the aftermath of IVF. I then move from a discussion of reproductive substances to reproductive activities, concentrating on research that addresses (trans)national bioeconomies of labor. Finally, I offer an overview of research that focuses on genomic technologies, primarily those developed alongside reproductive processes and technologies. Throughout these sections I am guided by the question: what does a continued emphasis on reproduction and gender in the bioeconomy make visible today? I argue that a continued focus on reproduction and gender is as necessary to research on the bioeconomy as a focus on the market was to earlier studies of biotechnology that neglected the role of capital accumulation in technological development.

Substantial beginnings

As mentioned, Franklin and Lock's 2003 edited volume *Remaking Life and Death: Toward an Anthropology of the Biosciences* offered one of the first theorizations of biocapital. Contributions to this edited volume opened up the theoretical grounds for future ethnographic research on how biotechnology intertwines with evaluations, ethical configurations, and (re)definitions of life itself. The book made way for future work on gendered bioeconomies in particular by suggesting that "biocapital is driven by a form of extraction that involves isolating and mobilizing the primary reproductive agency of specific body parts" (Franklin and Lock, 2003, p. 8). Similarly, Charis Thompson's 2005 work on the role of reproductive substances in the bioeconomy proposed that while Marx's analysis of capitalism stressed a mode of production that operated through the alienation of labor, today's "biomedical mode of (re)production" operates through the alienation of bodily substances (Thompson, 2005, pp. 11–12).

From this perspective, one might argue that assisted reproductive technologies, especially *in vitro* fertilization, offer a paradigmatic expression of gendered bioeconomies today. Though technologically assisted reproduction such as intra-uterine insemination has long been practiced in a variety of animal populations, since the 1978 success of human IVF in the United Kingdom, ARTs have become commonplace and highly profitable, resulting in what some call an IVF industry (Spar, 2006). IVF relies on the isolation and movement of reproductive substances out of a body (in one form) and into a body (in another form). This isolation of eggs, sperm, and other cells from their surroundings – in order to be analyzed, evaluated and eventually combined – opens up the possibility for each of these substances to be mobilized through the market (Cooper, 2007). While others have focused on the key role of the embryo to IVF and its associated transfers (Franklin, 2013a; Morgan, 2003, 2009), in this section I concentrate on the impacts of IVF on the conceptualization, circulation, and economization of egg and sperm.

Social scientists such as Emily Martin have investigated the ways in which evaluations of egg and sperm by scientific and medical communities often occur in a gendered manner, with stereotypical female and male characteristics attributed to substances (Martin, 1991). Taking this observation into an assessment of the economy of egg and sperm "donation," Rene Almeling's (2011) analysis of the now multi-billion dollar human egg and sperm market argues that cultural ideas about men and women create gendered economic conditions for the collection and dissemination of sex cells. Almeling shows that, in the United States, differential values attached to egg and sperm lead men who "donate" sperm to be understood as employees performing work, while women who donate eggs are viewed as altruistic gift givers (Almeling, 2011, p. 11).

Almeling argues that a seemingly symmetrical act of donation is interpreted through gendered understandings of women and men's reproductive value(s). While one might argue that gender is not symmetrical (Strathern, 1988), nor are sex cells mirror images of one another, Almeling's sociological analysis is useful in that it shows how reproductive biovalue is incalculable without considerations of gender. Additionally, both sperm and egg "donation" continue to problematically operate through the premise that certain desired traits (intelligence, musical ability, etc.) are to some extent contained in donor's genes and transferable via germ cells, making certain ethnic or educated "stock" more valuable on the cellular marketplace (Almeling, 2011; Tober, 2001).

Others researching sperm in particular have argued for a more situated understanding of semen donation. Diane Tober's ethnographic research on semen as "gift and goods" shows that while men at a California sperm bank were only secondarily motivated by altruistic intentions, the paying recipients of sperm frequently described the substance as an altruistic gift (Tober, 2001). More recently Sebastian Mohr has intentionally sidestepped the question of donor motivation in order to instead focus on how morals and masculinity are shaped through the experience of being a sperm donor in Denmark (Mohr, 2014). Similarly foregrounding national contexts of masculinity, but simultaneously broadening her argument to account for global similarities, Marcia Inhorn points to differential valuations of reproductive substances and acts as a by-product of persistent patriarchy (Inhorn, 2012). In summary, this ethnographic literature shows that systemic asymmetries between women and men, as well as their associated germ cells, create national and transactional nuances in interpretations and experiences of donation.

In recent years another gendered sector of the market in sex cells has grown: egg freezing or *oocyte cryopreservation*. This technique was initially developed for human use by cancer patients in an effort to preserve their fertility while undergoing harmful radiation treatments. Today IVF clinics increasingly offer "social egg freezing" or "fertility preservation" services, primarily to highly educated women in their mid-30s who wish to find a suitable partner prior to pursuing motherhood (Baldwin et al., 2015). Critically medicalized as "anticipatory infertility" (Martin, 2010), scholars of reproduction such as Lucy van de Wiel have argued that the function of egg freezing goes beyond fertility preservation, into the realm of calming anxieties about retaining the potential for future motherhood (van de Wiel, 2015). Such anxieties, one might add, have proven to be profitable domain for fertility clinics who are increasingly banking (on) eggs with uncertain futures.

Egg freezing joins other ARTs such as IVF in being an "'ambivalent topic' for feminism" (Franklin, 2013b, p. 186). Some scholars contest that while not an ideal choice, egg freezing provides a kind of "fertility insurance policy" for those women caught in economic and social structures that make them choose between career and children (Inhorn, 2013). Others have criticized this interpretation (Lockwood and Johnson, 2015), pointing to the nearly experimental standing of egg freezing, as well as the commercialization it brings to yet another area of female reproductive life (Morgan and Taylor, 2013). Regardless of one's take on the liberating or exploitative potential of egg freezing in particular or ARTs in general, it is clear that the increased use of IVF has given way to emergent human germ cell markets in the United States, United Kingdom and beyond.

Reproductive activities

Today ARTs are commonly used across the globe (Inhorn and Balen, 2002). IVF in particular is increasingly global (Franklin, 2007), yet national histories and regulations of IVF vary greatly (Franklin and Inhorn, 2016). These historical and regulatory differences have created new markets in what has been called "reproductive tourism" (Nahman, 2016), reproductive travel (Hudson and Culley, 2011), and "cross-border reproductive care" (CBRC). Though motivated to engage in CBRC by a variety of complex reasons (Inhorn and Gürtin, 2011), fertility patients often travel in order to access services or technologies that are prohibited, expensive or otherwise unavailable at home. The widespread increase in reproductive travel to and from stratified medical markets around the world has led to research on how configurations of economy, ethics, and infrastructure create differentiated technological capabilities, regulatory apparatuses and types of patient (in)accessibility.

Stressing how economic stratification and ethical differentiation plays out in the increased commodification of reproductive activities, Melinda Cooper and Catherine Waldby's book *Clincial Labor* (2014) describes a post-Fordist landscape where reproduction has moved out of the household and into the labor market. This characterization of the role of (re)production in the (bio)economy is somewhat familiar (Rapp, 2015; Thompson, 2005). However, Cooper and Waldby analyze reproductive labor, in the form of surrogacy, alongside other types of "in vivo labor." Pointing to similarities between surrogacy, clinical trial research participation, and organ donation, Waldby and Cooper locate a more generalizable pattern of transnational precarious labor economies. These precarious bioeconomies are often gendered, and always dependent upon what Lawrence Cohen would call "bioavailable" populations (Cohen, 2005). Like Adriana Petryna (Petryna, 2009), Cooper and Waldby stress that the pursuit of biomedical progress surrounding ARTs, pharmaceutical development and tissue regeneration depends upon the participation of often-disenfranchised human populations in such global bioeconomies.

Kalindi Vora's (2015) book *Life Support* also connects studies of surrogacy to other forms of labor, offering a post-colonial reimagining of surrogacy as well as what others might call "emotional labor" (Hochschild, 2002). Focused on India, which is the classic site for research on transnational gestational surrogacy (Singh, 2014), Vora tries to expand attention to various kinds of supportive labor. Vora argues that the labor she investigates centers around the cultivation and transfer of "vital life essence," both in material *and* immaterial forms. Here gendered and racialized histories of British colonialism are important because, according to Vora, they have left a legacy of vital essence exchange, not only in biological substances such as blood and organs, but also in forms of affective labor. Vora writes, "any analysis of biocapital must engage its roots in colonial labor allocation as a project of the racialization and gendering of labor" (Vora, 2015, p. 3). Her work could be described as an intersectional critique of biocapitalism and its scholarship which, she argues, often fails to consider that the body is not only revitalized biologically, but also affectively.

While Kean Birch argues that those researching bioeconomies tend to focus on "the 'biological' or 'material'" rather than the "political-economic" (Birch, 2016), feminist scholars of reproductive labor tend to conduct research that blurs this dichotomy all together. The circulation and valuation of material *and* immaterial phenomenon are being analyzed in tandem, and the definitions of reproduction, labor, and regenerative potential are being stretched to include physical and emotional exchanges and processes. Taking the biological not as a given, but as a preoccupation rooted in a cultural fascination with the nature/culture dichotomy, research on gendered bioeconomies is increasingly stretching understandings of both reproduction and the economy. Such research reasserts a longstanding commitment of feminist science studies – to show how the biological is also cultural, the material inherently semiotic (Haraway, 1991).

Genomic technologies

In addition to germ cell and labor markets, reproduction has been increasingly commodified and commercialized through emergent genetic markets (Ettorre et al., 2006). Many of these markets have been created alongside ARTs, which not only produce babies, but also create many other gendered things: understandings of the body, reproductive expectations and, importantly for this chapter, pharmaceutical, medical, scientific, and labor markets (Franklin, 1997). This section emphasizes the peripheral gendered bioeconomies emerging in and through the reproductive and regenerative aspects of genomic technologies.

A variety of genomic technologies have developed alongside IVF including Pre-Implantation Genetic Diagnosis (PGD). PGD facilitates the genetic selection of embryos during the IVF process, after egg and sperm have been joined and before embryos are transferred. This technique allows physicians to screen embryos for some single-gene diseases, such as Tay-Sachs or sickle-cell anemia, or for "aneuploidy" which may result in genetic abnormalities leading, for instance, to Turner's Syndrome. PGD can be understood as an extension of IVF into a genetic domain, hence it is often associated with "designer babies" (Franklin and Roberts, 2006). While popular representations often give an impression that PGD creates a genetic marketplace in which elite patients access "reprogenetic" technologies to improve future children, Sarah Franklin and Celia Roberts' multi-sited ethnography of PGD presents a more complex picture. Instead of market economies, they stress the importance of economies of hope in clinical settings, where expectations of successful treatment outcomes are both managed and sustained through clinical interactions. These economies of hope have diverse functions and temporalities, depending on one's relationship to PGD. As Franklin and Roberts write, "PGD patients and clinicians consequently occupy different economies of hope, since hope can 'run out' for a couple but not for a clinic (or the science of PGD more broadly)" (Franklin and Roberts, 2006, p. 213).

Over the last decade, PGD has become more common, and the procedure is now increasingly used for sex selection or "family balancing," potentially heightening the reproductive burden of women to not only reproduce perfectly healthy children, but also perfectly gender-balanced families (Chadwick, 2009). Even more recent technological developments such as pronuclear transfer, often called "three-parent IVF," raise similar issues. Like PGD, mitochondrial replacement technologies are explicitly developed for the medical purpose of avoiding inherited conditions, namely mitochondrial disease. But public representations of the technology often stress the bioethical debates surrounding potential eugenic or "unnatural" uses. While this public debate persists, scholars such as Donna Dickenson argue that regardless of its perceived bioethical dilemmas the availability and use of "three-parent IVF" to avoid genetic disorders increases the physical and emotional burden of reproducing the best children, especially for women (Dickenson, 2016).

Such arguments show the persistent salience of early social scientific findings on reproductive technologies such as the influential work of Rayna Rapp, who showed that women often take on the burden of reproductive decision-making during prenatal testing (Rapp, 1999). Now routine prenatal diagnostic procedures such as amniocentesis and NIPT (non-invasive prenatal testing) not only offer pregnant women more "choice," but also further biomedicalize the burdens of pregnancy through calculations of genetic risk (Reed, 2012; Zhu, 2013). The use of such "selective reproductive technologies" often reinforces preexisting inequalities and hierarchies (Gammeltoft and Wahlberg, 2014). As with PGD, in the case of prenatal testing new genomic technologies both produce and potentially mitigate parental fear, creating not only economies of hope but also economies of risk.

A similar critique has been developed of genetic testing for the BRCA gene. Pointing to a significant risk of ovarian and breast cancer, the BRCA gene has been utilized as a marker of high risk. But beyond simply an indicator of potential cancer, Kelly Happe's work shows that BRCA has become its own risk factor. The gendered and ethnic impacts of the routinization of cancer susceptibility tests, she argues, result in a discourse of rational decision-making and risk with the result of "diseasing" women's reproductive organs (Happe, 2006). Much like prenatal testing, the routinization of BRCA tests is influenced by the commercialization of genomic technologies and, especially in the United States, by a commercialized medical marketplace where prevention is underemphasized except when it results in major medical procedures such as surgery. Again like prenatal testing, such genetically based "prophylaxis" recreates already established gender and ethnic hierarchies and inequalities (Gibbon, 2006; Hall and Olopade, 2006).

Conclusion

In 2007 Kaushik Sunder Rajan argued: "one can understand emergent biotechnologies such as genomics only by simultaneously analyzing the market frameworks within which they emerge" (Rajan, 2006, p. 33). What might bioeconomic research imperatives look like today? Perhaps what this limited review of gendered bioeconomies literature makes clear is that the demands of analysis have increased. Not only are biotechnologies intertwined with markets, but they are also entangled with gendered and racialized bodies and biological substances, emerging from histories and contemporary entrenchments of what Rayna Rapp has called "stratified reproduction" (Rapp, 2001). Scholars continue the effort of feminist social scientists like Rapp, continuing to drag reproduction to the center of, this time, economic theory, thereby reminding readers that economies do not function outside of gender, or ethnicity.

From the earliest and arguably most persuasive strain of biocapital scholarship to more recent efforts to rethink surrogacy and selective reproductive technologies, bioeconomies continue to be gendered, in both senses of the term. Through this dual frame of gender, as a feminized dismissal and an embodied analytic, scholars continue to draw attention to the potential alienation and exploitation of women and their reproductive substances and labor, but also the primary role that reproduction and reproductive technologies broadly defined play in the bioeconomy more generally. Perhaps in light of such research, it might be established that an understanding of emergent biotechnologies today requires an analysis of gender as much as an analysis of the market frameworks in which they emerge.

Bibliography

Almeling, R., 2011. *Sex Cells: The Medical Market for Eggs and Sperm*. University of California Press, Berkeley.

Baldwin, K., Culley, L., Hudson, N., Mitchell, H., Lavery, S., 2015. Oocyte cryopreservation for social reasons: demographic profile and disposal intentions of UK users. *Reproductive BioMedical Online* 31.

Birch, K., 2016. Rethinking Value in the Bio-economy: Finance, Assetization, and the Management of Value. *Sci. Technol. Hum. Values* 42(3), 460–490. doi:10.1177/0162243916661633.

Chadwick, R., 2009. Gender and the Human Genome. *Mens Sana Monogr.* 7, 10–19. doi:10.4103/0973-1229.44075.

Cohen, L., 2005. Operability, Bioavailability, and Exception, in: Ong, A., Collier, S.J. (Eds.), *Global Assemblages: Technology, Politics, and Ethics as Anthropological Problems*. Blackwell Publishing, Malden, MA, pp. 79–90.

Cooper, M., 2007. Life, Autopoiesis, Debt. Distinktion *J. Soc. Theory* 8, 25–43. doi:10.1080/1600910X.2007.9672937.

Cooper, M. and Waldby, C., 2014. *Clinical Labor: Tissue Donors and Research Subjects in the Global Bioeconomy*. Duke University Press, Durham.

Dickenson, D., 2016. *Feminist Perspectives on Hum Genetics and Reproductive Technologies*, John Wiley & Sons, Ltd, Chichester, UK, pp. 1–5.

Ettorre, E., Rothman, B.K., Steinberg, D.L., 2006. Feminism Confronts the Genome: Introduction. *New Genet. Soc.* 25, 133–142. doi:10.1080/14636770600855176.

Franklin, S., 2013a. Embryo Watching: How IVF Has Remade Biology. *Tecnoscienza Ital. J. Sci. Technol. Stud.* 4, 23–44.

Franklin, S., 2013b. *Biological Relatives: IVF, Stem Cells, and the Future of Kinship.* Duke University Press, Durham.

Franklin, S., 2007. Stem Cells R Us: Emergent Life Forms and the Global Biological, in: Ong, A., Collier, S.J. (Eds.), *Global Assemblages.* Blackwell Publishing Ltd, pp. 59–78.

Franklin, S., 1997. *Embodied Progress: a Cultural Account of Assisted Conception.* Routledge, London; New York.

Franklin, S., Inhorn, M.C., 2016. Introduction. *Reprod. Biomed. Soc. Online* 2, 1–7. doi:10.1016/j. rbms.2016. 05. 001.

Franklin, S., Lock, M. (Eds.), 2003. *Remaking Life and Death: Toward an Anthropology of the Biosciences,* 1 edition. ed. School for Advanced Research Press, Santa Fe: Oxford.

Franklin, S., Roberts, C., 2006. *Born and Made: an Ethnography of Preimplantation Genetic Diagnosis.* Princeton University Press, Princeton.

Gammeltoft, T.M., Wahlberg, A., 2014. Selective Reproductive Technologies. *Annu Rev Anthropol.* 43, 201–216. doi:10.1146/annurev-anthro-102313-030424.

Gibbon, S., 2006. *Breast Cancer Genes and the Gendering of Knowledge: Science and Citizenship in the Cultural Context of the "New" Genetics.* Springer, London.

Hall, M.J., Olopade, O.I., 2006. Disparities in Genet Test: Thinking Outside the BRCA Box. *J. Clin. Oncol.* 24, 2197–2203.

Happe, K.E., 2006. Heredity, Gender and the Discourse of Ovarian Cancer. *New Genet. Soc.* 25, 171–196. doi:10.1080/14636770600855226.

Haraway, D.J., 1991. *Simians, Cyborgs, and Women: The Reinvention of Nature.* Routledge, New York.

Helmreich, S., 2008. Species of Biocapital. *Sci. Cult.* 17, 463–478. doi:10.1080/09505430802519256.

Hochschild, A., 2002. Emotional labour. *Gender: A Sociological Reader.* Jackson, S., Scott, S., (Eds.), 192–196.

Hudson, N., Culley, L., 2011. Assisted Reproductive Travel: UK Patient Trajectories. *Reprod. Biomed. Online* 23, 573–581. doi:10.1016/j.rbmo.2011. 07. 00doi:4.

Inhorn, M.C., 2013. Opinion: Women, Consider Freezing Your Eggs [WWW Document]. CNN. URL www.cnn.com/2013/04/09/opinion/inhorn-egg-freezing/index.html (accessed 16 February 2017).

Inhorn, M.C., 2012. *The New Arab Man: Emergent Masculinities, Technologies, and Islam in the Middle East.* Princeton University Press, Princeton, NJ.

Inhorn, M.C., van Balen, F., 2002. *Infertility Around the Globe: New Thinking on Childlessness, Gender, and Reproductive Technologies.* University of California Press, Berkeley.

Inhorn, M.C., Gürtin, Z.B., 2011. Cross-border Reproductive Care: A Future Research Agenda. *Reprod. Biomed. Online* 23, 665–676. doi:10.1016/j.rbmo.2011. 08. 00doi:2.

Lockwood, G., Johnson, M., 2015. Having It All? Where Are We With "Social" Egg Freezing Today? – ClinicalKey. *Reprod. Biomed. Online* 31, 126–127.

Martin, E., 1991. The Egg and the Sperm: How Science Has Constructed a Romance Based on Stereotypical Male-Female Roles. *Signs* 16, 485–501.

Martin, L.J., 2010. Anticipating Infertility: Egg Freezing, Genetic Preservation, and Risk. *Gend. Soc.* 24, 526–545. doi:10.1177/0891243210377172.

Mohr, S., 2014. Beyond Motivation: On What It Means To Be A Sperm Donor In Denmark. *Anthropol. Med.* 21, 162–173. doi:10.1080/13648470.2014.914806.

Morgan, L., 2009. *Icons of Life: A Cultural History of Human Embryos.* University of California Press, Berkeley.

Morgan, L., 2003. Embryo tales, in: Lock, M., Franklin, S. (Eds.), *Remaking Life and Death: Toward an Anthropology of the Biosciences.* School for Advanced Research Press, Santa Fe: Oxford, pp. 261–291.

Morgan, L.M., Taylor, J.S., 2013. Op-Ed: Egg Freezing: WTF?*. *Fem. Wire.*

Nahman, M.R., 2016. Reproductive Tourism: Through the Anthropological "Reproscope." *Annu Rev Anthropol.* 45, 417–432. doi:10.1146/annurev-anthro-102313-030459.

Petryna, A., 2009. *When Experiments Travel: Clinical Trials and the Global Search for Human Subjects.* Princeton University Press, Princeton.

Rapp, R., 2015. Clinical Labor: Tissue Donors and Research Subjects in the Global Bioeconomy by Melinda Cooper, Catherine Waldby (review). *Bull Hist Med.* 89, 370–371. doi:10.1353/bhm.2015.0058.

Rapp, R., 2001. Gender, Body, Biomedicine: How Some Feminist Concerns Dragged Reproduction to the Center of Social Theory. *Med. Anthropol. Q.* 15, 466–477. doi:10.1525/maq.2001.15.4.doi:466.

Rapp, R., 1999. *Testing Women, Testing the Fetus: The Social Impact of Amniocentesis in America*. Routledge, New York.

Reed, K., 2012. *Gender and Genetics: Sociology of the Prenatal*. Routledge, Oxford; New York.

Singh, H.D., 2014. "The World's Back Womb?": Commercial Surrogacy and Infertility Inequalities in India. *Am. Anthropol.* 116, 824–828. doi:10.1111/aman.12146.

Spar, D.L., 2006. *The Baby Business: How Money, Science, and Politics Drive the Commerce of Conception*. Harvard Business Review Press, Boston.

Strathern, M., 1988. *The Gender of the Gift: Problems With Women and Problems With Society in Melanesia*. University of California Press, Berkeley.

Sunder Rajan, K.S., 2006. *Biocapital: The Constitution of Postgenomic Life*. Duke University Press, Durham, NC.

Thompson, C., 2005. *Making Parents: The Ontological Choreography of Reproductive Technologies*. MIT Press, Cambridge, MA.

Tober, D.M., 2001. Semen as Gift, Semen as Goods: Reproductive Workers and the Market in Altruism. *Body Soc.* 7, 137–160. doi:10.1177/1357034X0100700205.

van de Wiel, L., 2015. Frozen in Anticipation: Eggs for Later. *Womens Stud. Int. Forum* 53, 119–128. doi:10.1016/j.wsif.2014. 10. 01doi:9.

Vora, 2015. *Life Support: Biocapital and the New History of Outsourced Labor*. Univ of Minnesota Press.

Zhu, J., 2013. Projecting Potentiality: Understanding Maternal Serum Screening in Contemporary China. *Curr Anthropol.* 54, S36–S44. doi:10.1086/670969.

11

Genomic hope: promise in the bioeconomy

Paul Martin

The field of genomics has attracted massive public and private investment over the last 20 years. It has been surrounded by high expectations that it will transform medicine and healthcare. These hopes provided the foundation for the creation of the genomics and biopharmaceutical industry, the adoption of genomic tools across life sciences research, and the integration of genomic knowledge into the core processes of the bio/pharmaceutical industry. More recently, the growth of genomic knowledge has been directly linked to new discourses about the possibility of a bioeconomy based on the production and commodification of knowledge about all living systems. In particular, the bioeconomy involves the collection, extraction and analysis of biomaterials and bio-objects, and the combination of this with different forms of individual and collective data. In the context of human health this integration focuses on the sequencing of genomes and linking these to personal medical information on a population scale with the aim of developing new diagnostic tests and forms of personalised therapy. The notion of the bioeconomy only emerged in the 2000s and was largely absent from public and policy discourse before then, but has since become one of the most important justifications for public investment in genomics.

However, progress in translating this new knowledge into widely used clinical applications or large numbers of genome-based drugs and diagnostics has been slow (Manolio et al., 2013; Burke & Korngiebel, 2015). Several new drugs of this sort have now reached the market (e.g. imatinib [Gleevec] and dasatinib [Sprycel]), but there have been high attrition rates of many genomic-based drugs in clinical trials, with only a small number of successful new therapies. There has been more success with diagnostics, with gene sequencing techniques providing tests for many rare genetic conditions and the somatic genotyping of cancer (Gagan & van Allen, 2015). However, very few widely used DNA-based diagnostics for common diseases have been introduced into routine clinical practice in the last decade.

Two important questions therefore arise: how has it been possible to sustain high levels of public and private support and investment for genomics over a long period in the face of limited tangible benefits for healthcare? How has support for the idea of the bioeconomy gathered momentum in the context of this slow progress in realising the benefits of genomics? In an attempt to address these questions and better understand the dynamics of genomic hope this chapter will focus on a key area in which these expectations have been articulated, namely official health and science policy documents. These have become important sites for the articulation of specific visions for how genomics will help improve health as well as contribute to the bioeconomy. They also provide the locus that links particular hopes and visions to practical political and policy activities.

Analytically, the chapter examines the construction of discourses of promise associated with both genomics and the bioeconomy. It will argue that the creation and maintenance of high levels of hope around the *future potential* of genomics for economic growth and competition in the bioeconomy has been key to sustaining support and investment. In doing this, it should be stressed that the concept of the bioeconomy is still 'in the making', and whilst some aspects of commercial biotechnology are now well established, important applications of genomics have yet to be successfully commercialised. In this sense, genomic hope, the application of the technology in healthcare, and the creation of key aspects of the bioeconomy are being co-produced.

Drawing on work in the sociology of expectations (Borup et al., 2006) the emphasis will be on understanding the performative function of genomic hope in winning support for particular health futures and policy initiatives. In particular, the chapter will build on existing work which has examined the role of socio-technical expectations in constituting both the field of genomics and the idea of the bioeconomy. Previous studies have considered the dynamics of future making in genomics (Hedgecoe & Martin, 2008), the role of expectations in forming specific areas of application, such as pharmacogenetics and personalised medicine (Hedgecoe, 2009a; Tutton, 2014), how particular professional discourses of hope have shaped the emergence of genomics (Hedgecoe, 2009b), the mutual constitution of hope and the commercialisation of biotechnology (Martin et al., 2008; Gislera et al., 2011; Tutton, 2011) and the bioeconomy itself (Birch et al., 2014; Haasea et al., 2015; Petersen & Krisjansen, 2015).

The UK has been chosen as the site for this analysis because it has a long track record of policy initiatives in this area and has established one of the largest programmes of investment in genomic medicine anywhere in the world.

The Genetics White Paper (2003)

Following the completion of the first draft of the human genome in 2000 the UK government started developing a policy for genetics in the NHS which culminated in the publication of a White Paper in 2003 (DoH, 2003) by the then 'New Labour' Government led by Tony Blair. This was the first policy of its kind in the UK and set out an integrated programme of investment, although on a rather modest scale with £50M of new money over three years. In particular, it articulated a coherent set of expectations around the clinical application of genetics which have continued to provide the broad template for similar hopes in all subsequent official policy documents. These were nicely summarised by John Reed, Secretary of State for Health, in the introduction:

> Advances in human genetics will have a profound impact on healthcare. Over time we will see new ways of predicting and preventing ill health, more targeted and effective use of existing drugs and the development of new gene-based drugs and therapies that treat illness in novel ways. Above all, genetics holds out the promise of more personalised healthcare with prevention and treatment tailored according to a person's individual genetic profile.
>
> *(ibid., p. 5)*

Here the emphasis was on diagnosis, targeting therapeutics and personalising therapy, with the claim that 'the Human Genome Project will pave the way for a revolution in healthcare' (ibid., p. 7). In the first section of the White Paper a series of illustrative cases were presented which compared current practice with an imagined set of future applications and users. In these scenarios, genetic technology is imagined to provide early detection of colon cancer, presymptomatic risk assessment in heart disease, better treatment of diabetes, and improved pharmacogenetic management of the anti-coagulant drug warfarin. The White Paper also discussed hopes for gene therapy, something that largely disappears from subsequent official policy in this domain.

In terms of a policy programme to enable this imagined future, the emphasis in the White Paper was on strengthening existing specialist services including testing laboratories and IT infrastructure; building genetics into mainstream services, including 'all branches of medicine', but with a focus on cancer specific screening programmes and GP networks; spreading knowledge across the NHS; empowering staff and patients through education and training on genetics; creating new knowledge and applications through research; the creation of Genetic Knowledge Parks; ongoing support for UK Biobank; and measures aimed at ensuring public confidence. Specifically it claimed that:

> The Government has an ambitious vision for harnessing these potential benefits for patients. We want the NHS to lead the world in taking maximum advantage of the safe, effective and ethical application of the new genetic knowledge and technologies as soon as they become available. . . . We intend to support specialised services and to encourage and assist clinicians and managers throughout the NHS to incorporate genetic advances into everyday clinical practice.
>
> *(ibid., p. 22)*

The ambitions of the White Paper were therefore framed as largely incremental and evolutionary, but the UK was seen as having a number of historical and institutional advantages in realising the potential of genetics: possessing a strong science base and the NHS as a comprehensive and integrated service. Central to this was a discourse of national competitiveness, with a desire that 'the NHS should lead the world'. However, policy rested on traditional ideas about enabling the production and diffusion of knowledge, with no real discussion of the NHS itself as a source of biovalue.

The only significant consideration of the economic benefits of genetics within the White Paper was the creation of five Genetic Knowledge Parks with an investment of £15M over five years. These were designed to create centres of research excellence that could collaborate with the private sector (ibid., p. 61). The key role of intellectual property (IP) was also considered, with the aim of improving the management of IP and reviewing the social and ethical issues raised by gene patents.

In summary, the White Paper was significant in signalling the importance of genetics for UK health policy and creating a coherent set of expectations that would become standard tropes. At this stage in the development of the field there were few concrete examples of the successful application of genetics in the clinic other than well-established genetic tests for classical Mendelian conditions. However, potential applications were envisaged through scenarios and high hopes were raised about the long-term potential of genetics to transform health and patient care. In terms of performativity the White Paper linked the 'new' genetics to a discourse of national competitiveness and the modernisation of the NHS, two major concerns of the New Labour government. This was largely rhetorical as there was little additional investment in genetics, but existing policy commitments, such as Biobank and the creation of an NHS IT infrastructure, were drawn into a more coherent programme. Most significantly, there was no narrative about the bioeconomy, something that would later come to dominate UK policy imaginaries in this domain.

Over the next decade a number of policy reports and initiatives would continue to place bioscience in general and genetics in particular at the heart of a science-led programme for the modernisation of the health service, but would also start to articulate the importance of the NHS itself as a source of economic advantage and value within the bioeconomy.

A strategy for human genomics in the UK (2012)

Nine years after the Genetics White Paper, the cross-departmental Human Genomics Strategy Group (HGSG) chaired by Sir John Bell published a major report setting out an integrated policy for the application of genomics in UK healthcare (HGSG, 2012). This implicitly referenced the strategy for genetics established by the White Paper and in its vision the HGSG recapitulated almost all of these earlier expectations, claiming that genomics:

> can then be harnessed to provide a greater ability to determine disease risk and predisposition, to support more accurate diagnosis and prognosis, and to select and prioritise therapeutic options in a wider set of pathological disorders. It can be used in every branch of medicine … Already, it is beginning to move 'from bench to bedside' and, as it does so, the potential for NHS-wide adoption and diffusion of genomic technology is becoming increasingly clear.
>
> *John Bell, Foreword to HGSG Report (2012, p. 3)*

Here the field of genomics had replaced genetics, but was still presented as a largely promissory domain 'in the making' with talk of seeing early signs of it 'beginning to move' and its potential for adoption and diffusion 'becoming increasingly clear'. As with the White Paper the starting point for the strategy was a clear vision (ibid., p. 3) with the claim that genomics is putting healthcare 'on the cusp of a revolution' involving the reclassification of disease and the targeting of therapeutics. The benefits of this would include more effective treatment sooner, the better use of NHS resources and highly specific and personalised treatments (p. 16).

The HGSG also echoed the White Paper in seeing genomics as critical to national economic competitiveness stating that: 'By 2020, the NHS will be a world leader in the development and use of genomic technology in the areas of healthcare and public health' (ibid., p. 17) and stressing the international strength of the UK science base and the NHS as providing the 'perfect environment in which to realise the potential of genomic medicine' (ibid., p. 16).

As with the White Paper, the HGSG report argued that the pursuit of its ambitious vision for genomics required a comprehensive series of initiatives, including: rigorous processes for establishing the clinical validity and utility of genomic tests; clear commissioning standards for clinical genetic testing; a secure and robust bioinformatics infrastructure; a healthcare workforce with the knowledge to make effective use of genomic technology; development of the UK legal framework regulating genomic data; and a co-ordinated approach to public engagement. Despite the striking similarity these policies had with the White Paper they were rather more specific and detailed and laid the foundation for very significant investment (see below).

However, there were a number of important discursive shifts within the HGSG report compared to earlier policy narratives. First, there was extensive use of the language of innovation, with a focus on the development and application of new technologies by and within the NHS. In a section on 'Changing the R&D model' (ibid., p. 26) it is noted that the pharmaceutical industry has a clear interest in the development of stratified medicine because of its unsustainable R&D model, which has resulted in a decreasing number of novel drugs at the same time as rising R&D costs (the so called 'productivity crisis' in the industry). In this context, the development of new innovation models based on public–private partnerships is framed as part of the solution to this crisis.

Secondly, the report reflected frustration with the rate of progress in adopting genomics in the clinic and acknowledged that 'There is, and has long been, a significant gap between the worlds of cutting-edge biomedical research and everyday healthcare' (ibid., p. 16). In particular, it warned that if this programme of reform and investment was not adopted the lead that the UK currently holds in this field could be severely, and perhaps permanently, undermined. The consequences of this would be bad for entrepreneurial opportunities across industry, research and academia. What is striking here is the move to justify major organisational change and investment not only on grounds of improved patient care, but also in promoting innovation and entrepreneurship within the bioeconomy.

Third, in contrast to the White Paper and its construction of imagined futures, the HGSG report combined contemporary examples of the use of genomics in the clinic with predictions about future applications. In Chapter 2, 'The potential impact of genomic technology in the NHS and clinical care' a series of case studies were presented, including the stratification of cancer treatment using specific genomic tests in combination with drugs such as imatinib (Gleevec, approved 2006), dasatinib (Sprycel, approved 2006), and vemurafenib (Zelboraf, approved 2011). Other examples relating to pharmacogenetic testing for drug response were also given, including gefitinib (Iressa, approved 2003), carbamazepine (generic drug, genetic test introduced in 2011), and Abacavir (approved 1998, test for hypersensitivity 2008). However, whilst these cases demonstrate proof of principle for the idea of disease stratification and pharmacogenetic testing, with Gleevec and Sprycel having blockbuster status, little is said about the rate of adoption. In this respect, it should be noted that the majority of these examples were well-established by 2012 and had been used in numerous reports during the 2000s to promote genomic medicine. Relatively few new examples were given by the HGSG report and no genetic tests were referred to for diagnosis of common diseases, a key promise at the launch of the HGP.

Finally, the report stresses the scale of the transformation within the NHS that is required, but justified this on the grounds that the potential benefits are 'immense'. For the HGSG this would require very significant funding as 'mainstreaming will come at a cost', but the Group argued that the potential of genomics merits such investment. One of the main reasons for this is that genomics, and biomedical sciences more broadly, were vital growth opportunities for the UK economy. The widespread adoption of genomics in mainstream clinical practice would therefore accelerate the commercial translation of those innovations and help fulfil the economic potential of genomics for the UK (p. 31).

In summary, the HGSG reiterated what has become a canonical set of hopes for genomic medicine with expectations of revolutionary change for clinical practice and patient outcomes. The use of an imagined future articulated through a vision statement, promissory language and scenarios performed a compelling role in justifying a series of policy initiatives and demands for resources based on improving health. As with the White Paper this was also linked to the twin projects of national competitiveness and the modernisation of public services. However, new discourses associated with the growth of the knowledge-based bioeconomy were also mobilised in the name of innovation, support for industry and creating value from the NHS itself. These were based on a series of new hopes about the economic benefits of genomics.

There was also a sense of progress within the HGSG report which rested on the concrete examples of the use of DNA tests to target therapy. However, the report is ambivalent about this. On the one hand case studies of existing practice are used to provide grounds for further hope, but the lack of progress is also acknowledged and used as the basis for claiming more resources and radical organisational change. As the field has progressed hopes associated with some technologies and applications have been emphasised (drug/test combinations), whilst others were quietly forgotten (gene therapy and genetic tests for common diseases).

In terms of performativity the HGSG report provided a powerful narrative to justify a programme of major investment and reorganisation of the NHS in order to realise the benefits of genomics for health, wealth and national competitiveness. In this context it is striking that less than a year after its publication, Prime Minister David Cameron announced a commitment to sequence 100,000 whole genomes over the next three to five years (Prime Minister's Office, 2012). Following an initial investment of £100M, this project, under the auspices of a dedicated delivery organisation, Genomics England, would go on to become the largest biomedical research and development project ever funded by the UK government.

Genomics England and the 100K Genomes Project (2013)

Genomics England Limited (GeL) was founded as a government owned 'for profit' company established to run the 100K Genomes Project, which involves conducting 100,000 whole genome sequences (WGS) on patients (or their tumours) recruited through the NHS in England. This project and the work of Genomics England was surrounded by high hopes of it becoming a world leader in genomic medicine:

> The genome profile will give doctors a new, advanced understanding of a patient's genetic make-up, condition and treatment needs, ensuring they have access to the right drugs and personalised care far quicker than ever before.
>
> *Prime Minister's Office (2012)*

The Executive Chair of GeL, Sir John Chisholm, described the project as 'the most important medical step in the 21st century' and 'in the order of the Human Genome Project' (Genomics England 2013d). In particular, it was anticipated that the 100K Genomes Project will 'deliver benefit to the community at large, both in terms of health and future wealth, and at the same time feeding back appropriate insights to the clinicians treating participating patients' (Genomics England 2013e). GeL also claimed that it will enable the UK to become the first country in the world to introduce this technology into its mainstream health system (Genomics England 2013a). The economic benefits were anticipated to emerge from several sources. First, by 'going straight to the answer rather than spending 20 years wandering around the hospital wards' (Genomics England 2013c) GeL believed that the NHS will save money previously spent on erroneous diagnostics and treatments. Second, GeL argued that the pharmaceutical industry will come to Britain to use the project infrastructure and this will lead to the development of new diagnostics and treatments (Genomics England 2013b). The formal launch of GeL coincided with the 65th Anniversary of the founding of the NHS and was directly linked to the objective of delivering a more personal service.

The initial £100M was invested 'to stay ahead in the global race' and to train a new generation of British genetic scientists to lead on the development of new drugs, treatments and cures, build the UK as the world leader in the field, to pump-prime DNA sequencing for cancer and rare inherited diseases and build the NHS data infrastructure (Prime Minister's Office, 2012).

The emphasis here was on creating the infrastructure and environment to make genomic medicine possible. However, what is most striking about GeL is its governance structure and mission, which followed a private business model, as well as its commercial style of operation. Specifically, it stated that in addition to improving public health it will be flexible, agile, able to move quickly as the market changes; participate in the market as a business talking to businesses, and ensure the benefits of the investment flows from the company to a large range of companies and contractors including SMEs (Genomics England, 2014). Central to this approach was the belief that genomics holds great promise for both improving healthcare and building the national bioeconomy, and that this can be done most effectively through the adoption of public–private partnerships run along business lines.

In the four years since its launch, a further £300M was committed by government and GeL has developed a dense network of collaborations with both domestic and international genomics companies, recruited and started sequencing patients and built an operating model and IT infrastructure to deliver the project. The massive investment into Genomics England can therefore be seen as the concrete realisation of the policy programme established by the HGSG report.

The importance of this infrastructure to future UK prosperity is highlighted in a recent prospectus designed to promote international business collaboration with UK firms and the NHS entitled 'Genomics and personalised medicine: how partnership with the UK can transform healthcare' (Department for International Trade, 2016). This is confident in tone and asserts that 'Genomics has transformed our understanding of disease and our ability to deliver care …' (ibid., p. 3) However, despite giving some examples of progress in clinical translation, the prospectus returns to a more tentative narrative of a technology still in the making and claims that by establishing the sequence of an individual's genetic material it is possible to identify sequences or mutations which are specific to that person (p. 4) and how this '… opens up the shift towards personalised treatment…' (p. 4).

The link to national economic development and competitiveness is also central to this document, with it claiming that:

> There has never been a better time to develop genomics and personalised medicine services in partnership with the UK. . . . With a clear strategy in place to increase the use of genomic information in day-to-day clinical medicine, a stream of new programmes will be introduced in the UK over the coming decade.
>
> *(ibid., p. 5)*

Here are the first discursive signs of the heavy investment made by the UK in genomic infrastructure after 2012 bearing fruit, and how this might improve health and form the basis for new industries, global trade and inward investment. Whilst concrete progress has started to become visible in these official documents, expectations remain high and very future oriented, as it is still too early to point to large-scale successes outside a few clinical niches or to system-wide transformation of services and care. However, the massive investment in infrastructure provides a platform not only for clinical research and knowledge transformation, but also for the realisation of national ambition in the knowledge-based bioeconomy. We can therefore see the ongoing and central role of promises and hope in mobilising resources and support for genomics, for linking this closely to the notion of the bioeconomy, and embedding this in a discourse that seeks economic advantage through technological innovation. These were largely absent from public and policy discourses at the time of completion of the HGP and mark a major transformation in the construction of this domain.

In considering the role of genomic hope in the emergence of a genome-based bioeconomy in the UK it should be noted that similar hopes exist in other countries and across the developed and developing world (Staffas et al., 2013). In particular, the OECD has played a key role in promoting the idea that genomics will provide significant economic benefits in the future (OECD, 2009; OECD & HUGO, 2010) and this has been adopted as a policy priority in many countries, including the United States (White House, 2012), France (Inserm, 2016), Russia (BIO2020, 2012) and China (Boterman, 2011). What is striking about these reports is their future orientation – with them often containing a vision for the application of genomics in healthcare – and a very similar set of policy recommendations to those seen in the UK promoting economic growth. Similarly, these are also linked to projects of improving national competitiveness in a global race. For example, this is quite explicit in the French report:

> This ambitious plan, overseen and supported by the State, is aimed at positioning France as a leader among the major countries involved in genomic medicine within the next ten years. Although it responds to a public health challenge in diagnostic, prognostic and therapeutic terms, this plan is also aimed at encouraging the emergence of a national medical and industrial sector for genomic medicine, and exporting this expertise.
>
> *(Inserm, 2016)*

The emergence of a consensus around hopes for the role of genomics in building a knowledge-based bioeconomy is therefore a powerful factor shaping the wider international development of the field.

Genomic hope and the making of the UK bioeconomy

This chapter has presented an analysis of some of the main shifts in UK policy discourses around genetics and genomics over the 15 years since the sequencing of the human genome. From this a number of points stand out. First is the very significant continuity in hopes, promises and expectations, with the list of potential benefits given in 2003 looking very similar to those stated in 2016. These include the idea of a biomedical revolution based on improved diagnosis and the targeting of therapeutics, leading to greater personalisation. This framework has strong similarities to the overarching 'genetic imaginary' that has remained stable since the late 1970s in much of the scientific, policy and social science literature on geneticisation (Weiner et al., 2017). However, this should not distract from important shifts in the hopes attached to specific applications, such as gene therapy or genetic testing for common conditions, which have been forgotten as scientific discovery has failed to find single genes linked to disease and technical progress in gene delivery has stalled. Previous work in the sociology of expectations (Brown & Michael, 2010) has highlighted the way in which new hopes are built on the collective forgetting of previous (unfulfilled) promises.

A second important analytical point is the rise of narratives about the genomics-based bioeconomy over this period. In 2003 the White Paper made no direct reference to this, although the notion of a science-led modernisation of the NHS was already linked to improving the UK's competitive place in the emerging global knowledge economy. By the end of this period the two themes of improving health and creating wealth were completely intertwined and the NHS was constructed as an important source of economic value and national advantage. This can be seen as the culmination of the unfolding logic of a commitment to building a knowledge economy in which public services are constructed as sites of knowledge production that can in turn become sources of value. The rise of a narrative about the place of the NHS in an emerging bioeconomy links it directly to helping solve the productivity crisis in the pharmaceutical industry, providing (anonymised) patient data to be linked with sequence information for exploitation by a growing genomics industry, and as a major purchaser of scientific services. This has then been packaged as the basis for new international investment and trade opportunities.

Third, it is important to draw attention to the performative function of genomic hope and the expectations that surround it. There are several dimensions to this relating specifically to domestic politics and the mobilisation of support and resources for particular health futures. Politically, the idea of modernising the NHS has considerable currency and the White Paper must be understood in this context. For the New Labour government this had echoes of previous narratives linking national renewal with the 'white heat' of technology. For subsequent Conservative-led administrations the emphasis was on improving productivity, and the quality and speed of services for consumers. More significantly, discourses about the potential of genomics for national economic renewal and the growth of the bioeconomy proved to be powerful factors in mobilising resources for a particular vision of the future of health and healthcare. Initially, this genomic vision served a largely rhetorical purpose, but after 2012 provided the rationale for a massive expansion of genomic research based on WGS.

Finally, it is worth returning to the questions raised at the start of this chapter: How has it been possible to sustain high levels of public and private support and investment for genomics over a long period in the face of relatively limited tangible benefits for healthcare? How has the idea of the bioeconomy developed in this context? The tentative answer suggested by the analysis presented here is centred on the dynamics of genomic hope and its performative role. What is striking is the ability of the advocates of genomic medicine to keep hope alive and centrally positioned in official policy over 15 years. This was achieved by linking powerful scientific, political and commercial interests to the hope that genomics will help improve health, modernise the NHS, create new forms of wealth, and increase the standing of the nation. At the same time, preparing for the era of genomic medicine has also been framed in terms of a project of state transformation in which the public sphere becomes another source of value to be governed by business methods and the logic of the market. Specifically, this has focused on creating an infrastructure to deliver new forms of knowledge production (based on WGS), reorganising the NHS, retraining healthcare professions, and developing a powerful socio-technical niche around Genomics England.

This still leaves a number of questions to be addressed by further research. To what extent will the hopes of genomic medicine ever be realised and over what time period? How long can genomic hope be kept alive without more widespread progress in the clinic? What changes to service delivery, professional work and patient care will be introduced in the name of genomics? Will these developments lay the foundation for new industries and the creation of national wealth, or will we remain trapped in an economy of promise? These questions are not only relevant to the UK, Europe and North America, but as shown above, have implications for the emerging global bioeconomy and the application of genomics in newly industrialising and developing nations. It is therefore vital to understand the dynamics of genomic hope and adopt a critical stance towards the claims being made in order to ensure that the benefits of genomics are not oversold and ultimately result in gains for global public health.

References

BIO2020 (2012) *Summary State Coordination Program for the Development of Biotechnology in the Russian Federation until 2020.* BIO 2020; Moscow, Russia.

Birch, K., Levidow, L. & Papaioannou, T. (2014) Self-fulfilling prophecies of the European knowledge-based bio-economy: The discursive shaping of institutional and policy frameworks in the bio-pharmaceuticals sector. *Journal of the Knowledge Economy* 5(1): 1–18.

Boterman, B. (2011) *Bio-Economie in China*. Rathenau Instituut and TWA Netwerk: The Hague, NL.

Borup, M., Brown, N., Konrad, K. & Van Lente, H. (2006) The sociology of expectations in science and technology. *Technology analysis & strategic management* 18(3–4): 285–298.

Brown, N. & Michael, M. (2010) A sociology of expectations: Retrospecting prospects and prospecting retrospects. *Technology Analysis & Strategic Management* 15(1): 3–18.

Burke, W. and Korngiebel, D. M. (2015) Closing the gap between knowledge and clinical application: Challenges for genomic translation. *PLoS Genetics*, February 26, 2015. https://doi.org/10.1371/journal.pgen.1004978.

DoH (2003) *Our Inheritance, Our Future: Realising the Potential of Genetics in the NHS*. Department of Health; London, UK.

Department for International Trade (2016) *Genomics and personalised medicine – how partnership with the UK can transform healthcare*. DTI: London, UK.

Gagan, J. & van Allen, E. M. (2015) Next-generation sequencing to guide cancer therapy. *Genome Medicine* 7: 80. DOI: doi:10.1186/s13073–13015–0203-x.

Genomics England (2013a) *About Genomics England*. Available at: www.genomicsengland.co.uk/100k-genome-project.

Genomics England (2013b). *Genomics England town hall engagement event. Afternoon: For clinicians, scientists, healthcare professionals, funders and allied organisations. Part 1*. Available at: www.genomicsengland.co.uk/town-hall-engagement-event.

Genomics England (2013c) *Genomics England town hall engagement event. Afternoon: For clinicians, scientists, healthcare professionals, funders and allied organisations. Part 2*. Available at: www.genomicsengland.co.uk/town-hall-engagement-event.

Genomics England (2013d) *Genomics England town hall engagement event. Morning: for public, patients and patient support charities. Part 1*. Available at: www.genomicsengland.co.uk/town-hall-engagement-event.

Genomics England (2013e) *Genomics England town hall engagement event. Morning: for public, patients and patient support charities. Part 2*. Available at: www.genomicsengland.co.uk/town-hall-engagement-event.

Genomics England (2014). How we work. Available at: www.genomicsengland.co.uk/how-we-work.

Gislera, M., Sornette, D. & Woodarda, R. (2011) Innovation as a social bubble: The example of the Human Genome Project. *Research Policy* 20(10): 1412–1425.

Haasea, R., Michieb, M. & Skinner, D. (2015) Flexible positions, managed hopes: The promissory bioeconomy of a whole genome sequencing cancer study. *Social Science & Medicine*. 130: 146–153.

Hedgecoe, A. M. & Martin, P. A. (2008) Genomics, STS, and the making of sociotechnical futures. In Hackett, E.J., *et al.* (eds) *Handbook of Science and Technology Studies*. 3rd Edition. MIT Press: Cambridge, MA. Pp. 817–839.

Hedgecoe, A. M. (2009a) *The Politics of Personalised Medicine: Pharmacogenetics in the Clinic*. Cambridge University Press: Cambridge, UK.

Hedgecoe, A. M. (2009b) Bioethics and the reinforcement of socio-technical expectations. *Social Studies of Science* 40(2): 163–186.

HGSG (2012) *Building on Our Inheritance: Genomic Technology in Healthcare. A Report by the Human Genomics Strategy Group*. HGSG, London. Available from: www.gov.uk/government/publications/genomic-technology-in-healthcare-building-on-our-inheritance.

Inserm (2016) *Presentation of the French Plan for Genomic Medicine 2025*. Press Release, 23rd June 2016. Inserm, Paris. Available from http://presse.inserm.fr/en/presentation-of-the-french-plan-for-genomic-medicine-2025/24328/. Accessed: 16th May 2017.

Manolio, T. A., *et al.* (2013) Implementing genomic medicine in the clinic: the future is here. *Genet Med.* 15(4): 258–267.

Martin, P., Brown, N., & Turner, A. (2008) Capitalising hope: the commercial development of umbilical cord blood stem cell banking. *New Genetics and Society* 27(2): 127–143.

OECD (2009) *The Bioeconomy to 2030: designing a policy agenda*. OECD International Futures Programme. OECD: Paris.

OECD & HUGO (2010) *Genomics and the Bioeconomy Symposium Report and Policy Considerations*. Montpellier, France, 17–18 May 2010. OECD, Paris. Available from: www.oecd.org/sti/biotech/47634101.pdf. Accessed 16 May 2017.

Petersen, A. & Krisjansen, I. (2015) Assembling 'the bioeconomy': Exploiting the power of the promissory life sciences. *Journal of Sociology*. 51(1): 28–46.

Prime Minister's Office (2012) *DNA tests to revolutionise fight against cancer and help 100,000 NHS patients.* Press Release, Prime Minister's Office, London, 10th December, 2012.

Staffas, L., Gustavsson, M. & McCormick, K. (2013) Strategies and policies for the bioeconomy and bio-based economy: An analysis of official national approaches. *Sustainability* 5: 2751–2769.

Tutton, R. (2011) Promising pessimism: Reading the futures to be avoided in biotech. *Social Studies of Science* 41(3): 411–429.

Tutton, R. (2014) *Genomics and the Reimagining of Personalized Medicine*. Ashgate: Farnham.

Weiner, K., Martin, P.A., Richards, M. & Tutton, R. (2017) Have we seen the geneticisation of society? Expectations and evidence. *Sociology of Health and Illness*, First published: 8 March 2017. DOI: 10.1111/1467–9566.12551.

White House (2012) *National Bioeconomy Blueprint*. The White House: Washington, DC.

12

Neoliberalism on drugs

Genomics and the political economy of medicine

Edward Nik-Khah

Introduction

In recent years, literature on neoliberal science and medical neoliberalism has suggested a link between recent changes in the organization of science and medicine, and economic ideas justifying them. This chapter discusses recent developments in biomedical science in terms of this work. The purpose in doing so is to illuminate forces shaping not only local developments such as genomics, but also the larger political economy of science (Tyfield et al., 2017).

Because a motley jumble of definitions for "neoliberalism" can currently be found in circulation, it is imperative at the outset to bring some precision to the term. Here, we are assisted by a growing body of work in the history of economics that has devoted itself to sharpening understanding of neoliberalism (e.g., Mirowski, 2013; Nik-Khah 2011a, 2017; Van Horn and Mirowski, 2009). While neoliberalism turns out to be heterogeneous, it is not featureless. Quite apart from the blandly "pro-market" position neoliberals are often presumed to support, they advance a distinct and unprecedented view of markets. The most distinctive feature of neoliberalism concerns knowledge – where it is located and what grants access to it. Hence, epistemology lies at the heart of the neoliberal worldview. Neoliberal epistemology has instructed expansionary – indeed, imperialistic – activities across the disciplines, advancing a thoroughgoing critique of claims of regulators, elected representatives, and even the community of scientists to access knowledge. Accordingly, there is a direct link between neoliberalism and science.

Moreover, neoliberals have suggested radical reforms in the face of this widespread and insurmountable ignorance. Not content to confine their activities to those areas traditionally understood to be the province of economics, neoliberals have sought influence over the organization of science. They continue to do so: the promise of future developments in biomedical science, including genomics, personalized medicine, and translational medicine, has aroused considerable interest within neoliberal circles, drawing extensive (and influential) commentary from them on scientific practice. For example, Stuart Hogarth (2015) has shown how neoliberal appointees to the US Food and Drug Administration (FDA) launched the Critical Path initiative to reengineer pharmaceutical science for the purpose of bringing about a better match (in their view) between the process of drug development and post-genomic biomedicine. This chapter seeks to understand the reasons for neoliberals' interest in genomics, and the ends to their activities are directed. It finds an answer in neoliberal epistemology: genomics may facilitate the unobstructed operation of the marketplace of ideas.

The marketplace of ideas

Neoliberalism is a politico-intellectual project, first devised as a response to socialism, social welfare liberalism, and – importantly – classical (laissez faire) liberalism. It is a collective effort that spans academic departments, general-purpose think tanks, special-purpose institutions, and transnational institutions (initially, the Mont Pèlerin Society), and employs a variety of tactics to advance its proponents' views of the ideal market society.

One of the gravest mistakes made by critics of neoliberalism is to conflate it with laissez faire (classical) liberalism. Neoliberals differ from their classical liberal predecessors in arguing that markets should be praised not for their presumed superiority in allocating goods, or in enhancing productivity, but rather for their *epistemic* virtues: the market acts primarily as a method of information processing and conveyance, the most powerful information processor known to humankind (Lave et al., 2010; Mirowski, 2011, 2013; Mirowski and Nik-Khah, 2017; Nik-Khah, 2014; Tyfield, 2016).

A peculiar epistemology resides at the heart of the neoliberal worldview. The individual human can never match the epistemic power of markets; therefore, goes the argument, markets should assume primary responsibility for generating knowledge. Although this argument was originally developed to rebut socialist planning ambitions on the grounds that the state could never adequately supervise the market's operations, the neoliberals' stress on the market's epistemic virtues would eventually lead them to challenge the academy as the primary site of the production and ratification of knowledge. At the hands of neoliberals, the "marketplace of ideas," which for the scientist often serves as a stand-in for the ideal of subjecting scientific theories to rigorous competitive trials, transmogrified into a thoroughgoing and generalizable critique of academic science: the government could never know enough to plan a complex economy, and the scientist had no privileged access to truth outside of the market (Nik-Khah, 2017).

This view guided neoliberals as they set out in the 1970s to forge enduring relationships with the pharmaceutical industry (Nik-Khah, 2014), and continues to guide the phalanx of think tanks and special purpose institutions that is the fruit of this union. It includes the American Enterprise Institute, the Competitive Enterprise Institute, the Independent Institute, the Manhattan Institute (particularly its Project FDA), the Mercatus Center, the Becker Friedman Institute (especially its Health Economics Initiative), and the Center for the Study of Drug Development to name only the most influential in the United States. In recent decades, these institutions have assumed primary responsibility for generating, circulating, and ratifying critiques of the performance of drug regulation and the organization of pharmaceutical science.

Take, for example, the efforts of the Center for the Study of Drug Development (CSDD), which is well known as the primary purveyor of figures for the "costs" of drug development in the United States. The CSDD qualifies as a neoliberal institution in that it was created in partnership with neoliberals (mostly affiliated with the Chicago School of Economics) for the express purpose of cultivating and advancing neoliberal arguments to retask pharmaceutical regulation and restructure pharmaceutical science. It has maintained a close affiliation with other neoliberal institutions, due in part to sharing members with them as well as to its adoption of a philosophical orientation immediately recognizable to fellow neoliberals. For both of these reasons, studies on pharmaceuticals produced by neoliberals include CSDD studies as obligatory references in their own works. To wit:

> A chief source of information about drug development and approval is the Tufts Center for the Study of Drug Development. Their information is often mined and analyzed from a libertarian perspective by researchers at the Competitive Enterprise Institute.
>
> *(Klein, 2000, p. 100, n. 5)*

This observation, from an academic economist who is also affiliated with the Competitive Enterprise Institute provides something of an insider's perspective on the collective effort to mobilize neoliberal ideas.

Think tanks have generated and promulgated a variety of easily digestible numbers – in addition to the cost of drug development, the length of a purported "drug lag," the "consumer surplus" of using drugs, and tallies of new molecular entities approved. The important point to realize is that the primary function of these numbers was never to facilitate public scrutiny of the markup practices of pharmaceutical companies, nor to launch a robust debate over whether pharmaceutical firms put scientific resources to their best uses. Instead, the intended lesson was that it was futile to try to outperform the marketplace of ideas. High "drug development costs," or an elongated "drug lag," served as evidence of the dangers inherent in the regulator supplanting the market's rightful role in generating and promulgating knowledge.

Although these studies have played feature roles in policy debates, it would be a mistake to conclude the only intended purpose of such work was persuasion. Indeed, one of the most distinctive neoliberal beliefs is that because most people are predisposed against markets, efforts to persuade the public are futile. Consequently, neoliberals also use such measures to audit and monitor, and thereby control regulators, in order to use the regulatory body as a firewall against democratic overreach. This intention was first clearly articulated in the general case by the Chicago School economist George Stigler (1973), and the approach was developed for the specific case of drugs by his protégé Sam Peltzman (1973). Neoliberals have long argued that new technologies always supersede the ability of the state to engage in effective regulation, urged regulators that cost-benefit procedures should inform regulatory decisions, and then supplied them with sanctioned measures of costs and benefits. Due in large part to the influential participation of neoliberals in the policymaking process, as well as in running and staffing the FDA, regulators have come to embrace this view, and neoliberals are only too happy to make common cause with this form of regulation. This is yet another way that neoliberals depart from their liberal predecessors. They called not for *laissez-faire* but for *activism*; they developed methods to control the state and fostered think tanks, which would stand at the ready to help exert such control.[1] Hence, think tanks not only promulgated neoliberal views; they exemplified them.

No set of performance measures can fully eliminate the ever-present threat of the obstruction of the marketplace of ideas by the human element. Neoliberals have turned such seemingly intractable problems into opportunities for entrepreneurs. One recent project sponsored by the Becker Friedman Institute at the University of Chicago (Jørring et al., 2017) introduced the "FDA Hedge," a financial instrument designed to allow pharmaceutical firms to offload "FDA approval risk" onto investors for profit.[2] Hence, an important element of neoliberals' practical success is their maintenance of facilities to incubate novel "solutions" and vet them for deference to the marketplace of ideas.

Neoliberal science

A primary reason neoliberals have found the case of pharmaceuticals so compelling is that the epistemic position rules out not only regulatory supervision of the market, but also calls into question any unique ability to access knowledge by the community of scientists. For neoliberals, science may participate in generating knowledge only so long as it is organized correctly and its results are used in the right way. Science must endorse the epistemic superiority of the marketplace; if it fails to do so, it is illegitimate. In expressing this idea, CSDD founder Louis Lasagna put it more concretely:

> It is common for critics of the use of the marketplace as a criterion of efficacy to point to the misplaced confidences of the past – in bleeding, leeches, puking, and purging. But such practices long ago fell into disrepute, not because of the double-blind, controlled trials, but because obviously better treatments came along.
>
> *(Lasagna, 1978, p. 872)*

In decades past, the neoliberal view of science often informed complaints about the way *academic* science produced knowledge about drugs (it took too long, it was too expensive, it was too critical, and yet, it was also somehow prone to think itself into dead ends), followed by treatises extolling the virtues of the corporate control of scientific priorities, and ending with calls for the further commercialization of science. But in an era where commercialized science is a fait accompli, the tendency has been to take the next step and re-envision science as but one part of an all-encompassing "information economy," which comprises not only doctors and consumers, but also the internet, smartphones, apps, and the like:

> We are firmly entrenched in the information economy. Consumers can go online, engage in social media, and ask as many friends and followers as possible about cars, appliances, schools, child care, vacations, lawn mowers, kitchen gadgets, and electronics before buying these products and services. Doctors can also access unprecedented amounts of data, and they can do so faster than ever before. They don't have to wait for the next conference or the next edition of a professional journal – they can share observations and outcomes instantaneously. Patients benefit because the doctor can combine specific knowledge about the individual patient with data on how similar patients responded to treatment. The medical marketplace will never be the same.
>
> *(Gulfo et al., 2016, p. 7)*

This breathless appeal to the "unprecedented amounts of data" supplied by social media has resonated far outside neoliberal think tanks (Landa and Elliott, 2013). At the hands of neoliberals, it clearly involves an expression of mistrust toward academic scientific practices, and hence amounts to a call for collecting *less* information. The diminution of the academic journal's status follows a favorite neoliberal complaint, that the pace of academic science is far too slow. Ignorance plays a fundamental role in maintaining this position:

> There is no way that pre-approval studies of drugs and devices, in tightly defined patient populations under scripted medical management protocols, can produce the kind of evidence that is available through real-world data acquisition and the Internet of Things. What's more, in the post approval, real-world setting, data that will enhance the selection of therapy for an individual patient can be made in an unprecedented manner, which can truly drive personal medicine.
>
> *(Gulfo et al., 2016, pp. 7–8)*

By flatly denying the possibility of arriving at knowledge before the drug is marketed, the authors (affiliated with the Mercatus Center) clearly intend to undermine the scientific basis for a regulatory decision. Post-market surveillance should *replace* large-scale randomized clinical trials because the former has mostly superseded the latter. Since the 1970s, neoliberals have argued that a characteristic of the well-functioning marketplace of ideas is that purchasers will pay for all the information that is appropriate for them to have (Nik-Khah, 2017). Hence, the market, and not the regulator, must instruct scientists in the information they should seek. Where randomized controlled trials (RCTs) are still used, studies using surrogate endpoints should supplant those examining final outcomes on patients' health, because requiring the latter may usurp the market's rightful role in evaluating the significance of these outcomes. In this way, neoliberals have sought to influence not only the uses of science, but also to reform its conduct.

Neoliberalism and genomics

Neoliberal think tanks have taken an intense interest in recent developments in the biomedical sciences, and house some of their most enthusiastic boosters:

> Medicine is on the cusp of a radical transformation. New sciences and technologies are poised to allow physicians to personalize treatment for every cancer patient; arrest or prevent the development of Alzheimer's disease; and radically lower health care costs by reducing the prevalence of expensive chronic diseases.
>
> Project FDA believes the FDA can become a bridge for innovation, rather than a barrier to it, and that this can be achieved without sacrificing patient safety. For instance, advances in molecular medicine that allow companies to target specific sub-groups of patients, combined with electronic health records, should allow the FDA to streamline and improve time-consuming and expensive pre-market product testing that can take a decade or more, and implement vigorous post-market surveillance of "real world" patients after drugs or devices demonstrate safety and efficacy in early testing. This approach will not only accelerate access to innovative products; it should enhance efforts to safeguard public health.
>
> *(Manhattan Institute, 2017)*

Here we encounter several stock neoliberal arguments about pharmaceutical regulation, along with something new. The stock arguments involve complaining about the expense of drug development costs, pinning the blame for these costs on regulation, calling for shifting emphasis to post-market surveillance, contending that new technologies will always supersede the ability of the state to regulate them, and nevertheless claiming that regulators can facilitate innovation. What is new has to do with their hinging on seemingly specific scientific developments (ones that are as yet mostly unrealized). Scrutinizing neoliberal views of these developments can shed light on neoliberal goals: which ones are pursued unconditionally and which may be safely jettisoned.

Panning out to examine a wider range of neoliberal writings on genomics and personalized medicine reveals the goal of lowering the price of therapies as inessential. Whereas some (including the Manhattan Institute, cited above) cite lower health care costs as a potential benefit of scientific advance, American Enterprise Institute scholar Thomas Stossel (2016) argues genomics may actually lead to *increases* in drug prices. But whatever the future of genomics holds, it turns out not to matter: neoliberals have lately advocated creating "Health Care Loans" as an alternative to price regulation or reformation of patent law to promote access to remedies (Montazerhodjat et al., 2016).[3] Here, as in the case of unruly regulators and the high costs of drug development, an apparent problem with the proper operation of the medical marketplace presents an opportunity to entrepreneurial innovators.

Given that genomics is usually taken to involve the collection of large amounts of molecular information to enable the fitting of therapies for distinct patient profiles, it may not be immediately apparent why neoliberals should welcome it. The answer rests with identifying how they anticipate the information will be generated and what will be done with it. Clearly, pharmacogenomics would involve collecting genetic information, and in this limited respect *more* information. But neoliberals have successfully argued that the decision to provide such information to a regulatory body should be left to the pharmaceutical firm (Hogarth, 2015). Regulators should cease requiring much in the way of pre-market data; nor should they require information on clinical endpoints, such as survival rates or the number of adverse outcomes (Gulfo et al., 2016, p. 17). At the hands of neoliberals, then, pharmacogenomics represents a peculiar notion of scientific advance – one involving an explicit prohibition on sharing evidence that could inform a regulatory judgment.

The move to deemphasize large-scale RCTs appears to be *unconditional*. This is different from the neoliberals' arguments about escalating drug prices. Prioritizing RCTs over drug prices is a consequence of the neoliberal move to subordinate science to the market. Although not every neoliberal wants to eliminate the RCT *right now*, all neoliberals argue for decreasing the scope of their use. For example, Tomas Philipson has argued for removing the "all or nothing" approval judgment (based on RCTs), allow for "partial approval" of drugs, and allow pharmaceutical firms to partially roll out their products (Philipson and Lakdawalla, 2016). The "bad drug" judgment is always open to criticism that it is made prematurely. The partial approval response envisions assigning a grade of "incomplete" to the drug, and waiting for the medical marketplace to generate knowledge.

It is important to have a clear understanding of what neoliberals understand this medical marketplace to be. Notwithstanding the rhetoric surrounding "personalized medicine," which may suggest a product designed to fit the consumer's profile, neoliberals do not primarily evaluate markets in terms of efficiently allocating resources to meet patients' definite health needs. What gets them excited, instead, is the possibility of constructing markets to better generate knowledge. When neoliberals say things like, "the medical marketplace [will] determine whether and for whom a new product is a real innovation" (Gulfo et al., 2016, p. 6), it should be understood as no mere rhetorical flourish. Of course, one may find here and there a neoliberal author arguing on behalf of some policy in terms of its allocative efficiency. But when the efficient allocation view comes into conflict with the informational view, concerns expressed in the former idiom must give way to those expressed in the latter.

When the neoliberal view of markets is taken seriously, the "patient" becomes reconceptualized as a participant in the activity of knowledge discovery. No longer do the agents formerly known as "patients" stand apart from the market. No longer does the market exist primarily to serve them. Instead, they, along with doctors, regulators, computers, debt instruments, and lots of other technical things are enrolled within a medical marketplace devoted primarily to knowledge discovery. Within this marketplace, the medicine consuming agent serves as something like a subroutine involved in solving an immensely complex problem, one which only the market has the capacity to solve. But since the natural and normal state of the neoliberal agent is ignorance, the ultimate purpose of this market generated knowledge is not to enlighten future consumers of medicine. Its purpose is simply to provide the kind of knowledge that patrons are willing to purchase. Hence, facilitating the operation of the marketplace of ideas aims not to create *more* knowledge available for all, but instead the *right kind* of knowledge selected for its usefulness to (because it is demanded by) well-heeled patrons.

Neoliberals well recognize that selecting as their goal "enabling the marketplace of ideas to operate" may conflict with the pursuit of patient health. In the memorable words of Peltzman, "the problem is not that we have too many thalidomides, but too few" (quoted in Nik-Khah, 2014). And neoliberals have also acknowledged that most people would be reluctant to accept more thalidomides in exchange for helping the marketplace of ideas to function better. Since the neoliberal turn in regulation there has been both "more thalidomides" and calls of medical consumers, regulators, elected representatives, and academic critics for change. That such calls have failed to produce anything remotely resembling a thalidomide-type policy response serves as a testament to the practical effectiveness of the neoliberal politics of knowledge.

Conclusion

Let us now take stock of the distinctly neoliberal positions on the medical marketplace. Knowledge is located neither in the individual person, nor in an autonomous community of scientists, but instead in the market. Those who propose to lean on the knowledge of the scientific community, by calling for more information about drugs to inform a (pre-market) regulatory decision, find themselves accused of foreclosing use of the marketplace of ideas, the only method available of arriving at true knowledge about drugs. This case is bolstered by recent developments in biomedical sciences which, interpreted right, reinforce this belief.

Put into practice, neoliberalism shapes the knowledge about drugs that is created. It also shapes ignorance. Although neoliberals tend to portray the ignorance of the individual, the regulator, and the scientist relative to the market as an unavoidable natural state, studies in the field of *agnotology* draw attention to the *deliberate* production of ignorance. Initially, studies that devoted themselves to anthropogenic global warming and tobacco cancer denialism attracted the lion's share of attention (e.g., Oreskes and Conway, 2010), but recent contributions have also detected agnogenesis (the deliberate production of ignorance) in the neoliberal response to financial crises and – importantly – pharmaceuticals (Mirowski, 2011; Mirowski and Nik-Khah, 2013; Nik-Khah, 2014, 2016; Fernández Pinto, 2017a, 2017b). Indeed, neoliberals have endorsed the creation of structural conditions that have allowed the practices to take hold (for such practices, see, e.g., Michaels, 2008a, 2008b; Mirowski, 2011; Lexchin, 2012a, 2012b). Paying attention to the political economy of medicine may help to enrich our understanding of how these ideas and practices cohere and how they work. It may also help explain why subjecting science to the market, which is justified by neoliberals for its unsurpassed ability to reveal knowledge, ends up serving precisely the opposite goal.

Notes

1 This observation was first made, of course, by Michel Foucault in his 1978–1979 Collège de France Lectures (later published in translation as *The Birth of Biopolitics*).
2 On the neoliberal orientation of the Becker Friedman Institute (formerly called the Milton Friedman Institute), see (Nik-Khah, 2011b).
3 The Health Economics Initiative of the Becker Friedman Institute is emerging as the leading neoliberal institution exploring the promise of financial engineering to deliver health care. See https://bfi.uchica go.edu/events/health-sector-and-economy.

References

Fernández Pinto, M. 2017a. "Agnotology and the new politicization of science and scientization of politics." In *The Routledge Handbook of the Political Economy of Science*, In: Tyfield, D., Lave, R., Randalls, S., and Thorpe, C. (eds.) New York: Routledge.
Fernández Pinto, M. 2017b. "To Know or Better Not to: Agnotology and the Social Construction of Ignorance in Commercially Driven Research." *Science & Technology Studies* 30(2): 53–72.

Foucault, M. 2008. *The Birth of Biopolitics: Lectures at the Collège de France, 1978–1979.* New York: Palgrave Macmillan.

Gulfo, J., Briggerman, J., and Roberts, E. 2016. "The Proper Role of the FDA for the 21st Century." *Mercatus Research, Mercatus Center at George Mason University.* Available: www.mercatus.org/publication/proper-role-fda-21st-century.

Hogarth, S. 2015. "Neoliberal Technocracy: Explaining How and Why the US Food and Drug Administration Has Championed Pharmacogenomics." *Social Science & Medicine* 131: 255–262.

Jørring, A., Lo, A., Philipson, T., Singh, M., and Thakor, R. 2017. "Sharing R&D Risk in Healthcare via FDA Hedges." MIT Sloan School Working Paper, 5194–5117.

Klein, D. 2000. "Policy Medicine Versus Policy Quackery: Economists Against the FDA." *Knowledge, Technology, & Policy* 13(1): 92–101.

Lasagna, L. 1978. "The Development and Regulation of New Medicines." *Science* 200(4344): 871–873.

Landa, A., and Elliott, C. 2013. "From Community to Commodity: The Ethics of Pharma-Funded Social Networking Sites for Physicians." *Journal of Law, Medicine & Ethics* 41(3): 673–679.

Lave, R., Mirowski, P., and Randalls, S. 2010. "Introduction: STS and Neoliberal Science." *Social Studies of Science* 40(5): 659–675.

Lexchin, J. 2012a. Sponsorship Bias in Clinical Research. *The International Journal of Risk & Safety in Medicine* 24(4): 233–242.

Lexchin, J. 2012b. Those Who Have the Gold Make the Evidence: How the Pharmaceutical Industry Biases the Outcomes of Clinical Trials of Medications. *Science and Engineering Ethics* 18(2): 247–261.

Manhattan Institute. 2017. Project FDA Homepage. Available: www.manhattan-institute.org/projectfda.

Michaels, D. 2008a. *Doubt Is Their Product: How Industry's Assault on Science Threatens Your Health.* New York: Oxford University Press.

Michaels, D. 2008b. Manufactured uncertainty: contested science and the protection of the public's health and environment. In: Proctor, R. and Schiebinger, L. (eds) *Agnotology: The Making and Unmaking of Ignorance.* Stanford, CA: Stanford University Press: 90–107.

Mirowski, P. 2013. *Never Let a Serious Crisis Go to Waste.* New York: Verso.

Mirowski, P. 2011. *Science-Mart: Privatizing American Science.* Cambridge, MA: Harvard University Press.

Mirowski, P., and Nik-Khah, E. 2017. *The Knowledge We Have Lost in Information.* New York: Oxford University Press.

Mirowski, P., and Nik-Khah, E. 2013. "Private Intellectuals and Public Perplexity: The Economics Profession and the Economic Crisis." In: Mata, T., and Medema, S. (eds.), *History of Political Economy 45(Supplement), The Economist as Public Intellectual,* 279–311.

Montazerhodjat, V., Weinstock, D., and Lo, A. 2016. "Buying Cures Versus Renting Health: Financing Health Care With Consumer Loans." *Science Translational Medicine* 8(327): 1–7.

Nik-Khah, E. 2017. "The 'Marketplace of Ideas' and the Centrality of Science to Neoliberalism." In: Tyfield, D., Lave, R., Randalls, S., and Thorpe, C. (eds.), *The Routledge Handbook of the Political Economy of Science,* New York: Routledge: 32–42.

Nik-Khah, E. 2016. "Smoke and Thalidomide." *Whitlam Institute Perspectives,* no. 14, Whitlam Institute within Western Sydney University.

Nik-Khah, E. 2014. "Neoliberal Pharmaceutical Science and the Chicago School of Economics." *Social Studies of Science* 44(4): 489–517.

Nik-Khah, E. 2011a. "George Stigler, the Graduate School of Business, and the Pillars of the Chicago School." In Van Horn, R., Mirowski, P., and Stapleford, T. (eds.), *Building Chicago Economics: New Perspectives on the History of America's Most Powerful Economics Program,* New York: Cambridge University Press: 116–147.

Nik-Khah, E. 2011b. "Chicago Neoliberalism and the Genesis of the Milton Friedman Institute (2006–2009)." In Van Horn, R., Mirowski, P., and Stapleford, T. (eds.), *Building Chicago Economics: New Perspectives on the History of America's Most Powerful Economics Program,* New York: Cambridge University Press: 368–388.

Oreskes, N. and Conway, E. 2010. *Merchants of Doubt.* New York: Bloomsbury Press.

Peltzman, S. 1973. "The Benefits and Costs of New Drug Regulation." In Landau, R. (ed.), *Regulating New Drugs,* Chicago, IL: University of Chicago Center for Policy Studies: 113–211.

Philipson, T., and Lakdawalla, D. 2016. "All or Nothing? Rethinking the FDA Approval Framework." *Forbes.* Available: www.forbes.com/sites/tomasphilipson/2016/11/04/all-or-nothing-rethinking-the-fda-approval-framework.

Stigler, G. 1973. "The Confusion of Means and Ends." In: Landau, R. (ed.), *Regulating New Drugs,* Chicago, IL: University of Chicago Center for Policy Studies: 10–19.

Stossel, T. 2016. "Prescription Drug Pricing: Scam or Scapegoat?" *American Enterprise Institute Report.* Available: www.aei.org/wp-content/uploads/2016/02/Specialty-Drug-Pricing.pdf.

Tyfield, D. 2016. "Science, Innovation, and Neoliberalism." In: Simon Springer, Kean Birch, and Julie MacLeavy (eds.), *The Routledge Handbook of Neoliberalism*, New York: Routledge: 340–350.

Tyfield, D., Lave, R., Randalls, S., and Thorpe, C. (eds.) 2017. *The Routledge Handbook of the Political Economy of Science.* New York: Routledge.

Van Horn, R., and Mirowski, P. 2009. "The Rise of the Chicago School of Economics and the Birth of Neoliberalism." In: Mirowski, P., and Plehwe, D. (eds.), *The Road from Mont Pèlerin: The Making of the Neoliberal Thought Collective*, Cambridge, MA: Harvard University Press: 139–178.

13

The value of the imagined biological in policy and society

Somaticizing and economizing British subject(ivitie)s

Martyn Pickersgill

Introduction

Attending the World Economic Forum this past week, I was struck by two trends. The first was that brain research has emerged as a hot topic. Not only was brain science or brain health a new theme at the meeting, research on the brain emerged in discussions about next generation computing, global cooperation, and even models of economic development as well as being linked to mental health or mindfulness. In a meeting frequented largely by economists and business leaders, I was surprised by the number of non-scientists who have become enchanted by brain science. Clearly this is the era of the brain, with mental health now part of a much broader discussion.

(Insel, 2015)

Thomas Insel, former Director of the US National Institute of Mental Health, wrote the above following his recent attendance at the World Economic Forum. Part of his regular and public-facing 'Directors Blog' series, Insel's comments are illustrative of how the biological has come to be situated within a range of economic regimes – how it has been "economized." In particular, through talk of an "era of the brain" (or of genes, or of epigenetics), chains of value-generation are conjured, instantiated, and replicated in ways that present the (imagined) biological as pro-missory matter (Brown and Kraft, 2006). This presentation can in turn act as a platform for the production of symbolic and financial capital via careful sociotechnical tinkering and positioning within the laboratory, the clinic, and the market.

Scholarship in science and technology studies has often focused on the "lively capital" (Sunder Rajan, 2012) that can be generated through newly and re-configured relationships between life, science, and business. Martin and Nightingale (2004), for instance, have shown how an imaginary of a "biotech revolution" has powered investment and attention toward extracting value from the soma, in the UK and beyond. This is even in the absence of "outputs" that have kept "pace with [this] increased research and development spending" (ibid.: 564). Concurrently, a wealth of careful social scientific and historical analyses have attended to an expansion of private-sector innovation in the realms of pharmacology, biobanking, tissue engineering, and so on.

Yet, the economization of biology exceeds the development of products, patents, and markets. As Insel's comments suggest, in a range of nations value is also extracted from bodies via the brains within them; in particular, through psychological techniques and personalized social interventions aimed at promoting mental health, enhancing wellbeing, and augmenting responsible citizenship. On this basis, it is not a challenging task to characterize populations as having come to include – or be focussed on generating – increasingly prosocial and content (or at least, decreasingly unsatisfied) individuals. Within such analytic sketches, the governance and subjectification of these agents could be considered to better optimize them for the generation and accumulation of wealth.

Characterizations like the above are as compelling to some as they are disturbing to others. In this chapter, I seek to move beyond sweeping celebrations or castigations to illustrate some of the ways through which soma and economy are articulated. Following a broader overview, I focus on UK programs of governmental intervention into the early development of children. I note how the (purported) modulation of the soma through parental practices and state welfare is increasingly framed as linked to societal costs and value. The concerns of this chapter can perhaps best be characterized not as with the body proper (Lock and Farquhar, 2007), but rather with the *imagined* biological. A range of assumptions and anticipated understandings underlie articulations of genes, brains, and bodies in diverse cultural arenas (including within the sphere of biomedicine itself). Together, these configure a discursive "biology" that does not necessarily have a direct and linear relationship with its fleshy referent, and may at times seem quite distant to it.

Plastic fantastic

The creation of capital is often associated with the plasticity of the (imagined) biological. To put this in a straightforward manner: wealth is commonly realizable through the transformation of biology from one kind of thing into another, with the product made being more valuable than the substrate. Indeed, the biology of the workforce is itself very much subject to economistic attention. The notion of plastic bodies that are responsive to changing workplace conditions (temperature, humidity, light, and so on) has for quite some time featured within the calculations of those who design and manage the built environment of zones of production. An article in US business magazine *Fast Company*, for instance, communicates this clearly through its headline, "Want more productive workers? Adjust your thermostat" (Friedman, 2012). In the piece, we are informed that not only do offices that are too cold mean that workers have "less energy available for concentration, inspiration, and insight," but they are also more likely "to perceive others as less generous and caring."

Conceptions of psyche and society are often included within assemblages of capital and bodies, and in discourses upon both. In particular, they are constructed as exerting somatic effects, and in turn impacted by biological change. In a sense, subjectivity is somaticized, but the (imagined) biological too can be shaped by agentic action and processes of subjectification. Psychological techniques such as mindfulness meditation, for example, are regularly regarded as working on the psyche through the brain, affecting this changeable organ in ways that enable agents to better adapt and respond to stressful situations (as *Forbes* strikingly illustrates in their article, "7 Ways Meditation Can Actually Change the Brain"; Walton, 2015). Further, new meta-discourses are emerging – such as "neuromanagement" – where, for instance, research is being urged into the neural processes underlying operational decision-making, with a view to further enhancing workplace productivity (Bian et al., 2012).

The biological at play within straightforward economic contexts – sites of manufacture and assembly, and of the financial and legal services that enable these to expand – is amorphous and flexible (see also Martin, 1994). It is constituted through wide-ranging academic studies of, and popular tropes regarding, varied bodily attributes and limitations, as well as enduring accounts of dynamic brain–body relationships. The plastic brain is especially important: as the site of worker subjectivity, and as a modulator of (un)productive bodies in response to the environment in which they are situated. The somatic, the psychic, and the social interweave in such accounts in complex and compelling ways.

British brains

As an observed and assumed reality, the malleability of biology remains key to a wide variety of economized practices. Within a range of nations, the plastic biology commanding some of the most intense cultural attention is perhaps the brain. Epigenetics is also increasingly visible within popular discourse, and intrigue around genetics endures. It is not unusual for somatic narratives to converge, with popular discussion linking together themes from neuroscience and epigenetics to wider societal concerns; for example, around the (potential) intergenerational effects of illegal drug use.

In the UK, the neurological is frequently configured as plastic. Education, for instance, has been one key site where neuroscientific technologies and concepts can be found. This includes wider discussions of development, as well as in educational policy and the classroom per se – including through the use of brain training games. In the substantial press coverage that the latter technologies have attracted, the possibilities of change, optimization, and enhancement across the life-course are commonly emphasized (Pickersgill et al., 2017). Within popular culture, behavioral shifts (such as during the teenage years) have been linked to a mutable neurology that changes over time (Choudhury et al., 2012). This is in response to external entities such as stimulants, as well as intra-bodily processes. The imagined biological might often emphasize the brain, but shifts between other icons of somatic concern also feature prominently. In particular, hormones continue to find cultural traction and are often interweaved with neurological accounts – such as in news items about the effects of testosterone on the brain (see e.g. Rainey, 2017).

The assumed plasticity of children's (and others') brains has come to focus the minds of many policymakers. In a range of social policy settings (e.g. addictions policy, older adult care, and so on), the mutability of the imagined (neuro)biological has been configured as a source of societal and economic value (including through intervention as a cost-saving measure). Much policy attention – as well as debate within social work and beyond – relates to the "early years." Generally, this term is taken to mean the first three years of a child's life (or "the first thousand days," as these are more commonly referred to in the United States, for instance). As we will see, the policy praxis regarding the early years intertwines established tropes of British civil society – e.g. citizenship, the economy, and welfare reforms – with imaginaries of neurological plasticity assembled, in part, through the international scientific literature.

Early life

The early years have been deemed by successive UK governments as salient in determining children's futures. The 'Sure Start' policy program, introduced by Prime Minister Tony Blair's Labour government, was a key initiative that firmly fixed attention on the social and emotional development of children. This formed part of a wider policy agenda to support "vulnerable young people" (Secretary of State for Social Security, 1999: 5). Specifically aimed at families living in socio-economically deprived contexts, the Sure Start program featured centers with educational activities for children and parents. These included play facilities for the former and child care and other support – such as debt counselling – for the latter. In Westminster and beyond, the program was commonly deemed a success, with one House of Commons Committee charged with scrutinizing policy in this area describing Sure Start as "innovative and ambitious," as well as "solidly based on evidence that the early years are when the greatest difference can be made to a child's life chances" (House of Commons, 2010: 3). The initiative evolved into a more universal program of Children's Centres in 2003 (Lewis, 2011) and varying degrees of support for children's services have continued under Conservative Prime Ministers David Cameron and Theresa May.

While state social services continue to be deemed important by recent governments, policy discourse has come to be increasingly permeated with themes of individual responsibility for the rational choices citizens are believed to make. The production of responsibility – to oneself, one's child, and to others, all with the aim of optimizing individuals and communities – has been a feature of speeches and reports that often express an idiom of somatic plasticity (Broer and Pickersgill, 2015a). Such themes were exemplified in David Cameron's "Life Chances" speech (January 11, 2016):

> [O]ne critical finding [from neuroscience] is that the vast majority of the synapses[,] the billions of connections that carry information through our brains[,] develop in the first 2 years. Destinies can be altered for good or ill in this window of opportunity. On the one hand, we know the severe developmental damage that can be done in these so-called foundation years when babies are emotionally neglected, abused or if they witness domestic violence. [...] On the other hand, we also know – it's common sense – how a safe, stimulating, loving family environment can make such a positive difference [...] So mums and dads literally build babies' brains [...] I believe if we are going to extend life chances in our country, it's time to begin talking properly about parenting and babies and reinforcing what a huge choice having a child is in the first place, as well as what a big responsibility parents face in getting these early years right.

The comments above underscore the importance of the imagined biological, placing considerable onus on parents to "build babies' brains" the *right* way (and, indeed, introducing the capacity to do so). In part, the post-Blair services that continue to be provided to families appear aimed precisely at developing the capacity to enact such individual responsibilities.

At a policy level, the mechanisms through which intervention in family and social life is thought to enable economic enhancement often relates to (somewhat diverse and mutable) neurobiological processes. In general, the brains of developing children are judged susceptible to environmental insult; accordingly, the immediate environment of a child is presented as salient to modulate. The mother is seen as of paramount importance, with her parenting practices regarded as shaping (or even constituting) that environment. These practices commonly become the subject of the intervention, which targets the (imagined) biological with the aim of – ultimately – benefiting society.

Under the current (2017) Conservative government, relevant social services can involve professionals working closely with pregnant and post-partum women, helping them to access employment and support, to look after and interact with young children, and to make an assortment of "positive choices." Political drivers have often been around the savings to "the taxpayer" that early intervention is argued to make. As one House of Commons Library

Briefing paper pointed out, "in addition to the social rationale for intervention, advocates of early intervention policies and programmes often cite the economic advantages in terms of cost savings to the public purse. This is based on the premise that early, preventative interventions deliver results for significantly less money than later, reactive interventions" (Bate, 2017: 12).

The mutable somatic

The imagined biological at stake in early intervention policies and practices largely relates to the brain. Yet, reflecting the ontological mixing so characteristic of articulations of plastic biology, genes too are important players in enjoinments to invest in and expand services aimed at the early years. Within policy documents, the infant genome has, for example, been presented as a significant producer of adult (neuro)biology, and is often framed as somewhat "fixed" (though new discussions about epigenetics have begun to recast this contention, as discussed below).

These themes are strikingly conveyed within the (in)famous "Allen Report" – i.e., Member of Parliament Graham Allen's "Early Intervention: The Next Steps" (Cabinet Office, 2011). For example, the report quotes two paragraphs from a 2007 working paper produced by the National Scientific Council on the Developing Child (based at Harvard's Center on the Developing Child) to advance an understanding of brains as designed by – but not wholly attributable to – the genome:

> Just as in the construction of a house, certain parts of the formative structure of the brain need to happen in a sequence and need to be adequate to support the long-term developmental blueprint. And just as a lack of the right materials can result in blueprints that change, the lack of appropriate experiences can lead to alterations in genetic plans. Moreover, although the brain retains the capacity to adapt and change throughout life, this capacity decreases with age.
>
> *(Cabinet Office, 2011: 12)*

Environmental inputs, then, impact the brain initially patterned by the genome. Such inputs might be social or chemical, although exposure to chemicals (like nicotine or alcohol) is often reduced to the result of discrete choices made by contextless individuals who feature predominantly in children's lives (often, mothers). The metaphor of genome as blueprint is, of course, a familiar one within Anglophone society (Nerlich et al., 2002), and has helped to convey a notion of biological inevitability and fixity. In the quotation above, the metaphor has been skilfully elaborated, precisely to unsettle these meanings and to present biology as plastic. In this way, the (well-recognized) trope of the linear gene is intertwined with a new story about the importance of the brain in/and society. The result is a montage of familiarity and novelty that has stimulated investment (e.g. into early intervention services) in ways that make straightforward cultural sense.

Within both policy and services aimed at the early years, the stabilization of attachments between parents and children is commonly held to be salient. Drawing from work in developmental psychology (most notably Bowlby [1969], 1999), attachment theory underscores the necessity of physical and psychological closeness between adults and their offspring, in order to develop the emotional regulation of the latter. In contemporary social policy and programs aimed at enhancing parenting practices, attachment is frequently described through a neurobiological framework. As in the Allen Report, across the UK it is common to read and hear about what is apparently a "growing body of evidence that adverse emotional and social experiences in infancy alter the architecture of the brain itself" (Furnivall, 2011: 5).

Within Scotland, which in a post-devolution context has an enhanced capacity to forge its own policy conversation about the early years, discourse around attachment and young people resonates with the psychosocial discussions that took place within and beyond the UK Labour government through their Sure Start and Children's Centre initiatives. In particular, the promotion of a neurobiologically informed notion of attachment as a mode of addressing deprivation and its interpersonal effects seems often to be considered a goal in and of itself (e.g. Furnivall, 2011). Nevertheless, the focus of the Scottish government on antisocial behavior has shifted toward the responsibilization of the child and their individual behavior (Tisdall, 2006), and hence discourse around childhood development must be considered as part of a broader policy ecology. Within the wider UK, the emphasis at a policy level of promoting attachment and building better brains often foregrounds the economy and the wider social and capital costs of failing to intervene. Of course, practitioners themselves might well be focussed on similar social democratic ideals to those constitutive of many children's services in the 1990s and early 2000s. Yet, the aforementioned Allen Report, for instance, constructs a financial burden to encourage investment: "every taxpayer pays the cost of low educational achievement, poor work aspirations, drink and drug misuse, teenage pregnancy, criminality and unfulfilled lifetimes on benefits" (Cabinet Office, 2011: x). Consequently, we can see that the neuropsychosocial discourse of attachment has come to be articulated in diverging – if not necessarily distinct – ways, further underscoring the mutability of the imagined biological.

Toward epigenetics?

Although (shifting) ideas about a plastic biology are of evident significance to early intervention policy and practice, it is also striking that policy and service actors can be critical of how, and the extent to which, somatic notions impact discourse and initiatives (as discussed in more detail in Broer and Pickersgill, 2015b). Within social policy spheres more generally, there is likewise – on occasion – recognition of the partiality of, and contestation surrounding, much of the biological knowledge circulating therein. We might usefully speculate whether this is, in part, related to the introduction of the scientific vocabularies, idioms, and metaphors documented above that trouble taken-for-granted biological ideas (ibid.). In essence, the layering of new (neuro)biological truths on top of somatic narratives relatively recently regarded as novel and valuable themselves (such as the "gene X for behavior Y" discourse) could, perhaps, be potentially destabilizing of the epistemic credibility of biomedical research per se within particular contexts. After all, if the veracity of prior certainties is now to be called into question, it is not surprising that newer knowledge is somewhat reflexively adjudicated (especially if it calls into question existing professional practice; cf. Pickersgill, 2011).

More generally, the fact that terms and concepts from neuroscience and genetics exist within policy reports or the accounts of those who develop new services does not necessarily imply fundamentally new kinds of social praxis. Rather, novel articulations of the imagined biological potentially reify policy paths already mapped or trod. Of course, reification itself can be regarded as a new social process, and the marks a reified extant policy might leave on the communities and subjects it touches cannot be dismissed.

By suggesting limits on the import of biology in the context of early intervention, I have no desire to dismiss the significance of the development of – or, maybe, the return to – a conception of sociality that is in some sense capable of leaving a somatic trace (Pickersgill, 2014). Rather, the import of the (imagined) biological seems in some sites to be growing, not receding. In particular, in the case of early intervention, ideas associated with epigenetics are becoming gradually more visible. A 2010 report by the Centre for Excellence and Outcomes in

Children and Young People's Services, for instance, notes how epigenetics "is now suggesting the environment, especially during pregnancy and very early childhood, activates or silences good and bad genes crucial for mental well-being and social adaptation." Some scientists have themselves urged new connections between biology and social policy: biologist Michael Meaney, for example, has hypothesized that epigenetics research could benefit early intervention practices by showing how "different biology may require different levels of interventions" (Meaney, 2014). Epigenetics, then, provides a route back to notions of the importance of genetics, while still enabling a plastic narrative of development that allows for intervention in ways that highly deterministic framings of genetics have struggled to accommodate.

Some of the greatest excitement around epigenetics within British policy communities pertains to how heritable epigenetic changes might be, and especially modifications that have significant phenotypic effects. Within Scotland, epigenetics research has been cast as a means through which to shed light on the widening of gaps in the health and wellbeing of the citizens of Glasgow (for instance, by the influential former Chief Medical Officer and member of the Scottish Government Council of Economic Advisers, Harry Burns; Scottish Parliament, 2013). Such a framing again points to the political plasticity of the mutable imagined biological. This essay opened with straightforward descriptions of capital-extraction and value-generation, and progressed to discuss regimes of early interventions that are today associated with particularly individualistic political rationalities. However, neurobiological and epigenetic notions also have traction in arguments for redressing various kinds of inequality. The possibilities of plasticity seem to be endless, and a kind of pragmatic enthusiasm remains considerable (if not unending).

Conclusion

In this chapter, I have sketched out some of the ways that the imagined biological interacts with various kinds of economizing processes and economic calculations. We have seen how political discussion and early intervention policy instantiates and expands an economized understanding of infanthood and development (e.g. the costs of poor infant attachment are deemed to have clear and calculable socio-economic effects, such that social welfare can be presented as a cost-saving measure rather than an end in itself). Biological notions – particularly as emerging from neuroscience and (epi)genetics – can play key rhetorical and practical roles; for instance, they have been used by politicians and those working in social care to authenticate and grow various kinds of state interventions and personal responsibilities. The actions of individuals are regarded in such contexts as impacting aspects of the soma, with varying implications for the spending and saving of capital.

Within social policy and beyond, genes might sometimes be seen as fixed, with only the somatic structures and functions they coded for – such as the brain – subject to imposition by the environment. At the same time, popular narratives on epigenetics allow for the actions of genes themselves to be presented as malleable, providing new sites through which the social might exert its effects. As the biological levels upon which personhood is imagined to be constituted are increasingly pluralized – a plurality that relates to the proliferation of formal bioscientific vocabularies and their circulation within societies – so too do the varieties and ontogenies of plasticity that can be operationalized within policy and practice (e.g. that aimed at the early years). The multiplicity of perspectives ensure that the imagined biological is itself plastic: bodies are understood to shift and change, and those understandings are shifting.

References

Bate, A. (2017) Early Intervention. House of Commons Library Briefing Paper Number 07647, 26 June 2017. Available at: www.researchbriefings.files.parliament.uk/documents/CBP-7647/CBP-7647.pdf

Bian, J., Fu, H., Shang, Q., Zhou, X. and Ma, Q. (2012) 'The study on Neuro-IE management software in manufacturing enterprises: The application of video analysis software', *Physics Procedia*, 33, 1608–1613.

Bowlby, J. (1999 [1969]) *Attachment* (second ed.), New York: Basic Books.

Broer, T. and Pickersgill, M. (2015a) 'Targeting brains, producing responsibilities: the use of neuroscience within British social policy', *Social Science and Medicine*, 132, 54–61.

Broer, T. and Pickersgill, M. (2015b) '(Low) Expectations, legitimization, and the contingent uses of scientific knowledge: engagements with neuroscience in Scottish social policy and services', *Engaging Science, Technology, and Society*, 1, 47–66.

Brown, N. and Kraft, A. (2006) 'Blood ties: banking the stem cell promise', *Technology Analysis and Strategic Management*, 18, 313–327.

Cabinet Office (2011) Early Intervention: The Next Steps. An Independent Report to Her Majesty's Government. Graham Allen MP. Available at: www.gov.uk/government/uploads/system/uploads/atta chment_data/file/284086/early-intervention-next-steps2.pdf.

Cameron, D. (2016) 'Prime Minister's speech on life chances', 11th January 2016, www.gov.uk/governm ent/speeches/prime-ministers-speech-on-life-chances.

Centre for Excellence and Outcomes in Children and Young People's Services (2010) Grasping the Nettle: Early Intervention for Children, Families and Communities. Available at: www.family-action. org.uk/content/uploads/2014/06/early_intervention_grasping_the_nettle_full_report.pdf.

Choudhury, S., McKinney, K. A. and Merten, M. (2012) 'Rebelling against the brain: public engagement with the "neurological adolescent"', *Social Science and Medicine*, 74, 565–573.

Friedman, R. (2012) 'Want more productive workers? Adjust your thermostat', Fast Company, 17th September 2012, www.fastcompany.com/3001316/want-more-productive-workers-adjust-your-thermostat.

Furnivall, J. (2011) 'Attachment-informed practice with looked after children and young people', IRISS Insights 10, www.iriss.org.uk/sites/default/files/iriss_insight10.pdf.

House of Commons (2010) House of Commons Children, Schools and Families Committee; Sure Start Children's Centres; Fifth Report of Session 2009–2010, Volume I, www.publications.parliament. uk/pa/cm200910/cmselect/cmchilsch/130/130i.pdf.

Insel, T. (2015) 'The Ignorance Project', NIMH Director's Blog. Available at: www.nimh.nih.gov/about/ directors/thomas-insel/blog/2015/the-ignorance-project.shtml.

Lewis, J. (2011) 'From Sure Start to Children's Centres: an analysis of policy change in English early years programmes', *Journal of Social Policy*, 40, 71–88.

Lock, M. and Farquhar, J. (2007) *Beyond the Body Proper: Reading the Anthropology of Material Life*, Durham, NC: Duke University Press.

Martin, E. (1994) *Flexible Bodies: Tracking Immunity in American Culture from the Days of Polio to the Age of AIDS*. Boston: Beacon Press.

Martin, P. and Nightingale, P. (2004) 'The myth of the biotech revolution', *Trends in Biotechnology*, 22, 564–569.

Meaney, M. (2014) 'Epigenetics offer hope for disadvantaged children', Child and Family Blog. Available at: https://www.childandfamilyblog.com/social-emotional-learning/epigenetics-offer-hope-disadvantaged-children/

National Scientific Council on the Developing Child (2007). The Timing and Quality of Early Experiences Combine to Shape Brain Architecture: Working Paper #5. Available at: http://developingchild. harvard.edu/wp-content/uploads/2007/05/Timing_Quality_Early_Experiences-1.pdf.

Nerlich, B., Dingwall, R., and Clark, D. (2002) 'The book of life: how the human genome project was revealed to the public', *Health*, 6, 445–469.

Pickersgill, M. (2011) '"Promising" therapies: neuroscience, clinical practice, and the treatment of psychopathy', *Sociology of Health and Illness*, 33, 448–464.

Pickersgill, M. (2014) 'Neuroscience, epigenetics and the intergenerational transmission of social life: exploring expectations and engagements', *Families, Relationships and Societies*, 3, 481–484.

Pickersgill, M., Broer, T., Cunningham-Burley, S. and Deary, I. (2017) 'Prudence, pleasure, and cognitive ageing: configurations of the uses and users of brain training games within UK media, 2005–2015', *Social Science and Medicine*, 187, 93–100.

Rainey, S. (2017) 'They can't smell sweat, girls use twice as many words as boys and their brains really DO switch off when you nag them! Surprising scientific facts about stroppy adolescents', *Daily Mail*, 1st March 2017, www.dailymail.co.uk/femail/article-4272928/Surprising-scientific-facts-stroppy-adolescents.html.

Scottish Parliament (2013) Health and Sport Committee 22nd January 2013: Official Report, www.parliament.scot/parliamentarybusiness/report.aspx?r=7902&mode=html.

Secretary of State for Social Security (1999) Opportunity for All: Tackling Poverty and Social Exclusion, http://dera.ioe.ac.uk/15121/1/Opportunity%20for%20all%20-%20tackling%20poverty%20and%20social%20exclusion.pdf.

Sunder Rajan, K. (Ed.) (2012) *Lively Capital*, Durham, NC: Duke University Press.

Tisdall, E. K. M. (2006) 'Antisocial behavior legislation meets children's services: challenging perspectives on children, parents and the state', *Critical Social Policy*, 6, 101–120.

Walton, A. G. (2015) '7 ways meditation can actually change the brain', *Forbes*, 9th February, 2012, www.forbes.com/sites/alicegwalton/2015/02/09/7-ways-meditation-can-actually-change-the-brain/#24aa04001465.

14

Responsible research and innovation

Ulrike Felt

RRI: a concept with a pre-history

When the notion of responsible research and innovation (RRI) gradually arose in the arenas of European and national-level policy around 2010, it had not emerged *ex nihilo*. As numerous analysts have pointed out, RRI can be seen as yet another buzzword in a row of pre-existing concepts addressing the relation between technoscientific and societal developments (Rip, 2014, Guston and Stilgoe, 2017, Hilgartner et al., 2017). Its most important predecessors were the programs supporting the study of *Ethical, legal and social implications/aspects (ELSI for the US and ELSA for Europe)* of scientific and technological developments. ELSI programs were launched in connection with the Human Genome Project in the United States, with the idea to reflect upon potential societal impacts through complementary research and public deliberation on the uses of new knowledge. Taken over and adapted to the European context, ELSA research became an integral part of the European Commission's Research Framework Programs from 1994 onwards as well as in many national funding programs – particularly in the field of the life sciences (Zwart and Nelis, 2009, Hilgartner et al., 2017).

As in the United States, in the European context the label ELSA research was introduced in a top-down manner, with no clear meaning attached to it. ELSA research would initially also focus on the field of life sciences and biotechnologies, with research steered through more or less substantive funding possibilities.[1] From its inception, ELSA was largely tied to, and somewhat pre-formatted by, the scientific research programs that it was meant to accompany. As a consequence, researchers were expected to ask a rather narrow set of questions, mainly along the lines of prior bioethical concerns, related to individual autonomy, harm or risk.

These funding arrangements and their framing of ELSA research attracted substantial critique from the social sciences and humanities. ELSA research could, as had been the critique toward bioethics, run the danger of smoothing out societal concerns rather than addressing issues, which may be inconvenient to researchers or policy makers but essential from a societal perspective (Zwart et al., 2014). It was highlighted that ELSA research would "not cover the whole range of social and economic realignments that accompany major technological changes, nor their distributive consequences, particularly as technology unfolds across global societies and markets" (Jasanoff, 2003, pp. 241–242). Furthermore, concerns were raised that these "programmed reflections" would lead to pre-emptied and ritualized forms of reflexivity and, thus, introduce a specific "division of moral labor" into research and innovation processes. Knowledge and technology creation would be delegated to the researchers, while ELSA researchers care for the reflection on the ethical, legal and social concerns (Felt et al., 2013, Rip, 2016).

As the ELSA research label was appropriated and gradually filled with meanings by those engaging in these programs, shifts became apparent in the ways questions were posed and in the focus of the research performed. While ethicists started to work in more empirically grounded ways, social scientists increasingly addressed normative questions. Moreover, new scientific developments, such as nanotechnology and synthetic biology, attracted ELSA research; consequently, ELSA researchers gradually extended their research areas, while starting to engage with larger questions and to develop new methodological approaches. In particular, from the turn of the twenty-first century onwards, ELSA research was partly seen as synonymous with the flourishing public engagement exercises. These activities taken together started to give shape to an interdisciplinary community loosely gathering around the ELSA label, accompanied by the creation of new journals and conference venues. Yet, this new "critical interdisciplinarity" was not without conflicts, showed limits and sometimes led to open failures. By now a number of studies have provided insights into the possibilities and limitations of such experimental collaborations. They pointed at the micro-politics and the tacit epistemic hierarchies which constrained the potential of such exercises in mutual engagement (Rabinow and Bennett, 2012, Felt, 2014, Viseu, 2015).

In the same period, other efforts addressed societal concerns, such as various forms of technology assessment (e.g. Guston and Sarewitz, 2002), research under the label of mid-stream modulation (e.g. Fisher et al., 2006), or enhanced ethical integration in laboratories (e.g. van der Burg and Swierstra, 2013). Across and beyond these sites of engagement, we witnessed a growing call to open up policy processes to other types of expertise (e.g. Stirling, 2008). While these activities were all meant to realize the spirit of better integrating research and society, we simultaneously encountered a constant concern that under economic and political pressures for continued innovation these efforts "could easily be sacrificed in favor of a sham program that merely gives the impression of doing so" (Fisher, 2005, p. 322).

This was the situation when RRI became an integral part of European research funding within the most recent research framework program, HORIZON 2020, which started in 2014.

RRI: opening up new possibilities?

How was European society conceptualized in research policy when introducing RRI? While late twentieth-century European policy discourses markedly gravitated around the notion of a *knowledge society*, the turn of the millennium brings a shift toward *knowledge economy* (Felt et al., 2007), with *innovation* gradually replacing knowledge thereafter. The importance of a steady flow of innovations to assure international competitiveness, has become the core mantra in European policy discourse. At the same time, societal actors are requested to be more supportive toward innovations for assuring societal well-being. It therefore seems as if citizens are requested to fully embrace the idea of a "self-experimental society" through demonstrating a clear "willingness to remain open to new forms of experience" (Gross and Krohn, 2005, p. 63) brought about by the steady flow of innovations. However, allowing for such an open experimental approach would necessarily mean asking questions, such as who frames the issues at stake, how are broader forms of participation assured, who identifies and addresses in- and exclusions, distributive justice and vulnerabilities, or what values will guide assessments on whether any experiment is a success or failure. It is from this backdrop that we analyze the introduction of the explicit reference to responsibility in research and innovation.

In the early phase of discussions around RRI, we witnessed various attempts to settle for a basic definition and to clarify foundational principles (for an overview on the development of RRI see Guston and Stilgoe, 2017). In order to work across many different communities, the concept had to be sufficiently vague to allow for broad adherence while remaining concrete enough to utilize it as a reasonably well-functioning device for policy-making and research practice. RRI was expected to allow for both the avoidance of unintended consequences of an innovation, and the movement of governance away from "reactive forms [...] to proactive forms" (Ribeiro et al., 2017, p. 89). Thus, we witness numerous efforts by social scientists (e.g. Owen et al., 2012) to highlight that RRI was to be different from ELSA: it should not only entail reflection upon techno-scientific developments and identify potential problems, but also should proactively support new ways of developing innovations in line with societal futures, which citizens feel are worth attaining. This shift enabled researchers to avoid the critiques expressed toward ELSA approaches, but in turn necessitated new ways of asking questions.

Early efforts to give meaning to RRI would define it as being

> a transparent, interactive process by which societal actors and innovators become mutually responsive to each other with a view to the (ethical) acceptability, sustainability and societal desirability of the innovation process and its marketable products (in order to allow a proper embedding of scientific and technological advances in our society).
>
> *(von Schomberg, 2011, p. 9)*

Later von Schomberg (2013, p. 54) would clarify that market mechanisms should not be the leading force in deciding "the normative dimension of what counts as an 'improvement'". Instead, careful processes of deliberation, including citizens and civil society actors, should decide on the value of innovations. This potentially points to a shift from focusing on the (market) value *of* innovation, to the values embedded *in* innovations (Felt, 2017). Also, it means asking questions of direction and scale of innovations (Felt et al., 2007) and not simply caring for the flow of innovations.

We, thus, encounter clear calls for RRI to support "those prospective, forward-looking dimensions of responsibility, (notably *care* and *responsiveness*) which allow consideration of purposes and accommodate uncertainty, a defining feature of innovation" (Owen et al., 2013, p. 29). RRI is specified further through the identification of four key-dimensions, "anticipation, reflexivity, inclusion and responsiveness" (Stilgoe et al., 2013, p. 1570), all of which should be fostered throughout the research and innovation process. Anticipation refers to systematic thinking of the many different potential outcomes of innovation while concurrently admitting our limited foresight capacities. Reflexivity means critically questioning "one's own activities, commitments and assumptions, being aware of the limits of knowledge and being mindful that a particular framing of an issue may not be universally held" (ibid., p. 1571). Inclusion draws our attention to questions of power and who is (not) given voice. Finally, responsiveness highlights the need to adjust "courses of action while recognising the insufficiency of knowledge and control" (ibid., p. 1572). Together, these dimensions should successfully "provide a framework for raising, discussing and responding" (ibid., p. 1570) to the key questions relevant to any sociotechnical trajectory that contemporary societies aim to embark upon.

What are the similarities or differences to the prior programs engaging with societal issues? In many ways RRI meets some of the same challenges and critiques ELSA research previously encountered. Rip (2016) has very convincingly pointed to the danger that RRI discourse could simply be "like the new clothes of the emperor" (p. 290) and pointed to the many hurdles a serious implementation of RRI would imply. Furthermore, it is unclear whether the label will survive the transition to the next EC framework program or if it will be replaced by yet a new one. Much EC funding is still strongly tied to research themes from the natural sciences, which potentially frames the kinds of questions that can and will be asked. Analysts have highlighted two further challenges, one being that RRI might serve neoliberal ideas through depoliticizing debate and deliberation (Pellizzoni, 2015), and the other that values and assumptions embedded in RRI might potentially reproduce dominant structural global inequalities (Macnaghten et al., 2014).

However, programs are appearing that genuinely embrace the idea of RRI as engagement along the whole process of innovation. This might lead, in the end, to a different outcome and not solely to the reflection on the outcome (e.g. the Dutch RRI program is in line with such an idea).[2] Also, RRI has called for a stronger participation of citizens and civil society organizations in research and innovation, for example through the fostering of citizen science. This collective and cooperative nature of RRI activities, as well as the emphasis on asking wider questions of the purposes and direction of action to be taken, might distance it from more neoliberal thinking. "RRI demands a paradigm shift in that it requires a voluntaristic co-responsibility approach to ensure the convergence of differentiated responsibilities towards some common goals" (Arnaldi and Gorgoni, 2016, p. 16). Thus, RRI promises to realize the open-ended character of its intervention in the innovation process, and could potentially become itself a social innovation (Rip, 2014).

RRI in practice: between outcomes and processes

How will RRI, once implemented, potentially contribute to the creation of new *geographies of responsibility* (Akrich, 1992), i.e. of new distributions of responsibility amongst diverse actors and institutions? In what follows, RRI's potential and limitations are investigated from two distinct angles: one in which responsibility is related to the outcomes of research and innovation, the other where responsibility is perceived as an essential characteristic in the process of producing knowledge and innovation (Stilgoe et al., 2013, Glerup and Horst, 2014). Across both, questions of the purpose of any innovation need to be posed. Even though these two angles are evidently closely intertwined, looking at them separately allows one to identify specific hurdles in the implementation of RRI.

Responsible outcomes

When looking into life science related issues, we witness that as the activities integrating science and society are constantly relabeled, so too is research in the life sciences. As Zwart and coauthors (2014, pp. 2–3) would put it:

> [L]abels such as "genetics" came to be refashioned as "genomics" (and its various derivatives), in order to reset the general focus of the field, while currently genomics is giving way to systems biology, synthetic biology, personalised medicine and other "post-genomics" headings (not completely unlike genomics, but with slightly different emphases).

Relabeling is never an innocent activity and, hence, represents a key element in any politics of knowledge. It shifts our attention, thus reframing problems as well as potential solutions. Working with the field of synthetic biology, for example, Hurlbut showed how, during the process of claiming a new field, researchers could voice the promise that they would now be able "to achieve forms of human benefits that would otherwise be unattainable" (Hurlbut, 2015, p. 113). In one and the same move, the field thus "constructed itself [not only] as *able to respond* [...] to basic problems of human welfare and security", but also "as having *the right response*" (Hurlbut, 2015, p. 113). Being able to reframe the issues at stake and the values to be cared for, to project the future to be aimed for and to define whose problems and concerns are to be addressed, are all clear expressions of power relations. Furthermore, Hurlbut underlines that we still tend to ask societal-consequences-of-innovation-questions rather than questioning the very "imaginations of necessity, responsibility, and the good that inform practices of [innovation] governance" (Hurlbut, 2015, p. 113) and whose imaginations and values drive innovation (e.g. Rip, 2014). Hence, questioning needs to begin at a much deeper level and become broader. This oftentimes proves to be complex due to the predominant modus of RRI being co-financed by the exact projects that should be reflected upon; thus, requiring arrangements with researchers from other disciplines or even in other institutional contexts, like industry.

Indeed, in considering issues related to human bodies, health or the environment, these broader, more fundamental questions gain particular importance. In these fields questions of ethics and responsibility have been present for quite some time. Thereby we encounter specific challenges and resistances in the process of implementing RRI as existing procedures are questioned. The notions of ethics and responsibility have often been already appropriated and integrated into field-specific vocabulary, usually with a rather narrow practical meaning (e.g. Informed Consent procedures are quite frequently claimed to be already a moment of engagement). Frequently, we also encounter utilitarian perspectives on ethics, which means striving for the maximization of technology's positive contributions while minimizing potential negative outcomes. In a neoliberal version this runs the danger of being translated into: "it is enough if actors avoid causing harm" (Rip, 2014). As long as major hazards can be kept at bay, any new technology should be seen as unproblematic. This makes an opening-up of the processes of innovation and asking broader value questions difficult and, thus, RRI hard to implement. Still, when it comes to citizen engagement, these undertakings quite frequently aim to convince participants to be supportive rather than opening up the debate to agenda setting. Unfortunately, RRI related activities very often serve rather as a mechanism to grease the wheels of technological progress and foster the advancement of the bioeconomy than to scrutinize the values embedded in the endeavors.

Responsible processes

How does one translate the ideal of RRI into research practice? What would the need to engage with societal actors and their values concretely mean for researchers on an everyday level? And how could the research and innovation process be opened up both toward society as well as toward other kinds of knowledge (European Commission, 2013)?

Indeed, contemporary research systems have become highly competitive, mainly functioning along more or less rigid new public management structures with a strong logic of delivery (in terms of papers, innovations, students) and a reliance on output indicators as a measure of success. Efficiency, excellence and relevance have become guiding values. The project has become the core organizational entity structuring and tightly timing research practices. This, in turn, has led to a steep rise in the number of researchers in doctoral and post-doctoral positions, while tenured positions remained in comparison relatively scarce. As a climate of competition along a narrow set of values prevails (Fochler et al., 2016), it is essential to ask why researchers should engage proactively in RRI activities as this investment would not be recognized by classical reward structures. Similar concerns might be expressed for the area of industrial development. In a neoliberal logic, the opening up toward society would tend to be limited to asking "does-this-innovation-create-potential-harm" questions and not mean considering wider societal values when it comes to decide on the directions of innovation.

More generally, understanding RRI in terms of a process means looking at articulations between the individual and the institutional in research. What is expected from the single researcher and what from the collective when it comes to questions of responsibility? Making processes of research and innovation responsible thus calls for the creation and circulation of devices, such as ethical reviews considering wider sets of values, codes of good scientific practice or frameworks fostering engagement with societal actors (Meyer, 2015). They are all meant to assure "procedural responsibility," implicitly assuming that this automatically would lead to "responsible outcomes."

If RRI should be integrated widely into researchers' practices, it therefore is not enough to appeal to their duty to be responsible and engage with society, but much rather "responsibility conditions" (Felt, 2017) need to be created, i.e. the ways in which research is organized, valued, and timed need reconstruction. RRI will not be able to succeed without institutionally fostering researchers' willingness and capacity to explore value-sensitive responses to the complex questions that arise at the interfaces of science and society. Research environments shape researchers' capacities to be "response-able." If the conditions are not favorable for this kind of engagement, RRI activities run the danger of being translated into ritual performances at the beginning or the end of a project, or researchers might choose the division-of-moral-labor model, which means outsourcing reflexivity and anticipation to social sciences and humanities.

Conclusion

We started with a brief sketch of the pre-history of RRI, showing how deeply it is embedded in earlier efforts to better integrate science and society. Thus, the strength of the concept cannot be seen in its radical newness, but rather in its capacity to re-articulate longstanding claims and concerns about innovation and society in new ways. In particular, this means creating spaces for fostering more collective forms of engagement; experimenting in new constellations of actors, and reconsidering the diverse values at stake in innovation. However, despite the transition to RRI seeming smooth, attention is needed. RRI comes at a time when innovation is seen as *the* key driving force of societal developments and neoliberal framings of innovation could potentially endanger the very idea of opening up toward society. Therefore, what needs to be nurtured and cared for is the collective and cooperative nature of RRI activities, as well as the capacity to ask sometimes potentially uncomfortable questions concerning the purpose and the direction of actions to be taken.

In that sense it is also essential to make distinctions transparent, "between normative research enlisted in the service of agendas – public or private – in which the frame is not itself open to question, and research that affiliates with efforts to question the frames within which politics, markets or any other entities are disciplined" (Suchman, 2013, p. 157). It is the latter which RRI should aim to support, if we do not want to simply contribute to "clothing the emperor" (Rip, 2016), instead of making innovation more responsive to societal needs.

Zooming into the life sciences and health related topics it quickly becomes obvious there is not one kind of RRI, which simply could be implemented. Instead, some form of situated compromise needs to be negotiated by the actors involved on *why* RRI activities should be undertaken, *in which form* (e.g. debate, co-design) and *at what moment* during the innovation process. That such a compromise might be fluid and can become questioned, creating considerable frictions in the process, is one of the challenges to be met by RRI activities. This also means time and space must be given to all actors in the innovation process and activities need to accommodate the different knowledge cultures. Thus, these engagement efforts have to be embedded into and valued in the context of institutions of research and development. This might contradict the equally present new public management ideals, when it comes to assessing research, and the concerns for the bioeconomy, which should be supported and advanced by innovations.

It is also essential to consider the situatedness of the researchers, who are supposed to engage in RRI work, to better grasp their understanding of social responsibility. While they become, in the framework of contemporary research, highly skilled in playing the indicator game and being accountable (i.e. accountants of their work and life in science), it is essential to support researchers to become capable and willing to be response-able. However, this responsibility cannot be solely delegated to the individual researcher. Institutional structures need to be created in which these kinds of engagements with society are valued and therefore gain meaning (Balmer et al., 2015).

Finally, RRI could be understood and implemented as a "technology of humility" (Jasanoff, 2003), i.e. become an approach that acknowledges the complexities and uncertainties linked to research and innovation in contemporary societies. We thus should render "the possibility of unforeseen consequences" visible, "make explicit the normative that lurks within the [scientific and the] technical; and [...] acknowledge from the start the need for plural viewpoints and collective learning" (ibid., 240). This calls for making RRI an integral part of the knowledge generation process and as a consequence "developing, nurturing and valuing researchers' capacities of anticipation, reflexivity, inclusion and responsiveness" (Felt, 2017, p.35). However, this should not mean to put in place rigid and standardized procedures. RRI activities need to remain open and flexible "to encompass and address a greater diversity of innovation system agents and spaces if they are to prove successful in their aims" (Wickson and Forsberg, 2015, p. 1159). Turning RRI into a powerful "technology of humility" further calls for researchers being ready to open up research to cooperation both across disciplines and between societal actors and them. However, all this cannot be realized without clear institutional commitments to provide space and time for this type of engagement within the rather narrowly normative and highly competitive environment of contemporary research, and to foster researchers' readiness to engage in forms of innovation which include wider societal values and concerns.

Notes

1 See http://cordis.europa.eu/elsa-fp4/src/about.htm, accessed: 01.05.2017.
2 www.nwo.nl/en/research-and-results/programmes/responsible+innovation.

References

Akrich, M. (1992) 'The De-scription of Technical Objects'. In Bijker, W. E. and Law, J. (ed.) *Shaping Technology/Building Society – Studies in Sociotechnical Change*. Cambridge, MA: MIT Press, pp. 205–224.

Arnaldi, S. and Gorgoni, G. (2016) 'Turning the Tide or Surfing the Wave? Responsible Research and Innovation, Fundamental Rights and Neoliberal Virtues', *Life Sciences, Society and Policy* 12(1): 6.

Balmer, A. S., Calvert, J., Marris, C., Molyneux-Hodgson, S., Frow, E., Kearnes, M., Bulpin, K., Schyfter, P., Mackenzie, A. and Martin, P. (2015) 'Taking Roles in Interdisciplinary Collaborations: Reflections on Working in Post-ELSI Spaces in the UK Synthetic Biology Community', *Science & Technology Studies* 28(3): 3–25.

European Commission (2013) *Options for Strengthening Responsible Research and Innovation*. Luxemburg: Publications Office of the European Union.

Felt, U. (2014) 'Within, Across and Beyond: Reconsidering the Role of Social Sciences and Humanities in Europe', *Science as Culture* 23(3): 384–396.

Felt, U. (2017) '"Response-able Practices" or "New Bureaucracies of Virtue": The Challenges of Making RRI Work in Academic Environments'. In Asveld, L., van Dam-Mieras, M.E.C., Swierstra, T., Lavrijssen, S.A.C.M., Linse C.A. and van den Hoven J. (eds.) *Responsible Innovation 3: A European Agenda?* Cham: Springer, 49–68.

Felt, U., Barben, D., Irwin, A., Joly, P.-B., Rip, A., Stirling, A. and Stöckelová, T. (2013) *Science in Society: Caring for our futures in turbulent times, Policy Briefing 50.* Strasbourg: ESF.

Felt, U., Wynne, B., Callon, M., Gonçalves, M. E., Jasanoff, S., Jepsen, M., Joly, P.-B., Konopasek, Z., May, S., Neubauer, C., Rip, A., Siune, K., Stirling, A. and Tallacchini, M. (2007) *Taking European Knowledge Society Seriously.* Luxembourg: Office for Official Publications of the European Communities.

Fisher, E. (2005) 'Lessons Learned from the Ethical, Legal and Social Implications Program (ELSI): Planning Societal Implications Research for the National Nanotechnology Program', *Technology in Society* 27(3): 321–328.

Fisher, E., Mahajan, R. L. and Mitcham, C. (2006) 'Midstream Modulation of Technology: Governance From Within', *Bulletin of Science, Technology & Society* 26(6): 485–496.

Fochler, M., Felt, U. and Müller, R. (2016) 'Unsustainable Growth, Hyper-competition, and Worth in Life Science Research: Narrowing Evaluative Repertoires in Doctoral and Postdoctoral Scientists' Work and Lives', *Minerva* 54(2): 175–200.

Glerup, C. and Horst, M. (2014) 'Mapping "Social Responsibility" in Science', *Journal of Responsible Innovation* 1(1): 31–50.

Gross, M. and Krohn, W. (2005) 'Society As Experiment: Sociological Foundations For a Self-Experimental Society', *History of the Human Sciences* 18(2): 63–86.

Guston, D. H. and Sarewitz, D. (2002) 'Real-Time Technology Assessment', *Technology in Society* 24: 93–109.

Guston, D. H. and Stilgoe, J. (2017) 'Responsible Research and Innovation'. In Felt, U., Fouché, R., Miller, C. and Smith-Doerr, L. (eds.) *Handbook of Science and Technology Studies, Fourth Edition.* Cambridge, MA: MIT Press, 853–880.

Hilgartner, S., Prainsack, B. and Hurlbut, J. B. (2017) 'Ethics as Governance in Genomics and Beyond'. In Felt, U., Fouché, R., Miller, C. and Smith-Doerr, L. (eds.) *Handbook of Science and Technology Studies, Fourth Edition.* Cambridge, MA: MIT Press, 823–852.

Hurlbut, J. B. (2015) 'Reimagining Responsibility in Synthetic Biology', *Journal of Responsible Innovation* 2(1): 113–116.

Jasanoff, S. (2003) 'Technologies of Humility: Citizen Participation in Governing Science', *Minerva* 41(3): 223–244.

Macnaghten, P., Owen, R., Stilgoe, J., Wynne, B., Azevedo, A., de Campos, A., Chilvers, J., Dagnino, R., di Giulio, G., Frow, E., Garvey, B., Groves, C., Hartley, S., Knobel, M., Kobayashi, E., Lehtonen, M., Lezaun, J., Mello, L., Monteiro, M., Pamplona da Costa, J., Rigolin, C., Rondani, B., Staykova, M., Taddei, R., Till, C., Tyfield, D., Wilford, S. and Velho, L. (2014) 'Responsible Innovation Across Borders: Tensions, Paradoxes and Possibilities', *Journal of Responsible Innovation* 1(2): 191–199.

Meyer, M. (2015) 'Devices and Trajectories of Responsible Innovation: Problematising Synthetic Biology', *Journal of Responsible Innovation* 2(1): 100–103.

Owen, R., Bessant, J. and Heintz, M. (eds.) (2013) *Responsible Innovation. Managing the responsible emergence of science and innovation in society.* Chichester, UK: Wiley.

Owen, R., Macnaghten, P. and Stilgoe, J. (2012) 'Responsible Research and Innovation: From Science in Society to Science for Society, With Society', *Science and Public Policy* 39(6): 751–760.

Pellizzoni, L. (2015) *Ontological Politics in a Disposable World.* Farnham: Ashgate.

Rabinow, P. and Bennett, G. (2012) *Designing Human Practices. An Experiment in Synthetic Biology.* Chicago: University of Chicago Press.

Ribeiro, B. E., Smith, R. D. and Millar, K. (2017) 'A Mobilising Concept? Unpacking Academic Representations of Responsible Research and Innovation', *Science and Engineering Ethics* 23(1): 81–103.

Rip, A. (2014) 'The past and future of RRI', *Life sciences, society and policy* 10(1): 17.

Rip, A. (2016) 'The Clothes of the Emperor. An Essay on RRI in and Around Brussels', *Journal of Responsible Innovation* 3(3): 290–304.

Stilgoe, J., Owen, R. and Macnaghten, P. (2013) 'Developing a Framework For Responsible Innovation', *Research Policy* 42(9): 1568–1580.

Stirling, A. (2008) '"Opening Up" and "Closing Down". Power, Participation, and Pluralism in the Social Appraisal of Technology', *Science, Technology & Human Values* 33(2): 262–294.

Suchman, L. (2013) 'Consuming Anthropology'. In Barry, A. and Born, G. (eds.) *Interdisciplinarity: reconfigurations of the social and natural sciences*. London/New York: Routledge, 141–160.

van der Burg, S. and Swierstra, T. (2013) *Ethics on the Laboratory Floor*. Basingstoke: Palgrave Macmillan.

Viseu, A. (2015) 'Caring for Nanotechnology? Being An Integrated Social Scientist', *Social Studies of Science* 45(5): 642–664.

von Schomberg, R. (2011) 'Prospects for Technology Assessment in a framework of responsible research and innovation'. In Dusseldorp, M. and Beecroft, R. (eds.) *Technikfolgenabschätzen lehren: Bildungspotenziale transdisziplinärer Methoden*. Wiesbaden: VS Verlag.

von Schomberg, R. (2013) 'A Vision of Responsible Research and Innovation'. In Owen, R., Heintz, M. and Bessant, J. (eds.), *Responsible Innovation: Managing the Responsible Emergence of Science and Innovation in Society*. Chichester, UK: Wiley, 51–74.

Wickson, F. and Forsberg, E. M. (2015) 'Standardising Responsibility? The Significance of Interstitial Spaces', *Science and Engineering Ethics* 21(5): 1159–1180.

Zwart, H., Landeweerd, L. and van Rooij, A. (2014) 'Adapt or Perish? Assessing the Recent Shift in the European Research Funding Arena From "ELSA" to "RRI"', *Life Sciences, Society and Policy* 10(11).

Zwart, H. and Nelis, A. (2009) 'What is ELSA genomics?', *EMBO Reports* 10(6): 540–544.

Part 3
Governance of medical genomics

15

Introduction

Stephen Hilgartner

The rise of genomics and its deepening entrenchment in a variety of practices and sites has been accompanied by significant epistemic and sociopolitical change. Since its earliest days, genomics has been a site where new forms of knowledge and new regimes of control have been co-produced. In a variety of sites, actors have contested control over the knowledge objects that genomics produces, the jurisdictions in which it is practiced, and the roles and relationships among agents that it instantiates. The stakes in these struggles have been diverse, including efforts to shape science policy, win scientific priority, distribute credit, spread information, capture property, establish authority, save face, protect human subjects, and maintain national security. New "knowledge-control regimes" (Hilgartner 2017) have taken shape in laboratories and databases, in research programs and funding agencies, in private property and novel forms of "open science," and in modes of governing relationships between science and its publics. Increasingly, as genomics has grown more significant in health and medicine, regimes have also been adjusted in medical practice and clinical research. Regulatory mechanisms and frameworks have been reconfigured; new ontologies, clinical guidelines, and informal standards have developed; and jurisdictional boundaries, both formal and informal, have shifted.

The six chapters that make up this section on governance each address different sites and/or modes of governance: from ethics programs, to biotechnology patents, to biomedical platforms, to genome diagnostics, to genomic information, to genome editing. They also present some of the variety in the ways social scientists have investigated modes of governance of genomics and health.

The section begins its examination of governance by considering the ethics programs established during the Human Genome Project (HGP). As the United States was consolidating plans to launch its genome program, James D. Watson, the newly appointed director of the National Institutes of Health's (NIH) genome office, announced that he planned to spend 3 percent of the genome project's $3 billion budget on studies of ethical and social issues. Soon the NIH and Department of Energy (DOE), the agencies that ran the U.S. genome project, had established a program focused on "ethical and social implications" (ELSI) of genome research. The resulting program was an innovative one: not only was ELSI's funding comparatively lavish, it was also focused on the single scientific domain of human genetics and genomics and placed under the management of the very NIH and DOE programs charged with running the HGP. Funding agencies in Europe developed similar programs under the rubric of ELSA, with the "A" standing for "aspects." The chapter by Stephen Hilgartner reviews the history of the U.S. ELSI program, examining the founding of the program and controversy about its institutional design and achievements. It also briefly discusses the ethics programs attached to major scientific initiatives in such fields as nanotechnology and synthetic biology. He argues that ELSI and ELSA programs underwrote the promise that the genomics revolution would be an orderly and beneficial one that would preserve human values, and concludes that the institutionalization of the concept of attaching "ethics" programs to major scientific initiatives is a significant development in science policy and, more broadly, in the politics of contemporary societies.

Patents and intellectual property are another important mode of governance in genomic medicine. Typically considered an element of innovation policy, patent law is generally regarded as an especially esoteric area of legal practice and scholarship. Blending cutting-edge science with arcane legal technicalities, it seems opaque not only to citizens but also to lawyers unschooled in its particularities. Yet the decisions made in this legal specialty are extremely consequential. Not only do intellectual property decisions distribute what economists refer to as "market power," but they also allocate what we might call "configuration power" – the ability to influence the specific configurations of social relations into which emerging technologies are integrated. Nevertheless, most legal scholarship and policy discussion of intellectual property focuses on questions about how to create efficient economic incentives to drive innovation (Boyle 2003) or on ontological issues about the definition of "patentable subject matter" (Winickoff 2015). Patent rules are generally justified in these terms, whereas specific cases are typically regarded as matters mainly of interest to the immediate parties. Rarely are patent decisions understood as quasi-constitutional decisions with far-reaching implications, the view advanced by a number of authors who examine intellectual property from a critical perspective (e.g., ibid.; Boyle 1996; Jasanoff 2005; Parthasarathy 2017). As a result, patent decisions grounded in a relatively narrow set of innovation-centric concerns end up granting actors significant powers to shape sociotechnical futures.

Biotechnology patents have been consistently controversial since the end of the 1970s, and citizens, activists, and other outsiders have challenged patents on life forms and genes, not only on the basis of innovation-centric arguments but also on the grounds of a wide variety of ethical, sociopolitical, environmental, and religious concerns. Yet as Shobita Parthasarathy observes in her contribution to this section, patent offices in the United States and Europe addressed these challenges in very different ways, raising the question of how to explain this diversity. Some observers have suggested that the difference stems mainly from the tradition in European intellectual property law of denying patents that threaten *ordre public*. But Parthasarathy contends that the variation stems from differences in history and political culture that have shaped the logics, institutions, and practices underlying patent policy. She also notes that the U.S. patent system left untouched the barriers that maintain the narrow techno-legal focus of patent policy. In contrast, the European Patent Office (EPO) took steps to increase its openness to citizen concerns.

As genomic technologies are being built in sites of clinical research and practice, modes of governance are simultaneously taking shape, and this dynamic process is the subject of the next chapter. Biomedical technologies, on the one hand, depend on governance mechanisms such as standardization, and on the other hand, operate as a form of governance, shaping the capacities and incapacities of actors in ways that constitute regimes of practice. These two aspects of governance are tightly intertwined, and are an inseparable part of the very process through which biomedical technologies emerge. Capturing such complexity poses challenges for analysts of genomic medicine. In the first edition of this *Handbook*, Cambrosio et al. (2009) applied the concept of "biomedical platforms" (Keating and Cambrosio 2003) to describe the hybrid formations taking shape as genomics is introduced into clinical research and practice. For these authors, the term *platform* refers, in the first instance, to a technological system capable of supporting a diversity of applications – a usage familiar from computers. But their more specialized definition focuses on the epistemic work that biomedical platforms enable by functioning as a nexus of stabilized interconnections between (a) biomedical entities, such as genes; (b) equipment used to manipulate those entities; and (c) modes of regulation, such as quality control standards. As such, platforms are important to the governance of biomedical practices, and introducing new platforms entails reconfigurations experienced in many sites.

In their contribution to the new edition of the *Handbook*, Alberto Cambrosio, Etienne Vignola-Gagné, Nicole Nelson, Peter Keating, and Pascale Bourret review the literature on a key genomic platform: next-generation sequencing (NGS), focusing specifically on its use in oncology – the clinical specialty that has so far made most use of NGS data. The authors conclude that significant transformations in cancer clinical research and care are taking place in oncology as genomic platforms are introduced. The vision of governance that emerges from their chapter is one in which the machinery of governance blends organizational, regulatory, and epistemic dimensions and is integrated into a process of mutual adjustment unfolding at many sites.

The next chapter – on genome diagnostics – addresses governance primarily as a matter of formal state regulatory policy and informal guidelines and "soft law" developed among professionals. Stuart Hogarth provides an informative review of how the frameworks for regulating diagnostic technologies in Europe and the United States are changing as new genomic technologies have entered research and clinical practice. The general tendency in the broad area of medical diagnostics has long been for innovation to occur in diffuse networks of public sector researchers and clinicians, with industry playing a secondary role. This structure, Hogarth argues, often leads new diagnostics to escape the scrutiny of formal regulatory processes. In this context, the boundary between the use of genomics in research and in clinical practice has remained indistinct, leading to questions about the standards for ensuring the efficacy of tests. One notable development was the codification of a framework for evaluation that considers the analytic validity of a test for a biomarker, its clinical validity for assessing the status of patients, its clinical utility for improving medical outcomes, and its broader ethical, legal, and social implications. Policymaking in this area has encountered obstacles, not the least of which is resistance, grounded in a neoliberal vision of the role of the state to regulatory solutions. Hogarth argues that future research should interrogate the gaps between policy recommendations and policy implementation, and suggests that more scholarly attention should be directed toward the dynamics of standard setting of even the most mundane of diagnostic devices and tests.

Since the early years of the Human Genome Project, the shape of the regimes intended to structure the production, circulation, and use of genomic data have frequently been contested. In his contribution, Edward Dove addresses the challenge of building regimes to protect personal and family privacy in a world in which genomic information is being routinely produced and collected. But for a variety of reasons, building data-privacy regimes that simultaneously permit "appropriate use" and prohibit "misuse" is problematic. Not only is the use/abuse divide a contested boundary that promises to remain so, but as Dove explains, the extant modes of regulating access to personal data fit the genomic context poorly. As a result, regimes that rest on the dominant "consent and anonymize" paradigm are insufficient to protect the privacy of individuals and families. Moreover, the category of "personal data" is unstable in this domain. Dove concludes that as increasing amounts of genomic data are collected, the task of protecting privacy is growing more important and more complex, and he calls for open discussion of these problems in the interest of developing collective expectations and definitions of rights and duties.

In the final chapter in this section, J. Benjamin Hurlbut provides a critical examination of the current debate about human germline genetic engineering. The recent development of the CRISPR/Cas9 technique, which allows precisely targeted editing of the genomes of virtually any organism, promises to greatly ease the intentional introduction of heritable changes into human germline, and it has inspired an intense and international debate about whether and under what circumstances such genetic engineering should be undertaken. Hurlbut analyzes how the current debate has been framed, bringing a historically informed perspective that makes visible the ways in which the questions asked have been circumscribed. Human germline engineering has been debated for more than 50 years, but in the 1980s a broad debate about human dignity and purpose was narrowed to a focus on safety and informed consent. This narrower focus has continued in the current debate surrounding CRISPR, which also echoes the calls for a moratorium on research and a process of careful deliberation familiar from the 1970s debate about recombinant DNA. More deeply, Hurlbut explores how the dominant framing of these debates allocate epistemic and political authority. In particular, he argues that the CRISPR deliberations have recapitulated the discredited linear model of innovation, presenting what is in effect a linear model of scientific responsibility that he terms the "imaginary of governable emergence." This view allocates epistemic authority to scientists, who are uniquely qualified to know "what risks are realistic, and thus what technological futures warrant ethical evaluation" (Hurlbut this volume). Society and its decision-making institutions are cast in a purely reactive role – even as they are charged with addressing the social and ethical implications. These demarcations, Hurlbut concludes, are then reified in the institutions and processes of governance. A good example of such demarcations in action – to bring this introduction back to the issues raised by the first chapter – is the HGP's ELSI and ELSA programs.

References

Boyle, J. 2003. 'Enclosing the Genome: What Squabbles over Genetic Patents Could Teach Us.' in: F. Scott Kieff (ed.) *Perspectives on Properties of the Human Genome Project*. The Netherlands: Elsevier.

Boyle, J. 1996. *Shamans, Software, and Spleens: Law and the Construction of the Information Society*. Cambridge, MA: Harvard University Press.

Cambrosio, A., Keating, P., Bourret, P., Mustar, P. and Rogers, S. (2009) 'Genomic Platforms and Hybrid Formations', in: P. Atkinson, P. Glasner and M. Lock (eds) *Handbook of Genetics and Society: Mapping the New Genomic Era*. London: Routledge; 502–520.

Hilgartner, S. 2017. *Reordering Life: Knowledge and Control in the Genomics Revolution*. Cambridge, MA: MIT Press.

Keating, P. and Cambrosio, A. 2003. *Biomedical Platforms: Realigning the Normal and the Pathological in Late-Twentieth Century Medicine*. Cambridge, MA: MIT Press.

Jasanoff, S. 2005. *Designs on Nature: Science and Democracy in Europe and the United States*. Princeton, NJ: Princeton University Press.

Parthasarathy, S. 2017. *Patent Politics: Life Forms, Markets, and the Public Interest in the United States and Europe*. Chicago: University of Chicago Press.

Winickoff, D. E. 2015. 'Biology Denatured: The Public-Lives of Lively Things', in: S. Hilgartner, C. A. Miller, and R. Hagendijk (eds.), *Science and Democracy: Making Knowledge and Making Power in the Biosciences and Beyond*. London: Routledge.

16

The Human Genome Project and the legacy of its ethics programs

Stephen Hilgartner

In the second half of the 1980s, an elite group of scientists set out to make entire genomes into tractable objects of biological analysis. Envisioning themselves as a scientific vanguard that would spearhead a revolution in the biological sciences and medicine, they sought to catalyze a "paradigm shift" in biology. The centerpiece of their scientific proposals was the Human Genome Project (HGP) – a concerted effort to map and sequence the genomes of the human and several model organisms; advance genome technology; and develop new tools for computational analysis. Their vision of revolutionary change captured imaginations and resources, and before the end of the decade, the U.S. Congress had committed $3 billion to the HGP, a joint project of the National Institutes of Health (NIH) and Department of Energy (DOE) (Cook-Deegan 1994). Smaller genome programs of varying configurations took shape in Europe and Japan (Jordan 1992). The prospect of transformational change – in biological research, medicine, and many aspects of everyday life – raised problems of governance (Hilgartner 2017). How could decision makers govern the "genomics revolution" in ways that realized its promised benefits while minimizing its potential for harm?

In 1988, James D. Watson, the newly appointed director of the NIH genome program, proposed an answer; namely, to devote 3 percent of the HGP budget to studying the ethical and social implications of genome research. Beyond ensuring that "society learns to use the information only in beneficial ways," Watson (1990, 46) warned about the possibility that failure to act could lead to "abuses," provoking a "strong popular backlash against the human genetics community." With the United States poised to spend some $3 billion over a fifteen-year period on the HGP, Watson's commitment promised to make what soon became the genome project's Ethical, Legal, and Social Implications (ELSI) program into the best funded bioethics enterprise in history. By 2014, the U.S. ELSI Program had provided $317 million in research support (McEwen et al. 2014).

The institutional design of ELSI contrasted sharply with other U.S. programs for addressing the societal dimensions of emerging technologies. Not only was ELSI's comparatively lavish funding targeted at a single scientific domain, but ELSI was placed under the management of the very NIH and DOE programs charged with running that research area's flagship project. In contrast, the National Science Foundation's relatively small ethics and values program (Hollander and Steneck 1990) and the Office of Technology Assessment (OTA), which Congress disbanded in 1995, both covered a wide range of topics. Although some observers perceived the potential for conflict of interest in this institutional design, others saw the idea of integrating analysis of ethical and social issues with the scientific research as a promising way to ensure relevance to emerging problems and thereby heighten policy impact.

The founding of ELSI thus introduced a new science policy concept, sometimes described as the "ELSI hypothesis," which proposed that "combining scientific research funding with adequate support for complementary research and public deliberation on the uses of new knowledge will help our social policies about science evolve in a well-informed way" (NIH 1993, 48). This vision proved attractive beyond the United States, and in Europe, various national and Europe-wide agencies also established well-funded "ELSA" programs (Stegmaier 2009; Zwart and Nelis 2009), with the "A" – short for *aspects* – chosen to avoid the connotations of *implications*, which suggested a focus on downstream "impacts." In the three decades since Watson's announcement, this science policy concept spread beyond genomics to such areas as nanotechnology and synthetic biology. Attaching programs on social and ethical issues to major scientific initiatives became a notable development in science/society relations in the twenty-first century.

This chapter examines the legacy of ELSI and ELSA programs, considering them as an instrument of governance, focusing mainly on the United States. (See Hilgartner et al. [2017] for a more extensive review of such programs, including those established by European national governments and the European Union.) The chapter begins by reviewing the history of the U.S. ELSI program, then turns to the rise of "post-ELSI" visions that emerged in the twenty-first century. After examining the debate about the purposes, achievements, and limitations of these programs, the chapter concludes with a brief discussion of institutionalized ethics programs as sites of "constitutional" change (Jasanoff 2004) in the governance of knowledge and polities.

The U.S. ELSI Program and the Human Genome Project

As a policy concept and institutional form, ELSI was forged amid ongoing debate about the wisdom of undertaking the HGP. Today, when the genomes of individual patients are increasingly being sequenced during clinical treatment (Cambrosio et al. *this volume*), it may be hard to imagine how ambitious the HGP seemed when it was first proposed in the mid-1980s. At the time, the laboratory techniques used to sequence DNA were slow and labor intensive. In 1988, when a National Research Council committee wrote a key report supporting the project, the longest continuous segment of DNA that had been sequenced was 150,000 nucleotides, an amount corresponding to only 0.005 percent of the 3 billion base pairs of the human genome (NRC 1988, 62, 65). One ambitious pilot project proposed to sequence 400,000 bases in its initial year – the equivalent of about 0.013 percent of the human genome. Many biologists remained unconvinced of the value of the HGP and some actively opposed it, arguing that a program to sequence the human genome could become a financial boondoggle and contribute to unproductive centralization that would draw resources away from more creative projects. Debate about the value of the program continued among scientists for several years after the U.S. project was officially launched in 1990.

The proposal to sequence the human genome also faced critics who worried about the problems that might emerge if the enterprise were successful. During the 1980s, the field of human genetics had been transformed by the development of predictive tests for several single-gene disorders inherited through Mendelian mechanisms. In the case of Huntington's disease, which often served as the archetypical example of the dilemmas raised by genetic testing, DNA analysis could predict whether an individual would get the disease decades before the onset of symptoms. Patient organizations that had formed around these single-gene diseases grew concerned about issues of genetic privacy, genetic discrimination, psychological effects, and the challenge of appropriately integrating genetic testing into clinical practice (Holtzman 1989; Nelkin and Tancredi 1989). The history of eugenics (Proctor 1988; Duster 1990) heightened concerns. To many, the applications of human genome research looked like a social and ethical minefield.

In this context, Watson's commitment to instituting an ethics program offered assurance that the HGP would not neglect the societal dimensions of genomics. To guide the program, the NIH and DOE established a Working Group, with representation from such fields as human genetics, bioethics, law, social sciences, and policy analysis. In seeking to "anticipate and address" emerging issues and "develop policy options" (NIH-DOE 1990, 66), this "ambitious experiment" in science policy (Juengst 1994, 121) expressed the hope of developing new, more effective means of governing emerging technologies. This aspiration also underlies more recent "post-ELSI" concepts (discussed below), such as those of "anticipatory governance" and "responsible research and innovation" (RRI).

A controversial program

ELSI's promise to "anticipate and address" the "implications" of genome research was sufficiently broad and ambiguous to provide encouragement to people who imagined it as many different things. The centerpiece of the ELSI program was providing funds to support research, mainly through investigator-initiated grants, but people also envisioned it functioning as an ethical watchdog, a quasi-regulatory body, a source of bioethical advice, and a public relations machine. Debate about ELSI began before its official launch and never completely disappeared. During the HGP, ELSI generally received favorable media coverage, and the existence of the program was often said to illustrate the genomics community's commitment to social responsibility. Yet many biomedical scientists, including some high-ranking NIH officials, resented devoting millions of dollars annually to funding what they regarded as "the vacuous *pronunciamentos* of self-styled 'ethicists'" (Juengst 1996, 64, 66). Meanwhile, a number of observers – including legal scholars (Annas, 1989), biologists (Hubbard and Wald 1993, 159), and historians of science (Weiner 1994) – saw ELSI as a strategy for diffusing political opposition to the HGP, in part by coopting potential critics. Even some people intimately involved in the program saw ELSI as a means of engineering acceptance. The first coordinator of the DOE's ELSI program, to take one prominent example, described ELSI in a retrospective commentary as a strategy to "deflect political challenges," win funding for the HGP, and address Congressional concerns (Yesley 2008, 3).

The proper design of the program was also a subject of debate. Some critics contended that an academic research program was the wrong mechanism for addressing ELSI issues, arguing for a national commission charged with making policy recommendations (Hanna 1995; Yesley 2008). Defenders of ELSI contended that it was unrealistic to expect any program or commission, no matter how it was constituted, to "resolve" the many issues raised by the rapidly advancing field of genomics. They also maintained that ELSI was achieving its main goals: creating the knowledge needed to anticipate issues and identify policy options, and building a cadre of ELSI researchers to advance the field (Juegnst 1996; ELSI Research Planning and Evaluation Group 2000). In the mid-1990s, the proper role of the ELSI Working Group became a subject of internal debate, with some arguing that the Working Group should offer policy recommendations and others saying its role should be limited to guiding the grants program. Eventually, conflict over such issues led Francis Collins, the second director of the NIH genome program, to disband the Working Group and consolidate control over the program.

As the HGP drew to a close in the early 2000s, the U.S. genome program presented ELSI, like the rest of the HGP, as a major success, and the government promised continued attention to ELSI issues as the genomics revolution continued. At the celebration of the completion of a "first draft" of the sequence of the human genome in 2000, President Bill Clinton stressed that "we must ensure that the new genome science and its benefits will be directed toward making life better for all citizens of the world, never just a privileged few…. And we must guarantee that genetic information cannot be used to stigmatize or discriminate against any individual or group." In 2003, the NIH published new goals for genomics, and these included a renewed commitment to ELSI research and education (Collins et al. 2003).

However, some observers concluded that ELSI had failed. McCain (2002) argued that the program had fostered among the public the misleading impression that the important issues raised by the scientific advances were being addressed. Philosopher Philip Kitcher (2001, 189), who had previously served on the ELSI Working Group, concluded that the program had been "doomed to fail" from the start because it falsely assumed that the problems raised by genomics could be resolved in "a politically neutral way."

In the years since the completion of the HGP, debate about the ELSI program has continued. Some biomedical scientists remained impatient with ELSI, and some bioethicists argued that ample ELSI funding has drawn attention away from more important problems in medicine and public health, such as inequality (Turner 2005). Other observers described ELSI's policy impact as minimal, suggesting it served as "a surrogate for policy making" (Fisher 2005).

Staff members at the National Human Genome Research Institute (McEwen et al. 2014) counter that ELSI has had significant, if subtle, influence. They point out that evaluating "impact" is difficult, explaining that much ELSI work does not have "a direct impact on policy or practice" but helps build a "foundation" for applied studies. They also contend that ELSI investigators "operating independently as scholars" influence policy through their work with policymaking bodies. Finally, they argue that ELSI's "most consequential impact" may stem from subtle and incremental changes in "the cultural milieu in which genomics research is conducted, genomic medicine is implemented, and genomic information is incorporated into decision-making in various areas of society more broadly" (ibid., 45).

Bioethical vision

The rise of ELSI, ELSA, and similar programs was part of the larger trend in the United States and other democratic countries of relying on "ethics" as part of the machinery of governance, a development especially evident in the management of moral problems related to medicine and biology. During the 1970s and 1980s, what became known as "bioethics" took shape as a domain of inquiry and decision making, growing from roots in moral philosophy, law, and social sciences to encompass many activities, performed in such sites as academic literature, institutional ethics committees, medical practice and education, private foundations, court decisions, and national commissions (e.g., DeVries and Subedi 1998; Evans 2011; Kleinman et al. 1999; Sperling 2013; Stark 2011). Despite ongoing contention, bioethics became an important source of authority. Its tendency to define problems as matters of values, rather than the result of (potentially changeable) allocations of power and privilege, contributed to its acceptance (Bosk 2008). Bioethics thus provided ontologies used to shape deliberation and "resolve" issues, which it generally framed as matters for moral reflection and judicious policy-making rather than political mobilization.

The U.S. ELSI program built on (and to some extent furthered) the successful institutionalization of bioethics – especially its American form, which tends to treat protection of the individual and the value of individual autonomy as cardinal values (Wolpe 1998; Jasanoff 2005). ELSI began from this starting point, which was the centerpiece of its initial research agenda (NIH-DOE 1990). More deeply, ELSI took on board the tendency in bioethics to focus on the social consequences of developments in biomedicine, rather than treating the content and authority of biomedicine as a topic for critical analysis. To be sure, ELSI did fund a few studies that subjected genome science to critical inquiry (Juengst 1996), but the program was largely restricted to the "implications" and "applications" of genomic knowledge and technology in downstream sites. What were defined as "ELSI issues" thus were understood as being distinct from "the science itself." The main exception – the ethics of research involving human subjects – had been a central concern of bioethics since the beginnings of its institutional emergence (Rothman 1991). Moreover, bioethics tends to frame human subjects concerns narrowly as a matter of the conduct of science rather than as entangled in hybrid practices that transcend a simple distinction between conduct and content (e.g., Keating and Cambrosio 2011).

Beyond genomics

Although it followed well institutionalized tendencies in U.S. bioethics, the visibility of ELSI created a collectively recognized policy direction for addressing the politics of emerging technologies: namely, the possibility of attaching "science and society" programs to major research initiatives. ELSI thus created a precedent that the promoters of emerging technologies drew on when advancing research programs in such fields as nano (Roco and Bainbridge 2007) and synthetic biology (Calvert and Martin 2009). But if ELSI became the exemplar of a new institutional form in science policy, it proved to be a flexible one. As actors proposed new programs to anticipate and address societal issues beyond genomics, ELSI was at times presented as something to emulate and at times as something to move beyond.

In discussions of the U.S. National Nanotechnology Initiative, for example, advocates of establishing a program to address societal concerns used ELSI as both a positive example to justify significant funding (e.g., Colvin 2003; Kurzweil 2003) and a negative one to propose fundamental changes in goals and design (e.g., Fisher and Mahajan 2006). As versions of the ELSI vision were taken up by new sociotechnical vanguards, a cascade of new "buzzwords" (Bensaude-Vincent 2014) entered science policy discourse. "ELSI" and "ELSA" were joined by "real-time technology assessment" (RTTA) (Guston and Sarewitz 2002), "upstream engagement" (Wilsdon and Willis 2004), "societal and ethical issues" (of nano), "anticipatory governance" (Barben et al. 2008; Guston 2014), "human practices" (Rabinow and Bennett 2012), and "responsible research and innovation" (RRI) (Felt 2016; Stilgoe et al. 2013; von Schomberg 2013). These new terms should not be appreciated only as discrete science policy concepts but also as "part of a lineage of visions" (Hilgartner 2015, 39) – with each new term reflecting a diagnosis of the limitations of ELSI (and other ELSI-like programs) and a vision of transcending those limitations.

Post-ELSI policy concepts – such as RTTA, upstream engagement, and anticipatory governance – all criticized ELSI's relative neglect of "upstream" issues, such as the shaping of research agendas or the normative assumptions embedded in biomedicine, and they sought to widen the range of social and ethical inquiry beyond "implications" and "applications." Advocates of replacing technocratic policymaking with more democratic forms of "public engagement" also found ELSI wanting. In keeping with its roots in bioethics, ELSI had been envisioned as a source of expertise: knowledge about ELSI issues would flow from experts to policymakers and the public. But as the new century began, an alternative vision emphasizing the importance of public dialogue about emerging technology grew increasingly prominent (Irwin 2006;

Nowotny et al. 2001), especially in Europe. Policymakers stung by the bovine spongiform encephalitis (BSE) crisis and the unexpected rejection of genetically modified crops and foods sought to learn to listen to a wide range of stakeholders, rather than merely speaking to them. New visions of engaging publics in various forms of dialogue and mutual learning also developed in the United States, especially in a program of "anticipatory governance" in nanotechnology (Barben et al. 2008). Recently, the concept of RRI has gained traction, especially in Europe (Felt *this volume*). It expands the focus of earlier institutionalized ethics programs, adding the fields of management and innovation studies and provoking some observers to worry that RRI portends a shift to a more promotional agenda (e.g., Zwart et al. 2014).

Predictably, these post-ELSI approaches encountered their own problems. Policymakers' commitment to truly two-way communication was often incomplete or ambivalent (Wynne 2006). Widespread acceptance of the linear model of innovation was shown to undermine some attempts to enact participatory deliberation (Felt et al. 2009). Social scientists who sought to engage with scientists (e.g., as "in-house" social scientists) encountered objections to their research agendas (e.g., Viseu 2015) or found themselves shifting roles like chameleons as they moved among contexts (Balmer et al. 2015). Rabinow and Bennett (2012) argue that their "experiment with synthetic biology," which aimed to build a truly collaborative engagement between social and natural scientists, failed owing to incompatible agendas and asymmetries of power. Joly (2015) also emphasizes the effects of power asymmetries, while also noting the difficulty of avoiding capture.

Advocates of various post-ELSI concepts do not deny the existence of challenges, but they argue that such problems do not negate the value of these institutional experiments. For example, Guston (2014), the leader of a Center for Nanotechnology in Society at Arizona State University, provides an extended discussion of the contributions and critiques of its program of "anticipatory governance," arguing that although such work cannot solve all the problems of governing emerging technologies, it can help to bend the "long arc of technoscience more toward humane ends" (ibid., 234).

Constitutional dimensions

Amid continuing debate about the extent to which various ELSI and post-ELSI approaches improve societal choices about emerging technologies, some scholars are asking broader questions about how these programs reflect and shape science/society relations (Irwin 2006) and, more broadly, the sociopolitical orders of the societies in which they operate (Jasanoff 2005). This work explores how emerging technologies and the machinery for governing them have "constitutional" dimensions, not in the narrow sense of constitutional law, but as a site of negotiations that reallocate authority and the capacity for action among experts, citizens, state institutions, and other actors (Jasanoff 2002). In a comparative study of biotechnology regulation, Jasanoff (2005) shows that Germany, the United Kingdom, and the United States approach governance in rather different ways. In particular, each country displays a different "civic epistemology" – that is, a patterned set of practices employed to "assess the rationality and robustness of claims that seek to order their lives" (ibid., 255). Even when these countries answered regulatory questions in broadly similar ways, the process and forms of public reasoning leading to these outcomes differed.

This constitutional perspective makes the empirical examination of the governance of emerging technologies relevant to central issues of democratic and political theory. Along these lines, Hurlbut (2017; *this volume*) argues that imagining science as a source of novelties to which society must respond positions scientists as uniquely capable of knowing what possible futures warrant ethical evaluation. In contrast, society is cast in a reactive role, even as it is made responsible for "addressing" the "social and ethical consequences" of technological change. Tallachini (2015) argues that the "soft law" of the European Union's ethics programs allowed extant regulatory frameworks to be extended more quickly than would be possible through a legislative process, contributing to a particular form of European integration and to the acceptance of emerging technologies. In a comparative study of Europe and the United States, Laurent (2017) provides an empirical examination of how nanotechnology became a site where democracy was at stake as it was problematized in ways that raised issues of constitutional significance. Hilgartner (2017, 217–219) argues that the scientific vanguard that launched the HGP, joined by state officials such as President Bill Clinton, promoted the imaginary of a collective "we" committed to appropriately guiding the applications of genomics. This imaginary underwrote the reassuring idea that the rise of genomics will be an "orderly and beneficial revolution – one that will preserve cherished human values in the midst of dramatic change" (ibid., 219).

Conclusion

The rise of ELSI and ELSA programs institutionalized an important development in the politics of science and the broader politics of contemporary societies: the attaching of "ethics" programs to scientific initiatives expected to lead to transformational change. As a policy concept, these programs have proved to be both durable and flexible. The cascade of new "post-ELSI" visions represents not only an effort to improve on earlier programs but also the challenge of revitalizing the broader imaginary of improving the capacity of societies to govern emerging technologies, such as genomics. Controversy about the value, the purposes, the achievements, and the possibilities of such programs will likely continue for the foreseeable future, as will adjustments in institutional designs, goals, and techniques of inquiry and engagement. But beyond debate about their objectives and institutional design, these programs have a constitutional significance. Perhaps most deeply their rise expresses the hope that a "we" capable of governing a world of normative disagreement and transformational change can be successfully constituted.

References

Annas, George J. 1989. Who's Afraid of the Human Genome? *Hastings Center Report* 19(4): 19–21.

Balmer, Andrew S., Jane Calvert, Claire Marris, Susan Molyneux-Hodgson, Emma Frow, Matthew Kearnes, Kate Bulpin, Pablo Schyfter, Adrian Mackenzie, and Paul Martin. 2015. Taking Roles in Interdisciplinary Collaborations: Reflections on Working in Post-ELSI Spaces in the UK Synthetic Biology Community. *Science & Technology Studies* 28(3): 3–25.

Barben, Daniel, Erik Fisher, Cynthia Selin, and David H. Guston. 2008. Anticipatory Governance of Nanotechnology: Foresight, Engagement, and Integration. In *The Handbook of Science and Technology Studies*. 3rd ed., edited by Edward J. Hackett, Olga Amsterdamska, Michael E. Lynch, and Judy Wajcman, 979–1000. Cambridge, MA: MIT Press.

Bensaude-Vincent, Bernadette. 2014. The Politics of Buzzwords At The Interface of Technoscience, Market and Society: The Case of 'Public Engagement In Science.' *Public Understanding of Science* 23(3): 238–253.

Bosk, Charles L. 2008. *What Would You Do? Juggling Bioethics and Ethnography*. Chicago: University of Chicago Press.

Calvert, Jane, and Paul Martin. 2009. The Role of Social Scientists in Synthetic Biology. *EMBO Reports* 10(3): 201–204.

Collins, Francis S., Eric D. Green, Alan E. Guttmacher, and Mark S. Guyer. 2003. A Vision for the Future of Genomics Research: A Blueprint for the Genomic Era. *Nature* 422(6934): 835–847.

Colvin, Vicki L. 2003. Statement: The Societal Implications of Nanotechnology. Hearing before the U.S. House Committee on Science, 108th Congress. First session, April 9, 2003. Serial no.108–113.

Cook-Deegan, Robert. 1994. *The Gene Wars: Science, Politics, and the Human Genome.* New York: W. W. Norton.

Duster, Troy. 1990. *Backdoor to Eugenics.* New York: Routledge.

DeVries, Raymond and Janardan Subedi, eds. 1998. *Bioethics and Society: Constructing the Ethical Enterprise.* Prentice-Hall, Upper Saddle River, NJ.

ELSI Research Planning and Evaluation Group. 2000. A Review and Analysis of the ELSI Research Programs at the National Institutes of Health and the Department of Energy. Final report, February 10, 2000. Available at: www.genome.gov/10001727/erpeg-final-report/ (Last accessed 29 November 2017).

Evans, John H. 2011. *The History and Future of Bioethics: A Sociological View.* Oxford: Oxford University Press.

Felt, Ulrike. 2016. "Response-able Practices" or "New Bureaucracies of Virtue": The Challenges of Making RRI Work in Academic Environments. Preprint. Published by Department of Science and Technology Studies, University of Vienna, November 2016. http://sts.univie.ac.at/publications, accessed June 16, 2017.

Felt, Ulrike, Maximillian Fochler, Annina Müller, and Michael Strassnig. 2009. Unruly Ethics: On the Difficulties of a Bottom-up Approach to Ethics in the Field of Genomics. *Public Understanding of Science* 18(3): 354–371.

Fisher, Erik. 2005. Lessons Learned from the Ethical, Legal and Social Implications Program (ELSI): Planning Societal Implications Research for the National Nanotechnology Program. *Technology in Society* 27(3): 321–328.

Fisher, Erik, and Roop L. Mahajan. 2006. Contradictory Intent? US Federal Legislation on Integrating Societal Concerns into Nanotechnology Research and Development. *Science and Public Policy* 33(1): 5–16.

Guston, David H. 2014. Understanding 'Anticipatory Governance.' *Social Studies of Science* 44(2): 218–242.

Guston, David H., and Daniel Sarewitz. 2002. Real-Time Technology Assessment. *Technology in Society* 24(1): 93–109.

Hanna, Kathi E. 1995. The Ethical, Legal and Social Implications Program of the National Center for Human Genome Res: A Missed Opportunity? In *Society's Choices: Social and Ethical Decision Making in Biomedicine*, 432–457. Washington, DC: National Academies Press.

Hilgartner, Stephen. 2015. Capturing the Imaginary: Vanguards, Visions, and the Synthetic Biology Revolution. In *Science and Democracy: Making Knowledge and Making Power in the Biosciences and Beyond*, edited by Stephen Hilgartner, Clark A. Miller, and Rob Hagendijk, 33–55. New York: Routledge.

Hilgartner, Stephen. 2017. *Reordering Life: Knowledge and Control in the Genomics Revolution.* Cambridge, MA: MIT Press.

Hilgartner, Stephen, Barbara Prainsack, and J. Benjamin Hurlbut. 2017. Ethics as Governance in Genomics and Beyond. In *Handbook of Science & Technology Studies*, 4th edition, 823–851. Cambridge, MA: MIT Press.

Hollander, Rachelle D., and Nicholas H. Steneck. 1990. Science- and Engineering-Related Ethics and Values Studies: Characteristics of an Emerging Field of Research. *Science, Technology, & Human Values* 15(1): 84–104.

Holtzman, Neil A. 1989. *Proceed with Caution: Predicting Genetic Risks in the Recombinant DNA Era.* Baltimore: Johns Hopkins University Press.

Hubbard, Ruth, and Elijah Wald. 1993. *Exploding the Gene Myth.* Boston: Beacon Press.

Hurlbut, J. Benjamin. 2017. *Experiments in Democracy: Human Embryo Research and the Politics of Bioethics.* New York: Columbia University Press.

Irwin, Alan. 2006. The Politics of Talk: Coming to Terms with the 'New' Scientific Governance. *Social Studies of Science* 36(2): 299–320.

Jasanoff, Sheila. 2002. In a Constitutional Moment: Science and Social Order at the Millennium. In *Science and Technology Studies: Looking Back, Ahead*, edited by Bernwald Joerges and Helga Nowotny, 155–180. The Netherlands: Kluwer.

Jasanoff, Sheila. 2004. *States of Knowledge: The Co-production of Science and Social Order.* London: Routledge.

Jasanoff, Sheila. 2005. *Designs on Nature: Science and Democracy in Europe and the United States.* Princeton, NJ: Princeton University Press.

Joly, Pierre-Benoît. 2015. "Governing Emerging Technologies? The Need to Think Out of the (Black) Box." In *Science and Democracy: Knowledge as Wealth and Power in the Biosciences and Beyond*, edited by Stephen Hilgartner, Clark A. Miller, and Rob Hagendijk, 133–155. New York: Routledge.

Jordan, Bertrand. 1992. *Travelling Around the Human Genome*. Montrouge, France: John Libbey Eurotext.

Juengst, Eric T. 1994. "Human Genome Research and the Public Interest: Progress Notes from an American Science Policy Experiment." *Am J Hum Genet* 54(1): 121.

Juengst, Eric T. 1996. "Self-Critical Federal Science? The Ethics Experiment within the Human Genome Project." *Social Philosophy & Policy* 13: 63–95.

Keating, Peter and Alberto Cambrosio. 2011. *Cancer Clinical Trials: Oncology as a New Style of Practice*. Chicago: University of Chicago Press.

Kitcher, Philip. 2001. *Science, Truth, and Democracy*. Oxford: Oxford University Press.

Kleinman, Arthur, Renee C. Fox, and Allan M. Brandt, eds. 1999. Bioethics and Beyond. *Daedalus* (special issue) 128(4): 1–325.

Kurzweil, Raymond. 2003. Statement: The Societal Implications of Nanotechnology. Hearing before the U.S. House Committee on Science, 108th Congress. First session, April 9, 2003. Serial no. 108–113.

Laurent, Brice. 2017. *Democratic Experiments: Problematizing Nanotechnology in Europe and the United States*. Cambridge, MA: MIT Press.

McCain, Lauren. 2002. Informing Technology Policy Decisions: The US Human Genome Project's Ethical, Legal, and Social Implications Programs as a Critical Case. *Technology in Society* 24(1): 111–132.

McEwen, Jean E., Joy T. Boyer, Kathie Y. Sun, Karen H. Rothenberg, Nicole C. Lockhart, and Mark S. Guyer. 2014. The Ethical, Legal, and Social Implications Program of the National Human Genome Research Institute: Reflections on an Ongoing Experiment. *Annual Review of Genomics and Human Genetics* 15(1): 481–505.

National Institutes of Health, National Center for Human Genome Research. 1993. *Progress Report: Fiscal Years 1991 and 1992*. NIH Publication No. 93–3550. Bethesda, MD: National Institutes of Health.

National Institutes of Health and Department of Energy. 1990. *Understanding Our Genetic Inheritance: The U.S. Human Genome Project: The First Five Years, FY 1991–1995*. Springfield, VA: National Technical Information Service.

National Research Council. 1988. *Mapping and Sequencing the Human Genome*. Washington, DC: National Academy Press.

Nelkin, Dorothy, and Laurence Tancredi. 1989. *Dangerous Diagnostics: The Social Power of Biological Information*. Chicago: University of Chicago Press.

Nowotny, Helga, Peter Scott, and Michael T. Gibbons. 2001. *Re-thinking Science: Knowledge and the Public in an Age of Uncertainty*. Cambridge, MA: Polity.

Proctor, Robert. 1988. *Racial Hygiene: Medicine Under the Nazis*. Cambridge, MA: Harvard University Press.

Rabinow, Paul, and Gaymon Bennett. 2012. *Designing Human Practices: An Experiment with Synthetic Biology*. Chicago: University of Chicago Press.

Roco, Mihail, and William Sims Bainbridge, eds. 2007. *Nanotechnology: Societal Implications I: Maximizing Benefits for Humanity*. Dordrecht: Springer.

Rothman, David J. 1991. *Strangers at the Bedside: A History of How Law and Bioethics Transformed Medical Decision Making*. New York: Basic Books.

Sperling, Stefan. 2013. *Reasons of Conscience: The Bioethics Debate in Germany*. Chicago: University of Chicago Press.

Stark, Laura. 2011. *Behind Closed Doors: IRBs and the Making of Ethical Research*. Chicago: University of Chicago Press.

Stegmaier, Peter. 2009. The Rock 'n' Roll of Knowledge Co-production. *EMBO Reports* 10(2): 114–119.

Stilgoe, Jack, Richard Owen, Phil Macnaghten. 2013. Developing a Framework for Responsible Innovation. *Research Policy* 42(9): 1568–1580.

Tallacchini, Mariachiara. 2015. To Bind or Not to Bind? European Ethics as Soft Law. In *Science and Democracy: Making Knowledge and Making Power in the Biosciences and Beyond*, edited by Stephen Hilgartner, Clark A. Miller, and Rob P. Hagendijk, 156–175. New York: Routledge.

Turner, Leigh. 2005. Bioethics, Social Class, and the Sociological Imagination. *Cambridge Quarterly of Healthcare Ethics* 14: 374–378

Viseu, Ana. 2015. Caring for Nanotechnology? Being an Integrated Social Scientist. *Social Studies of Science* 45(5): 642–664.

von Schomberg, René. 2013. A Vision of Responsible Research and Innovation. In *Responsible Innovation*, edited by Richard Owen, John Bessant, and Maggy Heintz, 51–74. London: John Wiley.

Watson, James D. 1990. The Human Genome Project: Past, Present, and Future. *Science* 248: 44–49.

Weiner, Charles. 1994. Anticipating the Consequences of Genetic Engineering: Past, Present, and Future. In *Are Genes Us? The Social Consequences of the New Genetics*, edited by Carl F. Cranor, 31–55. New Brunswick, NJ: Rutgers University Press.

Wilsdon, James, and Rebecca Willis. 2004. *See-through Science: Why Public Engagement Needs to Move Upstream*. London: Demos.

Wolpe, Paul Root. 1998. The Triumph of Autonomy in American Medical Ethics: A Sociological View. In *Bioethics and Society: Sociological Investigations of the Enterprise of Bioethics*, edited by DeVries, R. and Subedi, J. Prentice Hall, New York.

Wynne, Brian. 2006. Public Engagement as Means of Restoring Trust in Science? Hitting the Notes, but Missing the Music. *Community Genetics* 3(9): 211–220.

Yesley, Michael. 2008. What's ELSI Got to Do with It? Bioethics and the Human Genome Project. *New Genetics and Society* 21(1): 1–6.

Zwart, Hub A. E., Laurens Landeweerd, and Arjan van Rooij. 2014. Adapt or Perish? Assessing the Recent Shift in the European Research Funding Arena from 'ELSA' to 'RRI.' *Life Sciences, Society and Policy* 10(1): 1–19.

Zwart, Hub A. E., and Annemiek Nelis. 2009. What Is ELSA Genomics? *EMBO Reports* 10(6): 540–544.

17
Patenting

Shobita Parthasarathy

Biotechnology has challenged patent systems since the development of recombinant DNA (rDNA) in the 1970s (Hughes 2011; Parthasarathy 2017). Were genetically engineered organisms unpatentable discoveries of natural phenomena or patentable technologies? What about genes and other strings of DNA that could be identified, sequenced, and linked to major diseases? Did their status as pieces, or forms, of life matter? Were patent systems responsible for addressing ethical, social, and environmental concerns? Despite international agreements designed to harmonize patent law, countries have approached these questions in multiple ways. Focusing on the US and European patent systems, this chapter argues that this variability stems not merely from differences in law, but also in political culture, ideology, and history. As the US and Europe made different patent decisions, they developed different understandings of the patent system's role, as well as its appropriate knowledge and expertise. These differences will continue to matter and shape genomic futures.

The increase in patents in recent decades has stimulated concern about the commercialization of science and medicine, and the progressive enclosure of the public domain (Boyle 2010; Kleinman 2003; Mirowski 2011). This has attracted attention among legal scholars, humanists, and social scientists. Their work has revealed the philosophies that have guided the development of Western patent systems (Biagioli 2011; Bracha 2016), and explored the differences from how the developing world conceptualizes intellectual property (May and Sell 2005; Strathern 1996). Biotechnology patents, which challenge conventional notions of invention, have received much of this attention. Scholars have identified the values and assumptions that underlie legal definitions of nature, technology, discovery, invention, and even life (Jasanoff 2012; Bagley 2007; Hayden 2003). Others have shown how, as societies negotiate the ontological status of biotechnology, they are reordering legal and social orders (Pottage 2007; Winickoff 2015). Similarly, Daniel Kevles (2002a) and Mariachiara Tallacchini (2015) have argued that resolving questions about the patentability of biotechnology in both the United States and Europe has produced new spaces for ethics discussion.

Building upon this work, this chapter uses a historical and comparative approach to unravel the political configurations that underlie patent systems. Science and technology studies (STS) scholars have long argued that political culture and ideology shape how governments and their citizens approach appropriate knowledge and expertise for policy (Ezrahi 1990; Jasanoff 2005;

Porter 1996), particularly risk regulation (Daemmrich 2004; Davis and Abraham 2011). These epistemological differences are also evident across patent systems. Specifically, different understandings of the roles of and relationships between markets and government vis-à-vis innovation have led the United States and Europe to understand patents and patent systems differently (Parthasarathy 2017). The United States has what I call a *market-making* approach, assuming that an unfettered market will achieve overall social benefit and address moral concerns (Sandel 2013; Sarewitz 1996). Europe, by contrast, takes a *market-shaping* approach. While it is difficult to point to a pan-European political culture, most Western European countries have a strong welfare state and have traditionally seen clear responsibility for government in regulating innovation and shaping its moral, social, and economic implications (Gottweis 1998; Testa 2011). The state's role is to shape the marketplace and to maintain moral order (Metzler 2011; Sperling 2013).

These different approaches have influenced patent systems, including their approaches to life forms. The earliest European systems demonstrated concern about socioeconomic implications by passing laws that limited the term of a patent monopoly, requiring patent holders to commercialize inventions within a certain period of time, providing governments with authority to step in if patent holders set the prices of their goods too high, and prohibiting patents on certain essential goods including food and pharmaceuticals (Letwin 1954; Cassier 2008). Even as European patent systems evolved, they retained many of these provisions. Most notably for this chapter, patent laws across Europe contain *ordre public* provisions that allow governments to prohibit patents on inventions deemed contrary to public policy or morality (Sterckx and Cockbain 2012; Husserl 1938). This includes the 1973 European Patent Convention, which established the pan-European patent system.

By contrast, the US patent system has always treated inventors' interests and the public interest as essentially the same. The patent system's role is to provide an incentive to invent, which will produce economic growth and make more technologies available (Dobyns 1994; Walterscheid 1995). Early laws made it easy to apply for a patent, assuming that everyone was a prospective inventor (Biagioli 2011). While it limited patent length like its European counterparts, US legislators were generally hostile to the idea that patents ever hurt the public and were therefore unwilling to consider measures such as compulsory licensing (Oldfield Revision and Codification of Patent Laws 1914). Instead, they focused on ensuring procedural objectivity in the examination process by ensuring that bureaucrats had the technical training to properly evaluate the novelty and utility of inventions. And while the US Supreme Court identified a *moral utility* doctrine that suggested patents could be prohibited on moral grounds under exceptional circumstances, this prohibition has almost never been used (Bagley 2003; Lowell v. Lewis 1917).

A narrow legal question

Biotechnology provoked great excitement about the field's potential to address health and environmental problems, and a new industry quickly emerged to capitalize on financial opportunities (Rasmussen 2014; Wright 1994). However, scientists and civil society groups worried about ethical, social, and environmental implications, as did researchers and funding institutions, including the US National Institutes of Health (Krimsky 1984).

Patent systems did not escape these concerns (Jasanoff 2012; Kevles 2002a). In the late 1970s, Ananda Chakrabarty, a scientist at General Electric, tried to patent a genetically engineered microorganism designed to eat oil (Kevles 1994). The PTO argued that the microorganism could not be patented because it was a "product of nature," and the case landed at the US

Supreme Court (Winickoff 2015). The Court received amicus briefs from biotechnology companies, patent law associations, universities, and even individual scientists. All supported patentability, citing the importance of biotechnology to the United States' economic future (e.g., Genentech 1980). But the Court also received an amicus brief from an unlikely source, a coalition of civil society groups led by Jeremy Rifkin, who was also questioning the ethics and safety of rDNA research elsewhere (People's Business Commission 1980; Howard and Rifkin 1977). This coalition argued that life form patents would devalue living things by turning them into mere commodities, and would ultimately transform the relationship between humans and the ecosystem. The brief also questioned the Supreme Court's legitimacy to decide the case, arguing it was a political matter to be decided by democratically elected representatives in Congress.

The other briefs rejected these arguments, suggesting that the coalition simply didn't understand the patent system. It was a narrow legal domain, they argued, far from matters of life and morality. The Supreme Court ultimately agreed, ruling "anything under the sun made by man" was patentable (Diamond v. Chakrabarty 1980). As it did so it reinforced the narrowness of the US patent system: "The grant or denial of patents on micro-organisms is not likely to put an end to genetic research or to its attendant risks…patentability will not deter the scientific mind from probing into the unknown any more than Canute could command the tides." To the court and other amici, the coalition's concerns were irrelevant to the patent system; they were matters for other science and technology policy institutions to address. Congress seemed to agree, never taking up the issue. In fact, by the end of 1980 it had passed the Bayh-Dole Act, which allowed universities to hold patents on the fruits of federally funded research (Mowery et al. 2004).

The Chakrabarty decision and the Bayh-Dole Act did not end controversy. Rather, as the PTO began to approve patents on genetically modified plants and animals, environmental groups, animal rights activists, religious figures, and farmers joined the opposition (Kevles 2002b). Their concerns went unheeded. Traditional stakeholders and patent officials used the law, bureaucratic rules, and rhetoric to suggest that these outsiders lacked the appropriate expertise and background to participate. In effect, the US patent system erected an "expertise barrier": formal and informal rules that develop over the course of a policy domain's history and are shaped by the jurisdiction's political culture and ideology (Parthasarathy 2010). Policymakers also continued to emphasize that the patent system was not a regulatory domain, but rather simply a narrow arena concerned only with technical and legal questions of patentability.

Human gene patents in the United States

The practice of human gene patenting attracted broad attention in the 1990s, when the PTO began issuing patents on common diseases including breast cancer. Patient advocates and health care professionals predicted that such patents would lead to monopolies that would make genetic tests and treatments unaffordable (Andrews 2002; Cho et al. 2003). Bioethicists and long-term critics repeated their concerns about the commodification of life (Cahill 2001). Scientists and legal scholars argued that genes did not warrant patents because the discovery process was well-known, and worried that owners could refuse to license their patents or charge exorbitant fees, stifling research (Heller and Eisenberg 1998; Rai 1999). For example, when the NIH applied for patents on expressed sequence tags (ESTs) – pieces of DNA believed to be fragments of genes – scientists challenged the ingenuity underlying EST production, and argued that patents would wreak havoc in the emerging genomics field (Hilgartner 2017).

These concerns were realized over the next decade (Merz 1999; Scherer 2002), especially in the case of the BRCA genes, linked to inherited susceptibility to breast and ovarian cancer. Myriad Genetics, the biotechnology company who owned the patents, shut down other BRCA gene testing providers, charged high prices for its own test, and limited research dramatically (Parthasarathy 2007). Multiple policy committees reviewed the practice, but their reports did not lead to policy or legal change (Committee on Intellectual Property Rights 2006; Secretary's Advisory Committee 2010).

In 2007, the American Civil Liberties Union (ACLU) got involved. Although it had a long record of civil rights and civil liberties victories (Walker 1999), the ACLU was an outsider to the patent domain and to science and technology policy. But it had a science advisor with a family history of breast cancer who was concerned about BRCA gene patents, and she convinced the ACLU that rights to research and health were important civil liberties issues (Simoncelli 2013). The ACLU focused its attention on BRCA gene patents, motivating robust patient advocacy and support communities as well as public concern (Klawiter 2008; Stabiner 1997).

The ACLU sued Myriad Genetics and the PTO, challenging the patentability of human genes (Simoncelli and Park 2015). But it faced a long legacy of similar legal interventions that had been thrown out of US courts (Parthasarathy 2017). It identified two initial challenges: putting together its plaintiff list and deciding upon the lawsuit grounds. The ACLU could not establish "legal standing" on its own, as the courts had decided that plaintiffs needed to demonstrate direct physical or economic harm from the patent and to show that a favorable decision would redress the injury. These rules fit with the overall orientation of the US patent system: the public interest was embodied by inventors themselves. Thus, most patent cases were brought by industrial competitors, and focused on infringement. The ACLU handled this hurdle by assembling a group of plaintiffs that included medical geneticists and pathologists, research laboratories that Myriad had shut down, and women living with and at risk for breast cancer who were prevented by Myriad's monopoly to get a second opinion on BRCA gene testing (AMP v. Myriad 2009).

The ACLU then focused on whether human gene patents violated the product of nature doctrine (Winickoff 2015). Its driving concerns were research freedoms and access to health care. But, the lawyers knew that the courts would be unlikely to adjudicate them (Simoncelli and Park 2015). These strategies paid off. In June 2013, the US Supreme Court decided unanimously in favor of the plaintiffs (AMP v. Myriad 2013). Almost immediately, a number of providers began to offer BRCA gene testing and the horizons opened for genetics researchers as well.

This legal decision had important implications for science, medicine, and patients. However, the US patent system's understanding of patents or the public interest had not changed in any substantial way. While the decision certainly provided challengers with a primer on how to launch public interest cases in the patent system, the barriers to entry – in terms of legal standing and arguments deemed acceptable – remain restrictive. Indeed, the US patent system insists that patents are not to blame for the high costs of genetic tests or essential medicines (US Department of State 2016; US Patent and Trademark Office 2015).

Overall, the United States has built a governance approach that conceptualizes patents in techno-legal terms and at a deeper level, a market-making ideology. There is great resistance – rhetorical, legal, and institutional – to considering the socioeconomic or moral implications of human gene patents. The Myriad case succeeded precisely because it repackaged citizen frustrations about access to technology into a legal question about whether isolated human genes were products of nature. The underlying concerns about conflicts between the patent system and the public interest have not yet been addressed.

Life form patent policy in Europe

In 1988, at the biotechnology industry's request, the European Commission proposed the EU Directive on the Legal Protection of Biotechnological Inventions (BPD) to the European Parliament, expecting quick passage (Gold and Gallochat 2001). The law would harmonize European Union biotechnology patent law, which industry argued would attract investors and accelerate innovation. While the Commission saw life forms as technologies, EU Parliamentarians raised moral and socioeconomic concerns, just as US civil society groups had. But unlike US policymakers, the Parliamentarians saw life form patenting as a policy matter that required balancing public and private interests.

After ten years of debate and civil society pressure (Parthasarathy 2017), the Parliament and European Council passed a version of the BPD in 1998 (European Parliament and Council 1998). This final version differed substantially from the original Commission proposal. Changes included a new meaning of the *ordre public* clause, shaping the concepts of both "public" and "morality" in Europe with a focus on human and animal dignity, commodification of life, and bioethical concerns. Moreover, it created multiple exclusions from patentability, including uses of human embryos for industrial and commercial purposes, and allowed patents on animals only when the benefits outweighed the suffering experienced by the animal. And, it acknowledged the advisory role of the Commission's European Group on Ethics in Science and New Technologies. The BPD did allow human gene patents, a move that many would soon oppose.

Overall, Europe's guiding legislation followed an approach that went back to the early seventeenth century, but it also took into account bioethical concerns related to commodification and human and animal dignity. With this definition came a patent system in which citizens and their elected representatives, and ethics experts, had important roles to play.

Human gene patents in Europe

As in the United States, in Europe Myriad Genetics catalyzed debate over human gene patents when it submitted EPO patent applications covering the BRCA genes in the mid-1990s (Parthasarathy 2007; Gaudillière 2009). Citing impending patent protection, Myriad then tried to shut down European BRCA gene testing services. Scientists, physicians, and patients across Europe fought against it on multiple fronts. They were particularly shocked by Myriad's commercial orientation and practices, which conflicted with European public health approaches (Löwy and Gaudillière 2008; Parthasarathy 2007).

This included taking advantage of the European Patent Office's (EPO) opposition mechanism, which allows any third party to challenge a patent within nine months of its issue. By the early 2000s, civil society groups had used this mechanism dozens of times to express their displeasure about European patents (Parthasarathy 2017). In this case, twelve organizations – from hospitals to Greenpeace – filed EPO oppositions, questioning the novelty and inventiveness of the identification of the BRCA genes and suggesting that the patent application did not disclose enough information. But their primary concern was that Myriad's patents turned genes into commodities and limited access to health care, violating the EPC's *ordre public* clause (EPO 2005b). Indeed, European challengers adopted a strategy later used by their US counterparts; in addition to their primary concerns about the ethical and socioeconomic impacts of gene patents, they made traditional arguments that might be more palatable to the EPO.

During the opposition, the European Parliament issued a non-binding resolution agreeing that gene patents violated the *ordre public* clause (European Parliament 2001). It argued that public health would be hurt, tying the implications of patent-based monopolies to the patents themselves, a connection that US decision makers had been unwilling to make. In doing so, it envisioned a patent system that was responsible for both public health implications and for responding to public concerns. In fact, it explicitly rebuked the EPO on this basis, demanding "a review of the operations of the EPO to ensure that it becomes publicly accountable in the exercise of its duties." Although the US Congress had asked the PTO to review whether patents affected genetic testing in its 2011 America Invents Act, it had never challenged the accountability of its patent bureaucracy. In fact, its treatment of the PTO suggested that it saw accountability in terms of procedural objectivity and the issuance of patents, rather than in terms of balancing the interests of the inventor and the public.

Eventually, the EPO revoked one of Myriad's BRCA patents and substantially narrowed two others (EPO 2004), arguing that the invention was not novel, did not involve an inventive step, and was insufficiently described. It rejected *ordre public* concerns, arguing that the clause should be defined solely in terms of "legal and ethical values" (EPO 2005a). If the invention itself or the process of its creation was morally offensive, then commodifying it through patents would also be immoral and therefore a violation of the *ordre public* clause. In the BRCA gene patent case, however, opponents were arguing that the patent – not the invention – was immoral.

If the story stopped here, then we might conclude that US and European patent systems are not so different after all. After all, human genes are still patentable in Europe. But what happened next is particularly instructive. The EPO, recognizing that it was under growing scrutiny, took steps to make itself more open to the public. Meanwhile, France, Switzerland, and Belgium amended their compulsory licensing laws to restrict monopoly power related to gene patents (van Zimmeren and Van Overwalle 2010). Like the European Parliament, these national governments seemed to conclude that they should prevent the negative health and research implications of human gene patents. In the United States, decision makers and even some scholars had dismissed socioeconomic concerns as mere "anecdotes" (Caulfield et al. 2006) and defined the primary problem as a techno-legal one. But in Europe, where they defined the issue in distributional terms, compulsory licensing became an appropriate solution.

Conclusion

Analysts of patent law have pointed to the *ordre public* clause to explain the differences in how the US and European patent systems approach biotechnology. This legal explanation, however, is insufficient. In particular, it limits our ability to see the patent system in broader social and political perspective.

This brief chapter begins to unravel how political culture, ideology, and history shape patent system logics, institutions, rules, stakeholders, relationships, and practices. As a result, it allows us to observe the larger dimensions of the life form patent controversies. If we focused purely on patent decisions we would miss, for example, the legal and bureaucratic efforts by European patent system institutions to broaden their approach to knowledge and expertise and find ways to consider the implications of patent-based monopolies. We would also miss the expertise barriers erected and maintained by the US patent system and its stakeholders, to ensure that it remains a narrow techno-legal domain.

This analysis also helps us see how national and regional politics can shape our understanding of the moral and social implications of genomics. Not only did the two jurisdictions analyzed here see concerns about commodification differently, but they also viewed the patent system's role in shaping our genomic futures from very different perspectives.

References

Andrews, Lori (2002). "Genes and Patent Policy: Rethinking Intellectual Property Rights." *Nature Reviews Genetics*. 3: 803–808.

Association for Molecular Pathology et al. v. Myriad Genetics, Inc.: 659 US 12-398 (2013).

Association for Molecular Pathology et al. v. Myriad Genetics and PTO (2009), "Complaint," United States District Court Southern District of New York.

Bagley, Margo (2003). "Patent First, Ask Questions Later: Morality and Biotechnology in Patent Law." *William and Mary Law Review*. 45: 469–547.

Bagley, Margo (2007). "A Global Controversy: The Role of Morality in Biotechnology Patent Law." University of Virginia Law School: Public Law and Legal Theory Working Paper Series. Paper 57.

Biagioli, Mario (2011). "Patent Specification and Political Representation: How Patents Became Rights." In *Making and Unmaking Intellectual Property: Creative Production in Legal and Cultural Perspective*, edited by Mario Biagioli, Peter Jaszi, and Martha Woodmansee. Chicago: University of Chicago Press, 25–40.

Boyle, James (2010). *The Public Domain: Enclosing the Commons of the Mind*. New Haven, CT: Yale University Press.

Bracha, Oren (2016). *Owning Ideas: The Intellectual Origins of American Intellectual Property, 1790–1909*. New York: Cambridge University Press.

Cahill, Lisa Sowle (2001). "Genetics, Commodification, and Social Justice in the Globalization Era." *Kennedy Institute of Ethics Journal*. 11(3): 221–238.

Cassier, Maurice (2008). "Patents and Public Health in France: Exclusion of Drug and Granting of Pharmaceutical Process Patents Between the Two World Wars." *History and Technology* 24, 2: 153–171.

Caulfield, Timothy, Robert M. Cook-Deegan, F. Scott Kieff, and John P. Walsh (2006). "Evidence and Anecdotes: An Analysis of Human Gene Patenting." *Nature Biotechnology*. 24(9): 1091–1094.

Cho, Mildred K, Samantha Illangasekare, Meredith Weaver, Debra Leonard, Jon Merz (2003). "Effects of Patents and Licenses on the Provision of Clinical Genetic Testing Services." *The Journal of Molecular Diagnostics*. 5(1): 3–8.

Committee on Intellectual Property Rights in Genomic and Protein Research and Innovation, National Research Council (2006), *Reaping the Benefits of Genomic and Proteomic Research: Intellectual Property Rights, Innovation, and Public Health*, Washington, DC: National Academies Press.

Daemmrich, Arthur (2004). *Pharmacopolitics: Drug Regulation in the United States and Germany*. Chapel Hill, NC: University of North Carolina Press.

Davis, Courtney and John Abraham (2011). "A Comparative Analysis of Risk Management Strategies in European Union and United States Pharmaceutical Regulation." *Health, Risk & Society*. 13(5): 413–431.

Diamond v. Chakrabarty, 447 U.S. 303(1980).

Dobyns, Kenneth W. (1994). *The Patent Office Pony: A History of the Early Patent Office*. Fredericksburg, VA: Sergeant Kirkland's Museum and Historical Society.

European Parliament (2001), "European Parliament Resolution on the Patenting of BRCA1 and BRCA2 ('Breast Cancer') Genes," Joint Motion for a Resolution, October 3. Online at www.europarl.europa. eu/sides/getDoc.do?type=MOTION&reference=P5-RC-2001-0633&language=EN.

European Parliament and Council (1998), Council Directive (EC) No. 98/44 of 6 July 1998, 1998O.J. (L 213) 13.

European Patent Office (EPO) (2004). "Myriad/Breast Cancer Patent Revoked After Public Hearing" Press Release, May 18, on file with author.

EPO (2005a) "Interlocutory Decision in Opposition Proceedings (Article 106(3) EPC)," Patent No. 0705902, Application No. 95305605.8, June 9, 16.

EPO (2005b), "Minutes of the Oral Proceedings Before the Opposition Division."

Ezrahi, Yaron (1990). *The Descent of Icarus: Science and the Transformation of Contemporary Democracy*. Cambridge, MA: Harvard University Press.

Gaudillière, Jean-Paul (2009). "New Wine in Old Bottles? The Biotechnology Problem in The History of Molecular Biology." *Studies in History and Philosophy of Science Part C: Studies in History and Philosophy of Biological and Biomedical Sciences*. 40(1): 20–28.

Genentech (1980) Amicus Curiae in Diamond v. Chakrabarty, 447 U.S. 303.

Gold, E. R. and Gallochat, A. (2001) 'The European Biotech Directive: Past as Prologue,' *European Law Journal*, 7, 3: 331–366.

Gottweis, Herbert (1998). *Governing Molecules: The Discursive Politics of Genetic Engineering in Europe and the United States*. Cambridge, MA: MIT Press.

Hayden, Cori (2003). *When Nature Goes Public: The Making and Unmaking of Bioprospecting in Mexico*. Princeton, NJ: Princeton University Press.

Heller, Michael A. and Rebecca S. Eisenberg (1998), "Can Patents Deter Innovation? The Anticommons in Biomedical Research," *Science* 280, no. 5364: 698–701.

Hilgartner, Stephen (2017). *Reordering Life: Knowledge and Control in the Genomics Revolution*. Cambridge, MA: MIT Press.

Howard, Ted and Jeremy Rifkin (1977), *Who Should Play God?: The Artificial Creation of Life and What It Means for the Future of the Human Race*, New York: Dell.

Hughes, S. S. (2011) *Genentech: The Beginnings of Biotech*. Chicago, IL: University of Chicago Press.

Husserl, Gerhart (1938). "Public Policy and Ordre Public." *Virginia Law Review* 25, 1: 37–67.

Jasanoff, S. (2005) *Designs on Nature: Science and Democracy in Europe and the United States*. Princeton, NJ: Princeton University Press.

Jasanoff, Sheila (2012). "Taking Life: Private Rights in Public Nature." In *Lively Capital: Biotechnologies, Ethics, and Governance in Global Markets*, edited by Kaushik Sunder Rajan, Durham, NC: Duke University Press, 155–183.

Kevles, Daniel (1994). "Ananda Chakrabarty Wins a Patent: Biotechnology, Law, and Society, 1972–1980." *Historical Studies in the Physical and Biological Sciences*. 25(1): 111–135.

Kevles, Daniel (2002a). *A History of Patenting Life in the United States with Comparative Attention to Europe and Canada*. Report for the European Group on Ethics in Science and New Technologies, European Commission.

Kevles, Daniel (2002b). "Of Mice & Money: The Story of the World's First Animal Patent." *Daedalus*. 131(2): 78–88.

Klawiter, M. (2008) *The Biopolitics of Breast Cancer: Changing Cultures of Disease and Activism*. Minneapolis, MN: University of Minnesota Press.

Kleinman, Daniel Lee (2003). *Impure Cultures: University Biology and the World of Commerce*. Madison, WI: University of Wisconsin Press.

Krimsky, Sheldon (1984). *Genetic Alchemy: The Social History of the Recombinant DNA Controversy*. Cambridge, MA: MIT Press.

Letwin, William L. (1954) "The English Common Law Concerning Monopolies." *University of Chicago Law Review* 21, 3: 355–385.

Lowell v. Lewis, 1 Mason. 182; 1 Robb, Pat. Cas. 131 (Circuit Court, D. Massachusetts. May Term. 1817).

Löwy, Ilana and Jean-Paul Gaudillière (2008). "Localizing the Global: Testing for Hereditary Risks of Breast Cancer." *Science, Technology & Human Values* 33, 3: 299–325.

May, Christopher and Susan K. Sell (2005). *Intellectual Property Rights: A Critical History*. Boulder, CO: Lynne Rienner Publications.

Merz, Jon F. (1999), "Disease Gene Patents: Overcoming Unethical Constraints on Clinical Laboratory Medicine," *Clinical Chemistry* 45, no. 3.

Metzler, Ingrid (2011). "Between Church and State: Stem Cells, Embryos, and Citizens in Italian Politics." In Sheila Jasanoff, ed., *Reframing Rights: Bioconstitutionalism in a Genetic Age*. Cambridge, MA: MIT Press.

Mirowski, Philip (2011). *Science-Mart: Privatizing American Science*. Cambridge, MA: Harvard University Press.

Mowery, David C., Richard R. Nelson, Bhaven N. Sampat, and Arvids Z. Ziedonis (2004). *Ivory Tower and Industrial Innovation: University-Industry Technology Transfer Before and After the Bayh-Dole Act*. Palo Alto, CA: Stanford Business Publications.

Oldfield Revision and Codification of Patent Laws: Hearing Before the Committee on Patents, House of Representatives, 63d Congress 31-32 (1914).

Parthasarathy, S. (2007) *Building Genetic Medicine: Breast Cancer, Technology, and the Comparative Politics of Health Care*. Cambridge, MA: MIT Press.

Parthasarathy, S. (2010) "Breaking the Expertise Barrier: understanding activist strategies in science and technology policy domains" *Science and Public Policy*, 37, 6: 355–367.

Parthasarathy, S. (2017) *Patent Politics: Life Forms, Markets, and the Public Interest in the United States and Europe*. Chicago, IL: University of Chicago Press.

People's Business Commission (1980) Amicus Curiae in Diamond v. Chakrabarty, 447 U.S. 303.

Porter, Theodore (1996). *Trust in Numbers: The Pursuit of Objectivity in Science and Public Life*. Princeton, NJ: Princeton University Press.

Pottage, Alain (2007). "The Socio-Legal Implications of the New Biotechnologies." *Annual Review of Law and Social Science*, 3: 321–344.

Rai, Arti (1999). "Regulating Scientific Research: Intellectual Property Rights and the Norms of Science," *Northwestern University Law Review* 94, no. 1: 77–152.

Rasmussen, Nicholas (2014). *Gene Jockeys: Life Science and the Rise of the Biotech Enterprise.* Baltimore, MD: Johns Hopkins University Press.

Sandel, Michael J. (2013) *What Money Can't Buy: The Moral Limits of Markets.* New York: Farra, Straus, and Giroux.

Sarewitz, Daniel (1996). *Frontiers of Illusion: Science, Technology, and the Politics of Progress.* Philadelphia, PA: Temple University Press.

Scherer, Frederic (2002). "The Economics of Human Gene Patents." *Academic Medicine.* 77(12): 1348–1367.

Secretary's Advisory Committee on Genetics, Health, and Society (2010), *Gene Patents and Licensing Practices and Their Impact on Patient Access to Genetic Tests, Report 2010.* Online at https://osp.od.nih.gov/sa cghsdocs/gene-patents-and-licensing-practices-and-their-impact-on-patient-access-to-genetic-tests-rep ort-of-the-secretarys-advisory-committee-on-genetics-health-and-society/

Simoncelli, Tania (2013). "Amp V. Myriad: Preliminary Reflections," *GeneWatch* 26(2–3): 7.

Simoncelli, Tania and Sandra Park (2015). "Making the Case Against Gene Patents." *Perspectives on Science.* 23(1): 106–145.

Sperling, Stefan (2013). *Reasons of Conscience: The Bioethics Debate in Germany.* Chicago, IL: University of Chicago Press.

Stabiner, K. (1997) *To Dance with the Devil: The New War on Breast Cancer.* New York: Delacorte Press.

Sterckx, Sigrid and Julian Cockbain (2012). *Exclusions from Patentability: How Far Has the European Patent Office Eroded Boundaries?* New York: Cambridge University Press.

Strathern, Marilyn (1996). "Potential Property: Intellectual Rights and Property in Persons." *Social Anthropology.* 4: 17–32.

Tallacchini, Mariachiara (2015). "To Bind or Not Bind? European Ethics as Soft Law." In Stephen Hilgartner, Clark Miller, and Rob Hagendijk, eds. *Science and Democracy: Making Knowledge and Making Power in the Biosciences and Beyond.* New York: Routledge.

Testa, Guiseppe (2011). "More than just a Nucleus: Cloning and the Alignment of Scientific and Political Rationalities." In Sheila Jasanoff, ed., *Reframing Rights: Bioconstitutionalism in a Genetic Age.* Cambridge, MA: MIT Press.

US Patent and Trademark Office (2015), Report on Confirmatory Genetic Diagnostic Test Activity, September.

US Department of State (2016). "U.S. Disappointed Over Fundamentally Flawed Report of the UN Secretary-General's High-Level Panel on Access to Medicines." September 16. https://2009-2017.sta te.gov/r/pa/prs/ps/2016/09/262034.htm

van Zimmeren, E. and Van Overwalle, G. (2010) 'A Paper Tiger? Compulsory License Regimes for Public Health in Europe," *International Review of Intellectual Property and Competition Law (IIC)*, December 1, 2010, accessed November 24, 2014, http://ssrn.com/abstract=1717974.

Walker, S. (1999) *In Defense of American Liberties: A History of the ACLU.* 2nd ed. Carbondale, IL: Southern Illinois University Press.

Walterscheid, Edward C. (1995). "Inherent or Created Rights: Early Views on the Intellectual Property Clause," *Hamline Law Review* 19(1): 81.

Winickoff, David (2015). "Biology Denatured: The Public-Private Lives of Lively Things." In Stephen Hilgartner, Clark Miller, and Rob Hagendijk, eds. *Science and Democracy: Making Knowledge and Making Power in the Biosciences and Beyond.* New York: Routledge.

Wright, Susan (1994). *Molecular Politics: Developing American and British Regulatory Policy for Genetic Engineering, 1972–1982.* Chicago, IL: University of Chicago Press.

18

Genomic platforms and clinical research

Alberto Cambrosio, Etienne Vignola-Gagné, Nicole Nelson,
Peter Keating and Pascale Bourret

On (genomic) platforms

Scientists and clinicians increasingly use the term 'platform' to describe various facets of their work. Found on average in 23 titles/year of PubMed-indexed articles during the 1990s, this number rose to 248 during the first decade of the new century, and to 1,066 in the subsequent six years. A textual analysis of post-2000 titles shows that the term most frequently co-occurs with terms related to advanced, high-throughput ('omics') or biomedical imaging technologies. In this sense, platforms refer to technologies dependent upon sophisticated instruments, including computer equipment, bioinformatics tools, and biological reagents. The term is also used as a synonym for 'core facilities' – collections of equipment shared by researchers from one or more institutions.

Cambrosio et al. (2009) have outlined a more analytical definition of the term 'platform' that focuses on the epistemic work accomplished by these material and institutional configurations. In this view, biomedical platforms embody and engender stabilized interconnections between new biomedical entities (e.g., genes and mutations), the equipment and technologies necessary for the manipulation of those entities (e.g., reagents, sequencing machines), and the regulations that shape their proper use in clinical and laboratory settings (e.g., quality control standards). This definition of a platform links governance issues to the content of the activities performed on/by a platform and to the entities they (re)produce (Keating and Cambrosio 2003). Such an understanding emphasizes three related processes. First, it highlights the ongoing alignment of clinical research with biological research and its constitutive technologies, nowadays generically referred to as 'translational research'. This alignment does not entail a reduction of one sphere to the other; it is a two-way process of continual readjustment. Second, the use of a platform in laboratory or clinical research does not automatically entail its use in routine clinical practice. Articulation and regulation work is necessary to allow a platform to enter standard clinical use. Finally, platforms raise organizational and governance issues: to avoid technological obsolescence, platforms can only survive by becoming reflexive devices, i.e. by mutually adjusting their constitutive equipment to research questions in a two-way, evolutionary process.

This chapter reviews social science contributions to the analysis of genomic platforms deployed in bio-clinical activities. We focus on so-called next generation sequencing (NGS) technologies used for analyzing pathological material such as cancerous tissue, examining the manifold practices, institutions, and databases that make the interpretation and use of the ensuing results possible.

Platforms and the emerging contours of genomic medicine

At the beginning of the new century, prominent scientists argued that genomic technologies would revolutionize all areas of clinical practice, from diagnosis to prognosis and therapeutics, with oncology at the forefront of this transformation. Many argued that the identification of molecular markers of disease and the development of 'targeted' therapies (drugs specifically aimed at molecular abnormalities) would alter the field not only from a clinical viewpoint, but also from social, political, organizational and cultural perspectives (Conrad and Gabe 1999; Cunningham-Burley and Boulton 1999; Kaufert 2000). At the time, however, critics insisted that few *clinical* applications could be directly attributed to molecular genetics and that, consequently, post-genomic medicine remained a promissory note (Hedgecoe and Martin 2003). Even then, however, it was apparent that the number of genomic tests and applications was growing steadily and, more importantly, that aspects of genomics increasingly permeated thinking and activities in all major domains of contemporary biomedicine – albeit to varying degrees. Because oncology has played a pioneering role, it is a privileged site for analyzing the impact of genomics on medicine. But rather than asking whether the predicted revolution in oncology has been realized, we raise more open-ended questions: to what extent, under what guise, and with what consequences has genomics entered the clinic?

In research-intensive domains such as oncology, the nature of genomic initiatives has shifted in recent years from the diagnostic to the therapeutic domain, and has entailed new forms of collaboration between public and private actors. Two central notions – translational research and personalized medicine – have played a key role in this respect. *Translational research*, while sometimes criticized as the emperor's new clothes, has become the locus of several initiatives aiming at 'crossing the valley of death' (Kohli-Laven et al. 2011) that prevents the transformation of scientific discoveries emerging from laboratory, clinical, or population studies into clinical applications. *Personalized* (or *precision) medicine* refers to medical acts adapted to the individual (molecular) characteristics of a given patient (or, more exactly, increasingly smaller groups of patients). These terms are often freighted with calls for establishing clinical, scientific, and policy programs (Vignola-Gagné 2014), and their meaning has changed in recent years (Keating et al. 2016). It would nonetheless be foolhardy to dismiss them as rhetorical labels for they correspond to distinct spheres of biomedical activities subject to empirical investigation as such (Jones et al. 2011).

Social scientists have examined early developments in genomic medicine, especially regarding BRCA and hereditary susceptibility to cancer. In general, they characterized these activities as forms of predictive medicine focused not on *disease symptoms*, or even life-style risks, but on *molecular risk factors* (Bourret 2005, Gibbon et al. 2014), thus consolidating a shift in medical practice from clinically expressed to 'asymptomatic' diseases. However, new platforms have contributed to the emergence of a different side of genomic medicine, one that formulates novel plans of clinical action. In oncology, NGS results are made 'actionable' by linking them to clinical trials or off-label uses of approved drugs, and to new systems of classification and new venues for deliberating on 'actionability' (Nelson et al. 2013, see also Miller et al. 2016). This regime of genomic medicine produces different kinds of evidence and uncertainty than do genomic methods in the fields of hereditary cancer and medical genetics. The latter utilize a risk-based approach that categorizes the results of DNA sequencing as either (likely) pathogenic, (likely) benign, or Variants of Unknown Significance (VUS) (Timmermans et al. 2017, Skinner et al. 2016). In contrast, the domain of somatic cancers operates under a regime of actionability, and variants are generally categorized as diagnostic, prognostic, or predictive, in direct connection to therapeutic interventions. Even so, the definition of what constitutes a clinically useful or an actionable result remains deeply controversial, as different bio-clinical networks offer competing views and practices depending on their distance from the research and treatment poles of the translational spectrum.

While genomic platforms contribute to the (re)definition of clinical activities, the mobilization of their findings takes place within settings already shaped by pre-existing laboratory and clinical routines and models, both organizational and epistemic. Clinical judgment has long been deployed in a dense web of laboratory measurements, technoscientific mediations (Rheinberger 2011), and collective negotiations of their meanings (Atkinson 1995). Genomic medicine adds layers to these processes. Clinical sequencing, for instance, comes with its own path dependencies (Garcia-Sancho 2012), and exhibits certain traits of 'data-centric science' (Leonelli 2016). The clinical interpretation of genomic sequences is accomplished via a world of mediators, including databases and data processing algorithms. Accordingly, clinical teams increasingly interact with bioinformaticians and computational biologists. Genomic platforms have also led to the emergence of an entire sector of commercial providers for the production and the interpretation of genomic results (Curnutte et al. 2014). Understanding genomic medicine thus calls for continued attention to the resources and constraints created by new biotechnology and pharmaceutical industry practices.

In sum, by redefining clinical-laboratory interfaces, genomics is simultaneously redefining key features of medical work. The effects of this reconfiguration include the development of new ways to govern international endeavors that profoundly affect national and local practices; the development of new, hybrid connections between public research networks and commercial entities; ongoing attempts to align traditional clinical systems with the newest systems of evidence; a redrawing of the relations between medical specialists and patients; and a redefinition of the relations between research and routine practices.

Transforming the clinic, transforming healthcare

Featherstone et al. (2005) have argued that genetic platforms have not fundamentally transformed medical work. They even claim to have observed a revival of the clinical tradition. Will, Armstrong and Marteau (2010) also found that DNA tests for mutations underlying hereditary hypercholesterolaemia had little impact in lipids clinics specializing in prevention of cardiovascular disease. Well-established clinical measurements, such as cholesterol levels and more comprehensive risk profiles, often overrode DNA results. In this context, diagnosis was secondary to 'advice and treatment' (p. 915). In contrast, Rabeharisoa and Bourret (2009) found no evidence for a return to the clinical tradition in oncology and psychiatric genetics. They found changes in the settings, objects, and way in which clinical work was carried out, arguing that post-genomic clinical work is distributed 'in and between … "bioclinical collectives" … bring[ing] together researchers and clinicians from different disciplines and specialities, and strongly interconnect [ing] the clinic and the research' (p. 693). Moreover, they argued that the work taking place in these new collectives consisted of 'simultaneously producing the clinical relevance and the biological significance of mutations.' Finally, work in the 'clinic of mutations' involved 'constant interaction and feedback between the clinic and the research,' which transformed how medical judgments and decisions were elaborated.

Since these early studies, a new wave of STS investigations have provided clear evidence of change in clinical practice driven by the adoption of genomic technologies. For example,

Navon (2011) examined how genetic markers have been central to the delineation of new medical syndromes, a process he termed 'genomic designation.' The reconfiguration of biomedical practices does not stop at the doors of the clinic or the laboratory. Navon and Eyal (2016) show how genomic categories have a 'looping' effect on the patients so designated. Timmermans and Shostak (2016) review how genomic medicine participates in and is shaped by reimbursement arrangements, health delivery and administration schemes, state programs and regulations, professional groups, citizen initiatives and social movements. Hogarth (2012) analyzes how pharmacogenomics has transformed regulatory practices through the creation of 'new socio-technical spaces in the regulatory regimes for pharmaceuticals' and the development of 'novel regulatory experiments by a transnational network encompassing regulatory agencies, academic scientists and industry' (p. 441). The expansion of NGS activities also prompted the U.S. Food and Drug Administration to try to regulate this new domain of practice, in a departure from its tradition of restrained intervention in the diagnostic sector (Bourret et al. 2011; Curnutte et al. 2016). Aarden (2016) calls attention to how health care infrastructures shape the translation of genomic technologies, by comparing care pathways in the genetic testing of familial breast cancers in the United Kingdom, the Netherlands, and Germany. Despite the similarity of the diagnostic tests, he found different care trajectories in each country, modulated by national regulations, policies, institutional routines and guidelines. In the case of NGS, reimbursement issues have come to the fore and even constrain collective clinical decision-making. U.S. clinics have developed a range of makeshift yet elaborate solutions to problems such as dealing with generic, maladapted billing codes (Timmermans 2015). European countries have launched centralized initiatives to organize, finance, and control the development of sequencing and other high-throughput technologies while ensuring equal access to genomic medicine.

The issue is not simply whether genomic findings have redrawn disease categories or displaced clinical judgment. Rather, these platforms and processes interact in diverse and locally contingent ways. Hogan (2016a, pp. 138–140) argues, for example, that medical categories have also rewritten genomic theory, showing how the mapping of two distinct clinical disorders to the same location in the genome generated a new line of scientific inquiry into epigenetic imprinting in humans. Likewise, genetic findings have reshaped rather than supplanted clinical judgment, as researchers and clinicians have begun to mutually adjust and tentatively align patient case reports with genomic information. The regulatory landscape also continues to evolve as molecular technologies themselves evolve, creating an iterative process of regulatory and infrastructural change.

Interpretation, uncertainty, and clinical judgment

'Uncertainty,' in this context, can refer to two different processes. The first involves the use of microarrays and sequencing to resolve diagnostic uncertainty. Hogan (2016b) argues that the introduction of genomic platforms (microarrays) in the 2000s increased rather than reduced diagnostic uncertainty in prenatal testing. Turrini (2014) also shows how the (controversial) introduction of new gene-array technologies into prenatal diagnosis turned the interpretation of clinical data – an activity enabled by biobanks and databases linking molecular anomalies to many diseases and conditions – into an 'extremely complex and indeterminate' endeavor (p. 133). This is consistent with previous work by Hedgecoe (2003), who argued that using genetic entities to account for cystic fibrosis led to both an expansion of the boundaries of the disease and, in parallel, to an increase in the uncertainty concerning its classification (see also Timmermans and Buchbinder 2013). Hogan (2016b) claims, moreover, that some clinicians embraced these uncertain results, redefining them 'as starting points for research to improve the scope of prenatal diagnosis, and bring future certainty.' In short, genomic technologies have reduced diagnostic uncertainty in some ways while extending and/or displacing it in others.

The second meaning of uncertainty concerns the challenge of making decisions about patients when the information generated by high-throughput platforms is incomplete or unclear. Practitioners themselves acknowledge that 'interpretation of the clinical significance of genomic alterations remains the most severe bottleneck preventing the realization of personalized medicine in cancer' (Good et al. 2014). Especially in the field of hereditary diseases, sequencing technologies produce results categorized as Variants of Unknown Significance (VUS), meaning that practitioners have no evidence about their actual role in the aetiology of the disease. Timmermans et al. (2017) have described several strategies practitioners use to cope with the uncertainty related to VUS, ranging from developing standards to establishing expert collectives (a.k.a. 'molecular boards'). Standardization has taken the form of extensive efforts to prepare, mediate, and contextualize sequencing findings, drawing on infrastructures such as online databases of genomic variants, and bioinformatic tools such as algorithms and filters. Attended by 'bio-informaticians, laboratory analysts, laboratory directors, clinical geneticists, genetic counsellors, clinicians, and support personnel' (Timmermans et al. 2017), molecular boards review patient cases, analyze sequencing results, and make clinical recommendations. They provide a collective site for the (temporary) stabilization of interpretations linking platforms to clinical practices predicated on connections between drugs and mutations. The work of these bioclinical collectives is made possible by a data infrastructure that, in turn, necessitates substantial material and cognitive investments to ensure data interoperability (Leonelli 2013; Ribes and Polk 2015).

Skinner et al. (2016) have investigated the temporal instability of NGS findings in the clinical encounter and in patients' reception of sequencing information. They found that clinicians reported even negative results (no variant of interest found) to rare disease patients as provisional in light of quickly evolving technology. Patients and their families, who are often actively searching for a definitive diagnosis after encountering lack of knowledge about their condition, are especially sensitive to the possibility that future iterations of the technology might reveal a causal variant after all. This is consistent with Bourret's (2005) findings about the open-ended, ongoing character of hereditary cancer results. Patients also participate in clinical interpretation practices. In particular, Stivers and Timmermans (2016) have reported that, in clinicians' presentation of NGS findings to rare disease patients, the significance assigned to VUS is established through interaction with the patient. This interaction can downgrade or upgrade the causal significance attached to a variant along standard scales of evidence commonly used in the genomic clinic. Patients, on hearing about the potential effects of a certain variant judged not relevant to their case, might provide new observations about their disease that conform more readily with the expected effects of that variant. Stivers and Timmermans (2016), like Skinner et al. (2016), consider that reasoning with NGS findings is not an undiluted diagnostic practice, but contributes to a distinct form of care concerned with providing certainty and clarity about patients' conditions.

Genomic-driven clinical trials

The articulation of the laboratory and the clinic that takes place around NGS results also occurs further 'upstream' in clinical research. Indeed, we have argued that clinical trials are emerging as a crucial site for enacting genomic medicine (Nelson et al. 2014; Vignola-Gagné et al. 2017). Far from the simplistic description often provided by introductory textbooks, clinical trials are complex devices – complicated 'intervention ensembles' consisting of a 'coordinated set of materials, practices and constraints needed to safely unlock the therapeutic or preventive activities of drugs, biologics, and diagnostics' (Kimmelman and London 2015, p. 27). As evolving institutions, clinical trials need to establish the very practices they aim to evaluate, as well as the instruments and strategies of evaluation. Moreover, they jointly produce their objects of study and findings about those objects (Brives 2013; Keating and Cambrosio 2012).

The introduction of genomic platforms has disrupted the delicate equilibrium between the various components of the clinical trial machinery. In part, recent transformations in clinical research are due to sustained critiques of the trial system from trialists themselves and from a wide range of stakeholders (Will and Moreira 2010), but genomic technologies have also played an important role in these rearrangements. Uptake of NGS technologies in clinical trial work has been rapid across a broad swath of academic medicine and beyond, with oncology and rare and metabolic diseases standing at the cutting edge of these developments (Curnutte et al. 2016). NGS instruments increasingly allow researchers and clinicians to quickly identify and compare multiple molecular alterations found in tumor biopsies and other pathological samples. The study and evaluation of these alterations raise major organizational and sociotechnical issues that have catalyzed the reorganization of national clinical trial systems and the reconceptualization of the purpose of the trials themselves (Cambrosio et al. 2014).

The new forms of clinical trials organized around genomic platforms are no mere tests of therapies; they are complex clinical experimental systems. Their aim is no longer confined to proving safety and efficacy; rather, it includes learning about the biopathology of cancer and generating unexpected findings (Cooper 2012, Nelson et al. 2014). Such a reconceptualization of the very nature of clinical trials has profound consequences for how we understand their material organization and valuation (Helgesson et al. 2016). It also redraws connections between laboratory practices, clinical trials, and routine care. Crabu (2016) has introduced the notion of 'technomimicry' to describe how the practices of the clinic and the laboratory are reshaped to resemble each other in the context of clinical trials. Both the clinic and laboratory-based collectives attempt to recreate the practices and epistemic strategies of each other's sites, but do so by drawing on native instruments, entities and routines: 'the laboratory itself can be re-framed and adjusted to render laboratory facts and scientific phenomena congruent with the processes of care and the clinical management of patients,' and 'the clinic itself [...] reframed as a research site where patients are enroled not only for care, but also as participants in biomedical research activities' (p. 314).

The redesign of the clinical enterprise is not confined to local settings and single trials. The leaders of practice-changing, biomarker-driven clinical trials – along with state agencies, professional societies (Cambrosio et al. 2017), and such patient organizations as Friend of Cancer Research or Europa Donna – aim to reform the very framework of clinical research. What began in oncology will thus impinge on clinical research more generally. Previous research in other disease fields has shown how the configurations of practices and valuations stabilized by trials can be scaled up and used much beyond the narrow confines of a single clinical study project. Ribes and Polk's (2015) study of AIDS research infrastructures showed that the design and parameters of early cohort studies carried over into core resources and services and had a lasting impact on what researchers now can and cannot investigate. Petty and Heimer's (2011) study on HIV trials in developing countries similarly showed that these trials had lasting impacts on organizational routines, material environments, relations between professionals, and patient encounters. Such innovations must, however, contend with their necessary articulation with established clinical practices, a process that is a source of constant friction, making the translational pathway linking genomics and the clinic a rocky road.

STS scholars (e.g. Rabeharisoa and Callon 2008, Rabeharisoa et al. 2014) have also investigated the growing – albeit somewhat controversial – role of patient organizations in the redesign of the clinical enterprise. Generally speaking, patients are increasingly asked to participate in the design and performance of clinical trials as 'research subject, co-researchers, donors, campaigners, representatives and consumers' of novel genomic-based therapies (Kerr and Cunningham-Burley 2015). According to Haase et al. (2015), patients value participation in NGS projects not only for their own diagnostic interests but also for their contribution to the gift economies of future-oriented biomedical projects. The deployment of genomic platforms within innovative clinical trials thus also contributes to and is shaped by the redefinition of 'patienthood' that is associated with their staging.

Conclusion

The introduction of genomic platforms into clinical research and care practices is associated with substantial transformations. The nature and scope of genomic initiatives in cancer have expanded dramatically in recent years, raising several major organizational, regulatory, and sociotechnical issues. The often-controversial solutions to these issues include the redesign of a key component of clinical research – clinical trials – and their organizational infrastructure. Beyond their immediate investigative and therapeutic aims, a new wave of clinical trials is also testing new regulatory and organizational frameworks and exploring the large-scale transfer of genomic sequencing technologies with the explicit goal of improving public health. New forms of collaboration between public and private actors are being probed conterminously with the clinical development of targeted therapies. It is no exaggeration to claim that within these domains genomics is redefining the boundaries between research and care.

References

Aarden, E. (2016) 'Translating genetics beyond bench and bedside: a comparative perspective on health care infrastructures for "familial" breast cancer', *Applied and Translational Genomics*, 11 (December): 48–54.

Atkinson, P. (1995). *Medical Talk and Medical Work*. Sage: Thousand Oaks.

Bourret, P. (2005) 'BRCA patients and clinical collectives: new configurations of action in cancer genetics practices', *Social Studies of Science*, 35: 41–68.

Bourret, P., Keating, P. and Cambrosio, A. (2011) 'Regulating diagnosis in post-genomic medicine: re-aligning clinical judgment?', *Social Science and Medicine*, 73: 816–824.

Brives, C. (2013) 'Identifying ontologies in a clinical trial', *Social Studies of Science*, 43: 397–416.

Cambrosio, A., Keating, P., Bourret, P., Mustar, P. and Rogers, S. (2009) 'Genomic platforms and hybrid formations', in: P. Atkinson, P. Glasner and M. Lock (eds) *Handbook of Genetics and Society: Mapping the New Genomic Era*. London: Routledge; 502–520.

Cambrosio, A., Keating, P. and Nelson, N. (2014) 'Régimes thérapeutiques et dispositifs de preuve en oncologie: l'organisation des essais cliniques des groupes coopérateurs aux consortiums de recherche', *Sciences Sociales & Santé*, 32(3): 13–42.

Cambrosio, A., Bourret, P., Keating, P. and Nelson, N. (2017) Opening the regulatory black box of clinical cancer research: transnational expertise networks and "disruptive" technologies', *Minerva*, 55(2): 161–185.

Conrad, P. and Gabe, J. (eds) (1999) *Sociological Perspectives on the New Genetics*. Oxford: Blackwell.

Cooper, M. (2012) 'The pharmacology of distributed experiment: user-generated drug innovation', *Body and Society*, 18: 18–43.

Crabu, S. (2016) 'Translational biomedicine in action: constructing biomarkers across laboratory and benchside', *Social Theory and Health*, 14(3): 312–331.

Cunningham-Burley, S. and Boulton, M. (1999) The social context of the new genetics, in G.L. Albrecht, R. Fitzpatrick and S.C. Scrimshaw (eds) *The Handbook of Social Studies in Health and Medicine*. London: Sage, 173–187.

Curnutte, M.A., Frumovitz, K.L., Bollinger, J.M., McGuire, A.L. and Kaufman, D.J. (2014) 'Development of the clinical Next-Generation Sequencing industry in a shifting policy climate', *Nature Biotechnology*, 32: 980–982.

Curnutte, M.A., Frumovitz, K.L., Bollinger, J.M., Cook-Degan, R.M., McGuire, A.L. and Majumder, M.A. (2016) 'Developing context-specific Next-Generation Sequencing policy', *Nature Biotechnology*, 34: 466–470.

Featherstone, K., Latimer, J., Atkinson, P., Daniella, T. and Clarke, A. (2005) 'Dysmorphology and the spectacle of the clinic'. *Sociology of Health and Illness*, 27: 551–574.

Garcia-Sancho, M. (2012) *Biology, Computing, and the History of Molecular Sequencing: From Proteins to DNA, 1945–2000*. Chicago: University of Chicago Press.

Gibbon, S., Joseph, G., Mozersky, J., zur Nieden, A. and Palfner, S. (eds) (2014) *Breast Cancer Gene Research and Medical Practices: Transnational Perspectives in the Time of BRCA*. London: Routledge.

Good, B.M., Ainscough, B.J., McMichael, J.F., Su, A.I. and Griffith, O.L. (2014) 'Organizing knowledge to enable personalization of medicine in cancer', *Genome Biology*, 15(8): 438.

Haase, R., Michie, M., and Skinner, D. (2015) 'Flexible positions, managed hopes: the promissory bioeconomy of a whole genome sequencing cancer study', *Social Science and Medicine*, 130: 146–153.

Hedgecoe, A.M. (2003) 'Expansion and uncertainty: cystic fibrosis, classification and genetics', *Sociology of Health and Illness*, 25: 50–70.

Hedgecoe, A.M. and Martin, P. (2003) 'The drugs don't work: expectations and the shaping of pharmacogenetics', *Social Studies of Science*, 33: 327–364.

Helgesson, C.F., Lee, F., and Linden, L. (2016) 'Valuations of experimental designs in proteomic biomarker experiments and traditional randomized controlled trials', *Journal of Cultural Economy*, 9: 157–172.

Hogan, A. J. (2016a) *Life Histories of Genetic Disease: Patterns and Prevention in Postwar Medical Genetics*. Baltimore: Johns Hopkins University Press.

Hogan, A. J. (2016b) 'Making the most of uncertainty: treasuring exceptions in prenatal diagnosis', *Studies in History and Philosophy of Biological and Biomedical Sciences*, 57: 24–33.

Hogarth, S. (2012) 'Regulatory experiments and transnational networks: the governance of pharmacogenomics in Europe and the United States', *Innovation: The European Journal of Social Science Research*, 25: 441–460.

Jones, D., Cambrosio, A. and Mogoutov, A. (2011) 'Detection and characterization of translational research in cancer and cardiovascular medicine', *Journal of Translational Medicine*, 9: 57.

Kaufert, P. (2000) 'Health policy and the new genetics', *Social Science and Medicine*, 51: 821–829.

Keating, P. and Cambrosio, A. (2003) *Biomedical Platforms: Realigning the Normal and the Pathological in Late-Twentieth-Century Medicine*. Cambridge, MA: MIT Press.

Keating, P. and Cambrosio, A. (2012) *Cancer on Trial: Oncology as a New Style of Practice*. Chicago: The University of Chicago Press.

Keating, P., Cambrosio, A., and Nelson, N. (2016) '"Triple negative breast cancer": translational research and the (re)assembling of diseases in post-genomic medicine', *Studies in History and Philosophy of Biological and Biomedical Sciences*, 59: 20–34.

Kerr, A., and Cunningham-Burley, S. (2015) 'Embodied innovation and regulation of medical technoscience: transformations in cancer patienthood', *Law, Innovation and Technology*, 7(2): 187–205.

Kimmelman, J. and London, A.J. (2015) 'The structure of clinical translation: efficiency, information, and ethics'. *Hastings Center Report* 45(2): 27–39.

Kohli-Laven, N., Bourret, P., Keating, P. and Cambrosio, A. (2011) 'Cancer clinical trials in the era of genomic signatures: biomedical innovation, clinical utility, and regulatory-scientific hybrids', *Social Studies of Science*, 41: 487–513.

Leonelli, S. (2013) 'Global data for local science: assessing the scale of data infrastructures in biological and biomedical research', *BioSocieties*, 8: 449–465.

Leonelli, S. (2016). *Data-Centric Biology: A Philosophical Study*. Chicago: The University of Chicago Press.

Miller, F., Hayeems, R. and Hogarth, S. (2016) 'Informally regulated innovation systems: challenges for responsible innovation in diagnostics', in M. Boenink, H. van Lente, and E. Moors (eds), *Emerging Technologies for Diagnosing Alzheimer's Disease*. London: Palgrave-Macmillan, 227–244.

Navon, D. (2011) 'Genomic designation: how genetics can delineate new, phenotypically diffuse medical categories', *Social Studies of Science*, 41: 203–226.

Navon, D. and Eyal, G. (2016) 'Looping genomes: diagnostic change and the genetic makeup of the autism population', *American Journal of Sociology*, 121: 1416–1471.

Nelson, N., Keating, P. and Cambrosio, A. (2013) 'On being "actionable": clinical sequencing and the emerging contours of a regime of genomic medicine in oncology', *New Genetics & Society*, 32: 405–428.

Nelson, N., Keating, P. and Cambrosio, A., Aguilar-Mahecha, A. and Basik, M. (2014) 'Testing devices or experimental systems? Cancer clinical trials take the genomic turn', *Social Science and Medicine*, 111: 74–83.

Petty, J. and Heimer, C. (2011) 'Extending the rails: how research reshapes clinics', *Social Studies of Science*, 41: 337–360.

Rabeharisoa, V. and Bourret, P. (2009) 'Staging and weighting evidence in biomedicine: comparing clinical practices in cancer genetics and psychiatric genetics', *Social Studies of Science*, 39: 691–715.

Rabeharisoa, V. and Callon, M. (2008) 'The growing engagement of emergent concerned goups in political and economic life: lessons from the French association of neuromuscular disease patients', *Science Technology and Human Values*, 33: 230–261.

Rabeharisoa, V., Callon, M., Filipe, A.M., Nunes, J.A., Paterson, F., and Vergnaud, F. (2014) 'From "politics of numbers" to "politics of singularisation": patients' activism and engagement in research on rare diseases in France and Portugal', *BioSocieties*, 9: 194–217.

Rheinberger, H.-J. (2011) 'Infra-experimentality: from traces to data, from data to patterning facts', *History of Science*, 49: 337–348.

Ribes, D. and Polk, J. (2015) 'Organizing for ontological change: the kernel of an AIDS research infrastructure', *Social Studies of Science*, 45: 214–241.

Skinner, D., Raspberry, K. and King, M. (2016) 'The nuanced negative: meanings of a negative diagnostic result in clinical exome sequencing', *Sociology of Health and Illness*, 38: 1303–1317.

Stivers, T. and Timmermans, S. (2016) 'Negotiating the diagnostic uncertainty of genomic test results', *Social Psychology Quarterly*, 79: 199–221.

Timmermans, S. (2015) 'Trust in standards: transitioning clinical exome sequencing from bench to bedside', *Social Studies of Science*, 45: 77–99.

Timmermans, S. and Buchbinder, M. (2013) *Saving Babies? The Consequences of Newborn Genetic Screening.* Chicago: The University of Chicago Press.

Timmermans, S. and Shostak, S. (2016) 'Gene worlds', *Health*, 20(1): 33–48.

Timmermans, S., Tiebohl, C. and Skapderdis, E. (2017) 'Narrating uncertainty: variants of uncertain significance (VUS) in clinical exome sequencing', *BioSocieties* 12(3): 439–458.

Turrini, M. (2014) 'The controversial molecular turn in prenatal diagnosis', *Tecnoscienza: Italian Journal of Science & Technology Studies*, 5(1): 115–139.

Vignola-Gagné, E. (2014) 'Argumentative practices in science, technology and innovation policy: the case of clinician-scientists and translational research', *Science and Public Policy*, 41(1): 94–106.

Vignola-Gagné, E., Keating, P. and Cambrosio, A. (2017) 'Informing materials: drugs as tools for exploring cancer mechanisms and pathways', *History and Philosophy of the Life Sciences*, 39: 10.

Will, C., Armstrong, D. and Marteau, T. (2010) 'Genetic unexceptionalism: clinician accounts of genetic testing for familial hypercholesterolaemia', *Social Science and Medicine*, 71: 910–917.

Will, C. and Moreira, T. (2010) *Medical Proofs, Social Experiments: Clinical Trials in Shifting Contexts.* New York: Routledge.

19

Diagnostics

Stuart Hogarth

Introduction

Over the last two decades the heady expectations of a genomic revolution in biomedicine have been accompanied by much debate about whether the regulatory regimes that govern diagnostic tests are sufficiently robust to cope with the challenges that this brave new era might present. A persistent concern has been the pace of innovation and the paucity of the evidence base to inform clinical adoption of new genetic tests. As one diagnostics industry executive described the problem in 2003 (Winn-Deen 2003):

> [There has been] a noticeable lack of consensus within the genetics community about exactly when a test for a new marker was sufficiently validated for it to enter into clinical service. Some labs rushed to provide testing after the first publication, while others waited until the result had been replicated in multiple studies or multiple ethnic groups.

With the growing use of companion diagnostics and multi-marker prognostic signatures in oncology, and an increasing number of firms marketing pharmacogenetic tests or polygenic risk assessment for common, complex diseases, this issue is no longer confined to the field of rare disease genetics. Although the development of next-generation sequencing and non-invasive sampling techniques has sparked new concerns about the pace of DNA-based diagnostic innovation, the emerging fields of metabolomics and proteomics are broadening our definition of personalised medicine and further complicating the regulatory challenges. Given the pace of technological change, the field of molecular diagnostics continues to face significant challenges in validating the technical performance of new platform technologies, and in evaluating the clinical significance of new biomarkers.

Focusing primarily on Europe and the USA, this chapter describes how the regulatory frameworks for diagnostic technologies are changing, in part as a response to technological advances in (and with likely significant consequences for) personalised medicine. The chapter will provide an overview of how policy deliberation about the regulation of genomic diagnostics has evolved over the last two decades, and sets this policy debate in the context of broader concerns about the scientific rigour of diagnostic innovation. The chapter then describes policy developments in the statutory regulatory regimes that govern diagnostic tests as medical devices; the construction of a new regulatory regime for pharmacogenomics; and the growing role of health technology assessment as a gatekeeping mechanism.

A key argument advanced in this chapter is that it would be a mistake to simply characterise personalised medicine as another field where scientific advance has outstripped public policy. In practice, two decades of ELSI research, policy reports and public debate have anticipated the potential challenges of a scientific field that has advanced into the clinic at a scale and pace considerably less dramatic than many expected. If there is a lag, it is in moving from policy deliberation to policy implementation, and even in this respect the record is mixed.

A second key argument is the need to move beyond an approach that focuses solely on the challenges of the cutting-edge of technological change, a bias inherent in ELSI research funding. Concerns about diagnostic innovation are not restricted to molecular diagnostics and personalised medicine – a 2015 report from the National Academies of Science (NAS) suggested that in the USA, diagnostic errors are implicated in approximately 10 per cent of patient deaths, and 6–17 per cent of hospital adverse events. Amongst its many recommendations, the report emphasised the need for more robust evaluation of new diagnostic tests before they enter clinical practice (National Academies of Sciences, Engineering, and Medicine 2015). A 2015 report from the US Food and Drug Administration (FDA) similarly highlighted the risks to patients arising from poorly validated diagnostics (Food and Drug Administration 2015).

Finally, this chapter advances the argument that, in comparison with the burgeoning literature devoted to pharmaceutical regulation, the regulation of diagnostic tests has been a relatively neglected subject for social scientists, historians and legal scholars interested in bio-medicine. ELSI scholarship has done little to rectify this imbalance; there has been far more research on upstream governance topics such as tissue biobanking than downstream com-mercialisation of genomic diagnostics. In this chapter I will offer some suggestions for future research that might begin to fill this gap.

An evolving policy debate

Regulatory policy and ELSI scholarship have evolved over more than two decades, repeatedly shifting focus in response to technological innovations (such as next-generation sequencing), the emergence of new diagnostic modalities (such as companion diagnostics), and changes in the way that tests are delivered (such as direct-to-consumer [DTC] services). Alongside these dynamics, which we might characterise as endogenous to the field of genomics, are a broader set of exogenous factors that have shaped policy. Consider, for instance, the field of pharma-cogenomics, a term used to describe the application of genomic science in drug discovery and development, and the clinical use of pharmacogenetic tests to guide drug selection and dosage decisions based on information about how an individual's genetic make-up might influence drug metabolism. Here regulatory standards have developed through collaboration between regulatory agencies in the USA, Europe and Japan, reflecting the growing importance of transnational harmonisation in pharmaceutical regulation.

If one were to attempt a chronological sketch of the evolving policy debate, one might suggest that the earliest topic of debate was newborn screening for monogenic diseases (see for instance Nuffield Council on Bioethics 1993; National Research Council [U.S.] & Committee for the Study of Inborn Errors of Metabolism 1975). However, for a decade or so from the mid-1990s the topic attracting greatest attention was genetic risk prediction for common, complex diseases (CCDs). The discovery of the link between the APOE gene and Alzheimer's in 1992 and the discovery of the BRCA 1/2 genes linked to breast and ovarian cancer in 1994/5 fuelled much debate about the clinical utility of genetic risk factors for common complex diseases (CCDs), the psycho-social consequences of genetic risk assessment, and the dangers of premature clinical adoption when data on the predictive value of risk markers was still evolving. A succession of policy reports highlighted concerns about predictive genetic tests (sometimes referred to as susceptibility tests). ELSI scholarship in this period was primarily concerned with mapping the different regulatory actors and their respective powers and functions, comparative analysis of regulatory regimes in different countries, and proposals for regulatory reform (Javitt & Hudson 2006; Martin & Frost 2003; Hogarth et al. 2007).

Given the very limited number of predictive genetic tests that were available in this period, and the limited scale of their diffusion into clinical practice, this phase of policy deliberation might be viewed as a timely example of anticipatory governance; an effort to ensure that public policy did not lag behind technological change. Yet, in 2007 when a new wave of direct-to-consumer (DTC) genomics firms like Navigenics and *23andMe* began offering polygenic susceptibility tests for a range of CCDs, some ELSI scholars cautioned that regulatory action was still premature (Prainsack et al. 2008). DTC genomics subsequently became a popular topic for ELSI research, and concerns about DTC firms continue to frame much of the regulatory debate and, in the USA at least, have galvanised regulatory enforcement activity.

Moreover, by 2007 the attention of policymakers and regulatory agencies had diffused well beyond susceptibility testing. A wave of ELSI research and policy reports from around 2003 onwards had focused on pharmacogenomics (Hogarth et al. 2006; Hopkins et al. 2006; Melzer et al. 2003). Policymaking in this area was marked by a greater involvement of industry actors and a relatively minor role for the clinical genetics community (who were key actors in the debate about susceptibility testing). Other developments that have broadened the framing of the regulatory debate (but which this brief chapter lacks space to address), include the growing number of firms marketing proprietary multi-marker diagnostic signatures and the clinical application of two new technologies: next-generation sequencing and non-invasive prenatal testing.

Perhaps the most significant shift in framing might be very broadly characterised as a transition from 'innovation is too fast', to 'innovation is too slow'. This change was most clearly apparent in the USA, where it was linked to the deregulatory agenda of President Bush (Stewart 2002). In 2002 the new Bush administration disbanded the Secretary's Advisory Committee on Genetic Testing, a body which had focused its attention on regulatory reform, and replaced it with a new body: the Secretary's Advisory Committee on Genetics Health and Society, which began its work by explicitly stating that it would not address regulatory issues and then produced a series of reports that were chiefly concerned with barriers to innovation, such as reimbursement (SACGHS 2006) and gene patenting (SACGHS 2010). Concerns about how weaknesses in the regulatory regime for diagnostics lead to the clinical adoption of tests with limited evidence of safety and effectiveness have not been supplanted (Hayes 2015) – indeed, even SACGHS produced an important report on the topic in 2008 during the dying days of the Bush administration (SACGHS 2008) – but they are now supplemented by concerns about how fundamental failings in the diagnostic R&D process have resulted in disappointing progress in translating biomarker discoveries into clinical diagnostics (Wagner & Srivastava 2012; Poste 2011). This change in framing emerged in part from a growing consensus that major investments in genomic science have thus far had limited clinical impact (Evans et al. 2011; Green et al. 2011), but also from recognition that the limitations of the current diagnostic innovation process have a broader societal impact, in particular the opportunity costs of misdirected public and private funding in biomedical research and healthcare (Henry & Hayes 2012; Califf 2004).

A flawed system

Much diagnostic innovation has traditionally occurred at the interface of the clinic and the laboratory, in a 'hidden innovation system' which is rarely subject to formal regulatory controls (Hopkins 2006). The field of clinical genetics has exemplified the dynamics of this innovation model, in which clinicians and researchers based in the public sector play the lead role in the discovery and development of new biomarkers and new diagnostic techniques, and industry plays a supporting and secondary role. This model involves often very diffuse networks of actors operating without a clear co-ordinating central node; there is no single sponsor of a new technology, in the way that a pharmaceutical firm is the sponsor for a new therapeutic molecule. Hopkins has characterised this innovation system as 'hidden', both because it is often neglected in policy discourse, and because it has flown under the radar of formal oversight processes. Its diffuse and non-commercial character means that it generally evades both the statutory regulatory mechanisms that establish scientific standards for the safety and efficacy of medical devices by controlling market entry, and the soft law regulatory constraints of healthcare payors, who use Health Technology Assessment (HTA) to demand evidence of comparative clinical or cost effectiveness of new medical technologies.

In recent years, the hidden innovation model has been bolstered by enhanced funding for translational science programmes that encourage 'novel forms of clinical research designed to extend genomics into the clinic' (Kohli-Laven et al. 2011). If much of the policy debate around genetic testing oversight sought to create a clearer delineation of research and clinical practice (see for instance, the recommendations of the Task Force on Genetic Testing 1997), then arguably these recent developments have blurred the boundaries further. Such porous borders may provide valuable grist to the intellectual mill of STS scholars eager to map the heterogeneous networks of actors at work in this space, but from the perspective of public policy they serve only to heighten concern about a series of fundamental problems with diagnostic innovation, a field in which, critics suggest, researchers 'have often ignored … fundamental principles of the scientific process' (Hayes 2013). Advocates of what we might term *diagnostic reform* suggest that the failings in diagnostic R&D are multiple. Commonly cited problems include underpowered studies, various types of bias, insufficient research on clinical outcomes, over-fitting of data in retrospective analyses and a lack of prospective controlled studies (Feinstein 2002; Sox 1996). The porous boundary between research and clinical practice enables continued confusion over the appropriate scientific standards for diagnostic innovation.

Personalised medicine has not escaped these problems. For many years, research on the genetic epidemiology of CCDs was characterised by an overwhelming failure to replicate findings. A 2006 review by Ioannidis et al. described the field as dominated by small, underpowered studies, often with flawed designs and selective reporting of positive results (Ioannidis et al. 2006; Morgan et al. 2007). However, from 2007 a new wave of large-scale Genome-Wide Association Studies (GWAS), often organised as transnational collaborations, overcame the problem of small sample size and began to produce robustly replicated findings, but this development simply pushed the evaluative challenge downstream, presenting new challenges in how to validate the clinical validity and utility of polygenic susceptibility testing (McCarthy et al. 2008).

Again, these problems reflect broader challenges for the cause of diagnostic reform, in particular the need for regulators to establish and enforce clear scientific standards. As a recent commentary has suggested, in the absence of clear standards, uncertainty prevails:

... manufacturers, laboratory professionals, researchers and regulators are equally confused on what studies to do or accept as evidence for the clinical performance and effectiveness of medical tests.

(Horvath et al. 2014)

The changing terrain

In terms of statutory regulation, notable developments are the introduction of new regulations for diagnostic devices in the European Union (EU) in 2017 and in Australia in 2010, and efforts by the US Food and Drug Administration to establish a new regulatory framework for laboratory-developed tests (LDTs). Beyond these statutory regulatory frameworks, the last decade has seen initiatives to improve the evaluation of genomic diagnostics, including development of the ACCE evaluation framework (see Table 19.1) and the creation of evaluative bodies such as the UK Genetic Testing Network (Hogarth 2007). Broader diagnostic reform efforts have been promulgated by two interlinked communities: HTA and Evidence-Based Medicine (EBM), from which have emerged new standards such as the STARD framework for reporting diagnostic studies (Bossuyt et al. 2003). Such developments exemplify a broader process of regulatory expansion in the regimes governing healthcare technology adoption, in particular growing demand for evidence of comparative and cost-effectiveness by HTA agencies, and a greater role for clinical guidelines (Weisz et al. 2007; Timmermans & Berg 2010).

Regulating diagnostic devices

In the United States, the FDA was granted the legal authority to evaluate the safety and effectiveness of medical devices by the 1976 Medical Device Amendments to the Federal Food, Drug and Cosmetic Act (FDCA). According to the statutory definition, a medical device is deemed safe if the scientific evidence demonstrates 'that the probable benefits to health ... outweigh any probable risks' (US FDA: Code of Federal Regulations Title 21 § 860.7(d)(1)). Effectiveness is demonstrated when scientific evidence shows that 'in a significant portion of the target population, the use of the device ... will provide clinically significant results' ((US FDA: Code of Federal Regulations Title 21 § 860.7(e)(1)). How are these terms operationalised in the case of diagnostic devices? The FDA generally states that it is interested in two performance criteria: analytic validity and clinical validity. However, the term 'clinically significant results' is sufficiently ambiguous to merit closer attention. Social scientists could usefully investigate how this term has been operationalised in regulatory guidance documents and in regulatory decision-making, for instance, has the threshold for approval changed over time or does it vary significantly for different types of test?

Table 19.1 ACCE framework for test evaluation

ACCE framework for test evaluation	
Analytic validity	accuracy of test identifying the biomarker
Clinical validity	relationship between the biomarker and clinical status
Clinical utility	likelihood that test will lead to an improved outcome
Ethical, legal and social implications	

In the European Union, the IVD Directive, which came into force in 2003, is commonly understood as allowing manufacturers to validate their tests based solely on analytic performance although Hogarth and Melzer (2007) have disputed this interpretation. Such interpretive disputes notwithstanding, the new EU regulation, which will come into force in 2017, places far greater emphasis on the need for manufacturers to provide data on the clinical performance of diagnostic devices. Furthermore, a new risk classification system will ensure that a far greater proportion of diagnostic tests will be subject to regulatory scrutiny before market entry, including genetic tests, all cancer tests (the disease area where personalised medicine has advanced furthest) and all companion diagnostics. Specific provisions make explicit that the regulation will cover predictive genetic tests and DTC tests sold via the internet, two areas of ambiguity in the 2003 Directive. Despite a strengthening of regulatory controls, EU policymakers have been keen to present their approach as more flexible and less onerous than the USA's. Another area for future research would be to examine evidence of 'device lag', i.e. slower market entry in USA than EU. A more in-depth case study approach could examine which of the respective regulatory approaches offers the greatest protection against unsafe or ineffective medical devices.

Laboratory-developed tests

In the USA, the FDA's authority over genetic tests has been the subject of high-level policy reports, Congressional hearings and a number of different draft legislative proposals. The key regulatory loophole is not specific to genomics, but relates to what are now generally termed laboratory-developed tests (LDTs), i.e. a diagnostic test 'that is intended for clinical use and designed, manufactured and used within a single laboratory' (Food and Drug Administration 2014). The regulatory status of LDTs was first highlighted by the 1997 US Task Force on Genetic Testing (Hogarth et al. 2007). Since about 2003 the FDA has asserted its regulatory authority over LDTs on an ad hoc basis, but there has been resistance to proposals for a more systematic approach. Critics of FDA's move to regulate LDTs argue that such tests are already adequately regulated under the statute which governs clinical laboratories. However, the FDA has twice been given direction from Congress to address at least some portions of the LDT sector, and in the last decade two policy reports from government advisory committees have recommended that FDA regulate LDTs, arguing that the agency has the necessary legal authority, scientific expertise and regulatory experience to extend its oversight to this sector of the diagnostics market (SACGT 2000; SACGHS 2008).

FDA's earliest effort to progress this issue was stymied by the Bush administration political appointments to the agency in 2001. Limited progress was made under President Obama, in the shape of two draft guidances and continued ad hoc regulatory interventions against specific firms, but the failure to finalise guidance setting out a new framework is testament to the conservatism of the New Democrat politics of the Obama administration. Although draft legislation continues to be discussed in Congress, the prospects for progress under President Trump seem remote.

Other jurisdictions have begun to address the regulatory status of LDTs. Furthest advanced is Australia, where a new regulatory framework for IVDs was introduced in 2010, in part specifically to address oversight of the LDT sector by the Therapeutic Goods Administration (TGA). For low- and moderate-risk tests, laboratories must be ISO accredited, must register with the TGA and notify the agency about the tests they make. Test validation must meet TGA-endorsed standards, but will be carried out by the National Association of Testing Authorities (NATA) and the National Pathology Accreditation Advisory Council (NPAAC), the bodies responsible for assuring laboratory performance in Australia. Only high-risk tests will be subject to the same standards and processes as apply to test kits. This is an example of what regulatory theorists term 'responsive regulation': it strikes a balance between on the one hand relying wholly on either 'command-and-control' powers exercised by a central agency, or on the other hand focusing purely on mechanisms of self-regulation.

Although the precise details are somewhat different, the Obama administration–era draft FDA guidance had marked similarities: a risk-based approach with a variety of exemptions, the use of registries for tests exempt from full oversight, and a role for third parties in the governance process. The draft new EU IVD regulation retains a distinction between devices produced in health institutions (which are mainly exempt) and what it terms 'devices used in the context of a commercial activity to provide a diagnostic or therapeutic service' (Para 19 of Recital), which are not exempt. Although the regulation contains ambiguous wording and arguably contradictory clauses, and is likely to require additional clarifying guidance, even for health institutions, the new regulation will move closer to the TGA model and FDA's proposed model, with regard to high-risk LDTs produced in health institutions.

These areas of convergence suggest a policy space where a degree of transnational policy learning may be underway. An immediate priority for future research would be to investigate how the new Australian regulations are working in practice, a topic likely to be of interest in multiple jurisdictions that have yet to address this regulatory issue, for instance Canada, or where policy is still in a state of flux (as in the United States and the European Union).

Pharmacogenomics

Pharmacogenomics is the use of genomic science to study human variability in drug response. Proponents of pharmacogenomics suggest that it will lead to a fundamental transformation both in pharmaceutical R&D and in clinical practice, but uncertainty about the regulatory standards and processes for this emergent technological paradigm has been widely cited as an obstacle to more widespread and rapid adoption. However, in 2004 the FDA and the EMA each published reports suggesting that genomics would be a key component in a new scientific toolkit for drug development that would solve the pharmaceutical industry's longstanding productivity crisis; enable cheaper and faster drug development; and provide safer and more effective drugs (European Medicines Agency 2005; Food and Drug Administration 2004). Both agencies asserted a central role for themselves in the translation of genomics from bench to bedside. Pharmacogenomics thus presents an ideal case study of the role of regulators in the co-production of new biomedical technologies. Since 2004 a process of organisational restructuring, standard-setting and regulatory decision-making has constructed a new transnational regulatory regime for pharmacogenomics through an iterative process of regulatory experiment enacted by a network encompassing regulatory agencies in the USA, EU and Japan in conjunction with academic scientists and industry (Hogarth 2012). This process has been marked by the creation of new socio-technical spaces in the regulatory regimes for pharmaceuticals – a pre-regulatory space for the sharing of data outside the regulatory decision-making process and a pre-competitive space for the sharing of data between firms. It was marked also by the expansion of a transnational regulatory space for sharing data and setting standards across jurisdictional boundaries. The construction of this new regulatory paradigm was both driven by and reinforced the existing trend towards transnational governance in pharmaceutical regulation, and in particular it bolstered the growing authority of EMA as the *de facto* regulator for novel biomedical technologies in Europe.

Another dynamic worth noting is how the 2004 reports from FDA and EMA repositioned the agencies as enablers of innovation, the subsequent close interaction with industry in policy development and the repeated emphasis on faster and cheaper drug development as a key goal (Hogarth 2015). The construction of a regulatory regime for pharmacogenomics has thus been consistent with what Davis and Abraham term the era of *neoliberal corporate bias* in pharmaceutical regulation (Davis & Abraham 2013).

HTA for diagnostics

In the past, the decision on whether or how to use a test was a question for the clinical judgement of individual physicians, however, increasingly these decisions are made formally by healthcare funding or reimbursement systems, or health service provider organisations. The process of approval or recommendation is generally based on a review of the available evidence, a practice generally known as health technology assessment (HTA). One recent study suggests that new molecular diagnostics face heightened scrutiny by HTA agencies in the USA compared with more traditional diagnostic tests (Trosman et al. 2011), and the last decade has seen new initiatives to advance evidence-based evaluation of diagnostics, such as the diagnostics assessment programme established by the UK's National Institute for Health and Care Excellence, and the EGAPP process supported by the US Centers for Disease Control.

The growing importance of HTA for diagnostic innovation is illustrated by the Roche Amplichip, which in 2004 became the first pharmacogenetic microarray test to gain FDA approval. In the USA it was taken up by LabCorp, the country's second largest reference lab and was therefore widely available. However, a succession of negative HTA reports found insufficient evidence of clinical utility to support the use of the Amplichip to guide drug treatment (Blue Cross Blue Shield Technology 2004). This example illustrates the profound temporal and spatial bifurcation in the operation of power within the diagnostic regulatory regime. In the premarket space regulatory power accords with a traditional model of command-and-control regulation in which the sovereignty of the regulatory agency over the regulated industry is paramount, but in the postmarket space power is much more diffuse: not only is the balance of power between regulator and industry more equal but the role of gatekeeper is shared amongst multiple actors (Hogarth & Martin 2015). The changing power dynamics within this multi-level regulatory regime are a further topic for future research, for instance, is the trend towards convergence or divergence in the standards applied by regulatory agencies like FDA and HTA bodies?

Conclusion

This chapter began by suggesting that the regulation of diagnostic technologies has been a neglected research topic and in conclusion I will identify some ways forward. Concerns about the scale and pace of innovation in genomics are indicative of an unresolved tension between personalised medicine and evidence-based medicine. Examining the interactions and overlaps between the two distinct but related epistemic communities would be one fruitful approach for future research.

Whether flying under the banner of ELSI or RRI, researchers who believe that anticipatory governance mechanisms can play a useful role in steering biomedical innovation need a degree of reflexivity, and historical reflection on developments in this field may prove helpful. Well-characterised regulatory issues, highlighted over decades in a succession of policy reports produced in multiple jurisdictions have impelled administrative action and legislative reform but have yet to find final resolution and we might usefully interrogate the translational space between policy recommendation and policy implementation in search of lessons for the future.

There are strong drivers in ELSI funding mechanisms and in the intellectual orientation of STS scholarship for a focus on emergent technologies. The recent STS turn against innovation and towards maintenance (Russell and Vinsel 2016) offers hope for greater attention to the realm of everyday, established technologies and suggests an important direction for future scholarship on the regulation of diagnostics. The cutting-edge of personalised medicine will remain an important field of inquiry but future research should also investigate the dynamics of standard-setting and regulatory approval for mundane, but pervasive diagnostic technologies such as glucose meters.

References

Blue Cross Blue Shield Technology, 2004. Special Report: Genotyping for Cytochrome P450 Polymorphisms to Determine Drug-Metabolizer Status. *Assessment Program*, 19(9).

Bossuyt, P.M. et al., 2003. The STARD statement for reporting studies of diagnostic accuracy: explanation and elaboration. *Clinical Chemistry*, 49(1), pp. 7–18.

Califf, R.M., 2004. Defining the balance of risk and benefit in the era of genomics and proteomics. *Health Affairs*, 23(1), pp. 77–87.

Davis, C. & Abraham, J., 2013. *Unhealthy Pharmaceutical Regulation: Innovation, Politics and Promissory Science*, Basingstoke: Palgrave Macmillan.

European Medicines Agency, 2005. *The European Medicines Agency Road Map to 2010: Preparing the Ground for the Future*, London.

Evans, J.P. et al., 2011. Deflating the genomic bubble. *Science*, 331(6019), pp. 861–862.

Feinstein, A.R., 2002. Misguided efforts and future challenges for research on "diagnostic tests." *Journal of Epidemiology and Community Health*, 56(5), pp. 330 –332.

Food and Drug Administration, 2004. *Critical Path Initiative*, Rockville, MD.

Food and Drug Administration, 2014. *Draft Guidance for Industry, Food and Drug Administration Staff, and Clinical Laboratories: Framework for Regulatory Oversight of Laboratory Developed Tests (LDTs)*, Silver Spring, MD.

Food and Drug Administration, 2015. *The Public Health Evidence for FDA Oversighht of Laboratory Developed Tests: 20 Case Studies*, Bethesda, MD.

Green, E.D., Guyer, M.S. & National Human Genome Res Institute, 2011. Charting a course for genomic medicine from base pairs to bedside. *Nature*, 470(7333), pp. 204–213.

Hayes, D.F., 2015. Biomarker validation and testing. *Molecular Oncology*, 9(5), pp. 960–966.

Hayes, D.F., 2013. OMICS-based personalized oncology: if it is worth doing, it is worth doing well! *BMC Medicine*, 11(1), p. 221.

Henry, N.L. & Hayes, D.F., 2012. Cancer biomarkers. *Molecular Oncology*, 6(2), pp. 140–146.

Hogarth, S. et al., 2007. Closing the gaps: enhancing the regulation of genetic tests using responsive regulation. *Food & Drug LJ*, 62, p. 831.

Hogarth, S., 2007. From genomic research to public health practice – international policy implications. In B. M. Knoppers, ed. *Genomics and Public Health: Legal and Socio-Ethical Perspectives*. Leiden: Martinus Nijhoff, pp. 239–256.

Hogarth, S., 2015. Neoliberal technocracy: Explaining how and why the US Food and Drug Administration has championed pharmacogenomics. *Social Science & Medicine*, 131, pp. 255–262.

Hogarth, S. et al., 2006. Regulating Pharmacogenomics: An Overview of Developments in Various Countries and Industry Response to Regulatory Initiatives (A Report for Health Canada). Cambridge, UK. *A report for Health Canada, Cambridge, ON, Canada*.

Hogarth, S., 2012. Regulatory experiments and transnational networks: the governance of pharmacogenomics in Europe and the United States. *Innovation: The European Journal of Social Science Research*, 25(4), pp. 441–460.

Hogarth, S. & Martin, P., 2015. *The Ratio of Vision to Data: Promoting Pharmacogenetics Through Promissory Regulation in the USA*, London.

Hogarth, S. & Melzer, D., 2007. The IVD directive and genetic testing problems and proposals. In *A briefing presented to the 20th Meeting of Competent Authorities*. Lisbon.

Hopkins, M.M. et al., 2006. Putting pharmacogenetics into practice. *Nature Biotechnology*, 24(4), pp. 403–410.

Hopkins, M.M., 2006. The hidden research system: the evolution of cytogenetic testing in the National Health Service. *Science as Culture*, 15(3), pp. 253–276.

Horvath, A.R. et al., 2014. From biomarkers to medical tests: the changing landscape of test evaluation. *Clinica Chimica Acta*, 427, pp. 49–57.

Ioannidis, J.P.A. et al., 2006. A road map for efficient and reliable human genome epidemiology. *Nature Genetics*, 38(1), pp. 3–5.

Javitt, G.H. & Hudson, K., 2006. Federal neglect: regulation of genetic testing. *Issues in Science and Technology*, 22(3), pp. 59–66.

Kohli-Laven, N. et al., 2011. Cancer clinical trials in the era of genomic signatures: Biomedical innovation, clinical utility, and regulatory-scientific hybrids. *Social Studies of Science*, 41(4), pp. 487–513.

Martin, P. & Frost, R., 2003. Regulating the commercial development of genetic testing in the UK: problems, possibilities and policy. *Critical Social Policy*, 23(2), pp. 186–207.

McCarthy, M.I. et al., 2008. Genome-wide association studies for complex traits: consensus, uncertainty and challenges. *Nat Rev Genet*, 9(5), pp. 356–369.

Melzer, D. et al., 2003. *My Very Own Medicine: What Must I Know? Information Policy for Pharmacogenetics*, Cambridge, UK: Department of Public Health and Primary Care, University of Cambridge.

Morgan, T. et al., 2007. Nonvalidation of reported genetic risk factors for acute coronary syndrome in a large-scale replication study. *JAMA*, 297(14), pp. 1551–1561.

National Academies of Sciences, Engineering, and M., 2015. *Improving Diagnosis in Health Care*, Washington, DC: National Academies Press.

National Research Council (U.S.) & Committee for the Study of Inborn Errors of Metabolism, 1975. *Genetic screening : programs, principles, and research*, Washington: National Academy of Sciences.

Nuffield Council on Bioethics, 1993. *Genetic Screening: Ethical Issues*, London.

Poste, G., 2011. Bring on the biomarkers. *Nature*, 469(7329), pp. 156–157.

Prainsack, B. et al., 2008. Personal genomes: Misdirected precaution. *Nature*, 456(7218), pp. 34–35.

Russell, A. and Vinsel, L., 2016. Hail the maintainers. *Aeon*. Available at https://aeon.co/essays/innovation-is-overvalued-maintenance-often-matters-more.

SACGHS, 2006. *Coverage and Reimbursement of Genetic Tests and Services*, Bethesda, Maryland, USA.

SACGHS, 2010. *Gene Patents and Licensing Practices and Their Impact on Patient Access to Genetic Tests*, Bethesda, Maryland, USA.

SACGHS, 2008. *US System of Oversight of Genetic Testing: A Response to the Charge of the Secretary of Health and Human Services*, Bethesda, Maryland, USA.

SACGT, 2000. *Enhancing the oversight of genetic tests: recommendations of the Secretary's Advisory Committee on Genetic Testing*, Bethesda, Maryland, USA.

Sox, Harold C., M.D., 1996. The evaluation of diagnostic tests: principles, problems, and new developments. *Annual Review of Medicine*, 47(1), pp. 463–471.

Stewart, A., 2002. US advisory committee on genetic testing to be dismantled. *PHG Foundation Website*. Available at: www.phgfoundation.org/news/560/ [Accessed May 4, 2017].

Task Force on Genet Test, 1997. *Promoting Safe and Effective Genetic Testing in the United States*.

Timmermans, S. & Berg, M., 2010. *The Gold Standard: The Challenge of Evidence-Based Medicine and Standardization in Health Care*, Temple University Press.

Trosman, J.R., Van Bebber, S.L., Phillips, K.A., 2011. Health Technology Assessment and Private Payers' Coverage of Personalized Medicine. *Journal of Oncology Practice*, 7(3S), pp.18.

US FDA, *Code of Federal Regulations Title 21*.

Wagner, P.D. & Srivastava, S., 2012. New paradigms in translational science research in cancer biomarkers. *Translational Research*, 159(4), pp. 343–353.

Weisz, G. et al., 2007. The emergence of clinical practice guidelines. *Milbank Quarterly*, 85(4), pp. 691–727.

Winn-Deen, E., 2003. Fulfilling the promise of personalized medicine. *IDV Technology*.

20
Collection and protection of genomic data

Edward S. Dove

Introduction

However laudable the growing collection and use of genomic data may be from medico-scientific, economic, and social viewpoints, many people worry about the impact of this activity – if not about the nature of genomic data itself – on their privacy and the privacy of others. Arguably, privacy concerns have only intensified in recent years in the midst of deeper penetration of communication networks, advances in sequencing technology and information technology infrastructure (e.g. cloud computing), direct-to-consumer genetic testing, proliferation of systems such as artificial intelligence, and increasing commercial access to health-related patient data (Dove et al. 2017; Dove and Phillips 2015).

One of the central concerns in this discourse is that genomic data may be all too easily collected and used in ways that disrespect or disregard the fundamental rights and interests of individuals and society. A key 'right' within these is the right to respect for one's private and family life, which is enshrined in national and international human rights conventions (see also Bygrave 2014). For some people, however, privacy concerns associated with genomic data run deeper than with other types of personal data, including other types of health data. This is because genomic data are seen as private in a biologically interconnected sense that demands respect for both private *and* family life.

In this chapter, I reflect on key issues in the governance of the collection and protection of genomic data, with a focus on examples from European Union law. First, I discuss broadly how data privacy regimes regulate the collection and protection of genomic data. I examine how these regimes balance the needs of those wanting to collect and use genomic data for various purposes with the rights and interests of those from whom genomic data are taken and to whom they relate and implicate. In the genomic data context, those collecting and processing data include researchers (and their affiliated institutions), clinical organisations, and companies offering genetic testing. Following this discussion, I argue these regimes tend to generate or perpetuate several paradigms that can cause problems in an age of 'big data' analytics, direct-to-consumer genetic testing, and genomic data sharing initiatives. I then focus on two paradigms (see e.g. Kitchin 2014; Munns and Basu 2015; Sethi and Laurie 2013; Taylor 2012 regarding others): first, the tendency in regulation to create bounded, siloed categories of personal data;

and second, the 'consent or anonymise' paradigm. I conclude with a brief discussion of several emerging tools and initiatives that aim to strike a better equilibrium between robust exploitation of genomic data that promotes the public interest, while remaining consistent with the confines of existing law and preserving privacy as a fundamental value in biomedical research and clinical practice.

Data privacy regimes

Privacy in its informational dimension is a state of affairs whereby data relating to a person is in a state of non-access. It embodies a range of rights and values, such as the right to be let alone, intimacy, seclusion, and personhood. Data protection is a narrower 'set of legal rules that aims to protect the rights, freedoms, and interests of individuals, whose personal data are collected, stored, processed, disseminated, destroyed, etc. The definitive objective is to ensure fairness in the processing of data and, to some extent, fairness in the outcomes of such processing' (Tzanou 2013: 89). In this chapter I use the imperfect but relatively catch-all term 'data privacy regimes'.

Data privacy regimes such as the EU's General Data Protection Regulation (GDPR, Regulation (EU) 2016/679) tend to address the 'processing' of personal data, an umbrella term which includes any operation or set of operations performed on personal data (e.g. collection, storage, access or retrieval, disclosure, erasure). The regimes usually apply only to data that relate to an identified or identifiable *individual* person, known as the 'data subject' (hence the terms 'personal data' or 'personal information'). Of little surprise to the reader of this *Handbook*, this traditional definition is problematic in the genomics context given the inherently relational nature of much genomic data (Hallinan and de Hert 2017; Taylor 2012, Prainsack *this volume*).

As Tzanou observes, data privacy regimes seek to ensure fairness in the processing of data. One of the regulators' concerns is protection against informational harms. Potential informational harms that could arise from misuse or abuse of genomic data include psychological harms in the form of embarrassment, guilt, stress, anxiety, depression, and altered behaviour; social harms in the form of stigmatisation, embarrassment within one's social group, discrimination, and criminal prosecution; and economic harms in the form of loss of or change in employment (Laurie et al. 2014). These harms can affect not just individuals, but their families and communities as well.

Data privacy regimes generally fall within one of three types: (1) omnibus data privacy statutes often grounded in human rights principles (e.g. the GDPR); (2) sectoral laws or regulations that apply only to the demands of data privacy in a specific sector, e.g. The Health Insurance Portability and Accountability Act (HIPAA) Privacy Rule and the Genetic Information Nondiscrimination Act of 2008, both in the United States; and (3) other rules, principles, or standards that do not have the force of law, but that may nonetheless entail serious consequences when violated, such as professional or policy guidelines (e.g. American Society of Human Genetics' Code of Ethics, Royal College of Physicians guidance on genetic testing and sharing genetic information). Within each category there can be tremendous variation of scope and substance.

This variation means that approaches to regulating the processing of personal data differ across world regions. As Weber observes: 'From a comparative perspective, European regulations are quite advanced, setting a high level of data protection. In the United States and in Asia, the emphasis is more on self-regulatory approaches' (Weber 2013: 117). Some legal regimes operate on the scale of a regional grouping of countries, such as the EU; many apply only within a single country, while others are narrower still, for example, laws specific only to a single province or state within a country, or even to a specific type of industry. Laws may apply separately to public, private, or health/non-health institutions. It should be noted, though, that legal regimes can have a broad territorial scope. For example, the GDPR is especially far-reaching in its territorial scope, as it applies to organisations which have EU 'establishments' where personal data are processed 'in the context of the activities' of such an establishment. It also applies to non-EU established organisations where they process personal data about EU data subjects in connection with the 'offering of goods or services' (payment is not required) or 'monitoring' their behaviour within the EU. In the genomics context, this means the GDPR will apply to anybody who shares personal data within the EU area or collects and processes personal data from EU customers, research participants, or patients.

Despite these variations, there are several points of convergence. First, most data privacy regimes have two key objectives: (1) they aim to *enable* free flows of data necessary for the delivery of services (e.g. management of public health, biomedical research); (2) at the same time, they seek to establish appropriate privacy and security frameworks that *protect* personal data from misuse. Much of the discussion surrounding data privacy can be described in terms of the appropriate relationship between these two objectives. A second point of convergence is that data privacy regimes (particularly those within the first two types mentioned above) often impose specific duties on an individual or organisation who must determine the purposes and means of the personal data processing – the 'data controller' – as well as on an individual or organisation – known as the 'data processor' – who processes personal data on behalf of the data controller. A third point of convergence is that consent often operates as a primary legitimator of data processing, as does in its own way anonymisation – the dual impact of which I address below. A fourth and related point of convergence, and one that I wish to argue below, is that while some data privacy regimes are seen as providing robust protection, many also tend to perpetuate or breathe life into several problematic paradigms in the genomics context.

Problematic paradigms

Data as regulatory bounded objects

Data privacy regimes tend to draw distinctions between different categories of personal data and can include 'genetic data', 'medical data', 'health data', 'biometric data', and 'data relating to health'. Categorising data is a fraught process, not the least because the distinctions between the different types are nebulous and often artificial (Manson and O'Neill 2007).

For example, 'genetic data' is a sub-category of personal data for the purposes of data privacy regulation in the EU, but the distinction from (other types of) health data is far from clear. The EU's GDPR introduces the regulatory object of 'genetic data', which was absent in the previous data privacy regime: EU Directive 95/46/EC. 'Genetic data' covers 'personal data relating to the inherited or acquired genetic characteristics of a natural person which give unique information about the physiology or the health of that natural person and which result, in particular, from an analysis of a biological sample from the natural person in question'. This is an unduly narrow definition, given that much genetic data does not provide 'unique' information about a sole individual, but rather quite often the individual's genetic family members as well (Hallinan and de Hert 2017). This definition does align, however, with the law's general fixation with individual ('data subject') rather than familiar or group protections, whether for privacy violations, discrimination, or otherwise.

But how far might the notion of 'genetic data' under the GDPR extend? Regulators and policymakers in the Member States may have to consult scientists to determine how 'genetic data' is distinct from the category 'data concerning health' (discussed below), and how the emphasis placed on 'unique' information accords with scientific understanding of genes that suggests a more relational approach to what these data provide (Laurie 2002; Taylor 2012; Widdows 2013). It may be that those responsible for the collection and use of these data will err on the side of caution and assume that *all* genetic data should be treated as a category of personal data for the GDPR, even if they do not provide 'unique' information about the physiology or the health of an individual (though whole genome sequence data would qualify on uniqueness grounds). If this happens, the majority of genomic research data would be covered by the legal provisions speaking to 'genetic data' even when they might not be truly 'unique.'

The GDPR also introduces a definition of 'data concerning health' that was not included in Directive 95/46/EC. It is defined broadly, covering all 'personal data related to the physical or mental health of a natural person, including the provision of healthcare services, which reveal information about his or her health status'. Recital 35 of the GDPR makes clear that 'data concerning health' includes *information derived from* the testing of a genetic sample, which is subtly distinct from genetic data per se. But this is an unnecessary and perhaps unhelpful distinction because it is unclear whether it will always hold in practice. Consider, for example, the results of a genetic test that are recorded in a medical record. The data relating to one's genetic characteristics will be 'genetic data', while the information derived from the characteristics revealing information about one's health will be 'data concerning health'. Here, the distinction between 'genetic data' and 'data concerning health' would seem to collapse because either way, the recorded data will enable a healthcare professional to derive the necessary information about the patient's health.

This subtle but arguably unhelpful distinction aside, ultimately, both 'genetic data' and 'data concerning health' are treated as sensitive data for the purposes of the GDPR (along with other categories such as data revealing racial or ethnic origin, which triggers stricter data protection rules than those applying to other categories of personal data), but the legal separation between the two categories fuels concerns of persistent genetic exceptionalism in regulation. And indeed, the inclusion of genetic data as a stand-alone regulatory category reflects, I would argue, a belief held by some publics (not to mention governments and regulators) that genetic data have greater interpretive potential than other types of data (genetic data are characteristically viewed as more 'sensitive' than other categories of data because they are both inherently identifying and a source of familial information). In turn, this generates a belief that implications from failure to respect its collection and use are much greater than other types of data.

More crucially, the legal fictions arising from these nebulous categories of data reflect the tendency of regulatory instruments to categorise and bound objects of regulatory concern by bespoke legal definitions and their own sets of regulatory rules. A key resultant problem is the separation of the human data subject from her siloed and regulated – but still deeply valued – data parts. Laurie observes that: 'In its attempt to create "bounded objects" as the appropriate focus for regulatory attention, the law seems to overlook the experience for the data [...] *subject*. An alternative regulatory perspective would be to recognise and acknowledge the enduring connection between subject-object, and the potential identity-significant implications of this' (Laurie 2017: 65). Law has an inclination to categorise objects; the socio-political context of genomics that has given rise to recent categorisation and boundary creation compounds matters (Manson and O'Neill 2007). The creation of new categories of 'genetic data' and 'data concerning health' in the GDPR sees the triumph of arguments from proponents who have long-criticised the predecessor legislation for under-protection in data privacy regimes. However, what potentially results from the multiplication of regulated bounded objects is not so much greater protection for the data subject, but rather a fragmentation of the relationship between the data subject and the object vis-à-vis their personal data. In turn, this comes with an increased risk of regulatory failure to protect against the very harms that the reforms were designed to address. Contrariwise, if the proliferation of new bounded categories adds to confusion about what is regulated, this could drive a culture of caution for data controllers, thereby potentially thwarting the core objective of data protection law to promote data flow.

'Consent or anonymise'

The paradigm of 'consent or anonymise' means that, more often than not, researchers and genomic data controllers operate on a false assumption, perpetuated in law, that personal data (including genomic data) can be collected and used by researchers and others who have *either* obtained the consent of the data subject *or* plan to anonymise the data. To be clear, the benefit to data users is that many fewer legal restrictions apply to data that have been anonymised, while consent ostensibly allows them to engage in processing of personal data, potentially for an indefinite period in a research setting. Both paths are seen by many data controllers and regulators alike as providing a necessary and sufficient condition for the legal and ethical use of genomic data.

There are two key problems with this approach, however. First, there are severe limits to genomic data anonymisation; some would argue that genomic data simply cannot be adequately anonymised (Dove and Phillips 2015). Although a dataset may be properly anonymised according to conventional approaches (i.e. through randomisation or generalisation), its cross-linking with data available elsewhere (e.g. from another dataset) can make it possible to infer data subjects' identities. Large datasets, particularly those including extensive genomic data, cannot be completely safe from inferential exploitation and ultimately data subject re-identification (EAGDA 2013). For example, a research team re-identified five people selected at random from a DNA database – without using a reference sample. Using their DNA, ages, and the US states where they lived, the team identified the five individuals, as well as some of their relatives, identifying in total nearly 50 people (Gymrek et al. 2013). Another research team has found that relying on as few as 30 to 80 individual (statistically independent) single nucleotide polymorphism (SNP) positions can enable unique individual identifiability (Lin et al. 2004); and another team found that an individual's SNP profile could potentially be identifiable even when it is aggregated with 1,000 or more other samples (Homer et al. 2008).

For the sake of argument, even if anonymisation were possible for genomic data, there are two other significant problems. First, the methods and degree of anonymisation required to warrant fewer legal restrictions are almost always unspecified (or when specific, are inconsistent), causing legal uncertainty when it comes to working with genomic data (Knoppers and Saginur 2005; Phillips and Knoppers 2016). Second, anonymisation of genomic data is particularly unhelpful when researchers or clinicians want to link the data to other data sources over time; to re-contact donors to enhance research aims; to obtain additional information or invite participation in other research projects; to communicate a clinically actionable finding; to identify and correct errors or to amend genomic data when new information becomes available; or to allow donors to withdraw their genomic data from a study. Thus, while anonymisation may be championed as a means of achieving strong privacy protection, in the genomics context it offers very limited or no utility to (not to mention respect for) either researchers or patient-participants.

The other key problem with this paradigm is the perpetuation of the view that privacy concerns can be assuaged through a properly drafted information sheet and consent form:

> consent, whether specific or broad, is typically an all-or-nothing affair; it is hardly a locus of privacy control – and wrongly or rightly, many associate privacy with control. Potential

[research] participants tend to have two main choices when consenting: they can participate or not – and if they do, they can later withdraw (albeit to a varying extent). The space for negotiation over the terms of data access and use is virtually non-existent, an important issue for those who view privacy as a necessary dimension of autonomy. [And], it bears emphasizing that consent does not fully address privacy concerns, as obtaining participants' consent to share their data does not absolve data stewards (or custodians) and data users from their legal obligations to use data fairly and lawfully.

(Dove 2015: 680)

In sum, we should be wary of perpetuating a worldview that absolute anonymity is ever achievable, let alone desirable: no data exist in splendid isolation from their social context. When it comes to genomic data, data privacy regimes can do better than to encourage reliance on a false and ethically fraught binary of consent or anonymisation.

Emerging tools and initiatives

So far, I have discussed how data privacy regimes regulate the collection and protection of genomic data; I have also suggested that these regimes can generate or perpetuate problematic paradigms. Despite this, or perhaps because of this, in recent years, tools and initiatives have been developed that afford robust governance and data access authorisation mechanisms, while holding data users accountable for their actions. They also work both to promote and protect the public interest in privacy and to advance biomedical research and medicine by engaging publics and encouraging stakeholders to regulate data use rather than technology or types of data.

Examples of these initiatives include the conceptual development of 'processual regulation', a novel framework that requires a temporal-spatial examination of regulatory spaces and practices as these are experienced by all actors, including the relationship of actors with the objects of regulation (Taylor-Alexander et al. 2016), and principles-based approaches to the governance of health-related research using personal data (Laurie and Sethi 2013; Sethi and Laurie 2013), which moves beyond the 'consent or anonymise' paradigm. Another example is a data access approach called 'registered access', whereby genomic database applicants are approved for access by providing details of their identity for authentication and agreeing to terms and conditions of data use during the registration process. Though it does not intend to replace open or controlled access models, the registered access model is seen as providing a more efficient process for both researchers and data custodians; it is also easier to automate because the approval process is less demanding (e.g. it does not rely on the review of descriptions of research proposals) (Dyke et al. 2016a). Further, a 'data sharing privacy test' has been proposed, which distinguishes degrees of sensitivity within categories of data recognised as 'sensitive' under the law, and proposes guidance for determining whether to apply a heightened level of protection when sharing genomic and health-related data for international data sharing initiatives, namely based on: (1) the data's sensitivity (noting variation of definition and of protection within predefined categories of sensitive data in data privacy regulation); (2) the potential resulting harm from possible re-identification of the data; and (3) individuals' expectations with respect to the data being shared (Dyke et al. 2016b). Both the 'registered access' model and 'data sharing privacy test' aim to achieve robust data subject protection and use of genomic data in ways that do not lead to further bounded object creation or reliance on consent or anonymisation.

Conclusion

As more genomic data are collected and used for myriad purposes, protecting privacy is becoming an increasingly complex but vital task. As a result of the uptake of genomic information in research and clinical practice as well as in individual hobby (or interest), new questions are emerging that impact on extant laws and regulations governing the collection and protection of genomic data. These questions include whether, and if so, when, an individual can appropriately insist or expect a clinician to have regard to information generated through direct-to-consumer genetic testing and (be able to) place that information directly into his or her own medical record. They also include whether a researcher has a legal duty or ethical responsibility to feedback information such as incidental findings into a care pathway and/or perhaps to design a research study with this possibility in mind so that certain privacy-protecting techniques are *not* adopted. These questions do not yield easy regulatory responses. But, if we wish to initiate an open discussion about how best to regulate the collection and protection of genomic data, I would posit that expectations and duties may be best constructed not necessarily top-down through rules-driven law, but rather through the prism of flexible governance structures that reflect a publicly negotiated social compact on the responsible use of genomic data that serve the common good.

Acknowledgements

The author wishes to thank Graeme Laurie and Mark Taylor for their invaluable comments.

References

Bygrave, L.A. (2014) *Data Privacy Law: An International Perspective.* Oxford: Oxford University Press.

Dove, E.S. (2015) 'Biobanks, data sharing, and the drive for a global privacy governance framework' *Journal of Law, Medicine & Ethics* 43: 675–689.

Dove, E.S. and Phillips, M. (2015) 'Privacy law, data sharing policies, and medical data: a comparative perspective' in Gkoulalas-Divanis, A. and Loukides, L. (eds), *Medical Data Privacy Handbook.* Cham: Springer: 639–678.

Dove, E.S., Laurie, G. and Knoppers, B.M. (2017) 'Data sharing and privacy' in Ginsburg, G.S. and Willard, H.F. (eds) *Genomic and Personalized Medicine, 3rd Edition.* Waltham, MA: Elsevier: 143–160.

Dyke, S.O.M., Kirby, E., Shabani, M., Thorogood, A., Kato, K. and Knoppers, B.M. (2016a) 'Registered access: a "Triple-A" approach' *European Journal of Human Genetics* 24: 1676–1680.

Dyke, S.O.M., Dove, E.S. and Knoppers, B.M. (2016b) 'Sharing health-related data: a privacy test?' *npj Genomic Medicine* 1, article number: 16024: 1–6.

Expert Advisory Group on Data Access (EAGDA) (2013) 'Statement for EAGDA funders on re-identification' Available at: https://wellcome.ac.uk/sites/default/files/eagda-statement-for-funders-on-re-identification-oct13.pdf

Gymrek, M., McGuire, A.L., Golan, D., Halperin, E. and Erlich, Y. (2013) 'Identifying personal genomes by surname inference' *Science* 339: 321–324.

Hallinan, D. and de Hert, P. (2017) 'Genetic classes and genetic categories: protecting genetic groups through data protection law' in Taylor, L., Floridi, L. and van der Sloot, B. (eds) *Group Privacy: New Challenges of Data Technologies.* Cham: Springer: 175–196.

Homer, N., Szelinger, S., Redman, M., Duggan, D., Tembe, W., Muehling, J., *et al.* (2008) 'Resolving individuals contributing trace amounts of DNA to highly complex mixtures using high-density SNP genotyping microarrays' *PLoS Genetics* 4: e1000167.

Kitchin, R. (2014) *The Data Revolution: Big Data, Open Data, Data Infrastructures and Their Consequences.* London: SAGE Publications.

Knoppers, B.M. and Saginur, M. (2005) 'The Babel of genetic data terminology' *Nature Biotechnology* 23: 925–927.

Laurie, G. (2002) *Genetic Privacy: A Challenge to Medico-Legal Norms.* Cambridge: Cambridge University Press.

Laurie, G. and Sethi, N. (2013) 'Towards principles-based approaches to governance of health-related research using personal data' *European Journal of Risk Regulation* 4: 43–57.

Laurie, G., Stevens, L., Jones, K.H. and Dobbs, C. (2014) *A Review of Evidence Relating to Harm Resulting from Uses of Health and Biomedical Data*. London: Nuffield Council on Bioethics.

Laurie, G. (2017) 'Liminality and the limits of law in health research regulation: what are we missing in the spaces in-between?' *Medical Law Review* 25: 47–72.

Lin, Z., Owen, A.B. and Altman, R.B. (2004) 'Genomic research and human subjects privacy' *Science* 305: 183.

Manson, N.C. and O'Neill, O. (2007) *Rethinking Informed Consent in Bioethics*. Cambridge: Cambridge University Press.

Munns, C. and Basu, S. (2015) *Privacy and Healthcare Data: 'Choice of Control' to 'Choice' and 'Control'*. Farnham: Ashgate Publishing.

Phillips, M. and Knoppers, B.M. (2016) 'The discombobulation of de-identification' *Nature Biotechnology* 34: 1102–1103.

Sethi, N. and Laurie, G. (2013) 'Delivering proportionate governance in the era of eHealth: Making linkage and privacy work together' *Medical Law International* 13: 168–204.

Taylor-Alexander, S., Dove, E.S., Fletcher, I., Ganguli Mitra, A., McMillan, C. and Laurie, G. (2016) 'Beyond regulatory compression: confronting the liminal spaces of health research regulation' *Law, Innovation and Technology* 8: 149–176.

Taylor, M. (2012) *Genetic Data and the Law: A Critical Perspective on Privacy Protection*. Cambridge: Cambridge University Press.

Tzanou, M. (2013) 'Data protection as a fundamental right next to privacy? "Reconstructing" a not so new right' *International Data Privacy Law* 3: 88–99.

Weber, R.H. (2013) 'Transborder data transfers: concepts, regulatory approaches and new legislative initiatives' *International Data Privacy Law* 3: 117–130.

Widdows, H. (2013) *The Connected Self: The Ethics and Governance of the Genetic Individual*. Cambridge: Cambridge University Press.

21

In CRISPR's world: genome editing and the politics of global science

J. Benjamin Hurlbut

The advent of technologies of genome editing have opened a new chapter in the history of biotechnology, offering means to achieve precise, targeted, reliable, genetic modifications in virtually any organism, from microbe to mammal. The advent of CRISPR/Cas9 has made genome editing accessible to a wide range of researchers engaged in an enormous range of projects. The technique is "cheap, quick and easy to use" radically expanding access to powerful tools to modify genomes, organisms, and ecosystems (Ledford 2015a, 2015b; NASEM 2017a). It has sparked imaginations of a mushrooming menagerie of bioengineered creatures as humanity turns its editorial eye to life at all levels of the evolutionary ladder.

Visions of the technological potential and social implications of CRISPR offer a unique window into the contemporary politics of biotechnology, not least because certain projects in genome editing research touch on fundamental commitments to the protection of human life. This chapter explores the politics of emerging debates about ethical and governance challenges associated with genome editing, focusing in particular on applications of CRISPR to human germline genetic engineering (HGGE). As researchers have begun to develop techniques for editing human gametes and embryos, a significant international debate has taken shape about whether and under what circumstances these techniques should be used to make heritable, genetic alterations in human beings. At stake are imaginations of the right allocations of authority and responsibility between scientific and democratic institutions, and the processes through which those imaginations are made to shape actual practices of governance.

I offer an analysis of the international debate over human germline genome editing. I attend in particular to notions of what warrants public ethical deliberation, and in what terms and frames. I argue that the debate is informed by a longstanding imaginary in U.S. politics of biotechnology that positions science and technology as generative of sociotechnical futures, and therefore as most able to know what futures are possible and what warrants worry. This imaginary is not merely one of technological emergence, but of *governable emergence:* political institutions' capacities to govern technological futures are defined against science's putative capacity to generate, predict, and manage those futures. In this sense, an imaginary of technoscientific change is simultaneously an imaginary of the necessary and right allocation of responsibilities for governance.

The imaginary of governable emergence shapes patterns of normative deliberation, for instance when scientific experts seek to silence public ethical concerns on epistemic grounds by characterizing them as grounded in unrealistic technological scenarios – "science fiction" uninformed by scientific fact. The debate around human genome editing offers an example of a reverse move: one in which a longstanding ethical limit is declared untenable and unreasonable precisely because it has become technologically feasible to transgress it.

Early debates over human germline genetic engineering

Serious public discussions of the social and ethical ramifications of human genetic engineering date back more than half a century (Evans 2002). Early debate unfolded with advancing knowledge about genetics, evolution, and molecular biology, and alongside new technologies for shaping human reproduction like oral contraception and in vitro fertilization. In contrast to contemporary debate, these discussions tended not to focus on specific technologies but on a general expectation that the future would bring a range of biological capacities for controlling human reproduction and that it was therefore necessary to reflect upon the ends to which such capacity could be – and ought to be – used.

Some of the most prominent biologists of the day were leading figures in these discussions. They tended to focus not merely on mitigating genetic disease in individuals, but on the potential for enhancement, directed evolution, and technologically driven eugenics. For instance, Robert Sinsheimer, a molecular biologist whose research was foundational for the development of biotechnology, celebrated "defined genetic improvement of man" as a means "to carry on and consciously perfect" the human species (Sinsheimer 1969). In a paper for an influential 1962 conference on *Man and His Future,* Julian Huxley envisioned powers that raise – and must be guided by – the question: "what are people for?" (Wolstenholme 1963).

By the early 1980s, these expansive debates about the nature, purpose, and future of human beings had given way to more circumscribed concerns about risk and informed consent. At that time, somatic gene therapy experiments and the first successful germline alterations of mice elicited widespread concerns about HGGE. Following the call of three major religious organizations for greater attention to the "new era of fundamental danger triggered by the rapid growth of genetic engineering" and to associated questions of "the fundamental nature of human life and the dignity and worth of the individual human being," a U.S. federal bioethics body conducted a study of human genetic engineering (United States 1982, 95). This study, *Splicing Life,* marked a shift from the relatively wide scope of earlier ethical debate to a more narrow focus within mainstream bioethics on risks associated with limited medical interventions like correcting disease-causing genes (Evans 2002). Following *Splicing Life,* the U.S. National Institutes of Health authorized its Recombinant DNA Advisory Committee (RAC) to review all proposed human gene therapy experiments. From 1983, the RAC declared that it "will not at present entertain proposals for germ line alterations," a policy that remains in force more than thirty years later (National Institutes of Health 1985). From this point forward, mainstream bioethics scholars generally treated eugenic ambitions as speculative and beyond the scope of immediate bioethical concern, focusing on risk and individual autonomy even where they envisioned quite radical forms of genetic alteration (e.g. Buchanan et al. 2001).

The RAC's approach to gene therapy was in keeping with its general focus on technical risks in governance of recombinant DNA (rDNA) research. The RAC was institutionalized for the purpose of implementing the rDNA guidelines that were developed at the 1975 Asilomar conference. Following the advent in the early 1970s of new techniques for modifying DNA, prominent researchers called for a voluntary moratorium on recombinant DNA experiments out of concern that they might unwittingly create dangerous new pathogens. These scientists convened a meeting at the Asilomar conference center in northern California to establish guidelines for research. These guidelines, they hoped, would simultaneously limit biohazard risks and forestall federal regulation. The meeting, which is remembered as a crucial act of scientific responsibility that opened the way to the benefits of biotechnology, focused narrowly on issues of biosafety risk, setting aside broader social, ethical and security dimensions. The guidelines that were initially developed at Asilomar became the basis for governance of rDNA research in the United States, including human gene therapy (Hurlbut 2015).

As somatic gene therapy experiments emerged in the early 1990s, the RAC called for renewed attention to germline engineering but preserved a relatively narrow focus on issues of feasibility and risk. It continued to see germline therapy as both unnecessary and unrealistic, reinforcing the existing RAC policy. At the same time, some prominent figures argued that the widely accepted ethical bright line between somatic and germline gene therapy should be seen as reflecting a technical judgment about the state of technology, not a moral proscription (e.g. Cook-Deegan 1994). This approach discouraged engagement with the broad ethical questions of pre-1980s debate even as it tacitly treated the prohibition against HGGE as provisional and subject to revision as technology advances.

In Europe, by contrast, the same scientific developments in the early 1980s catalyzed attention to the implications of human genetic engineering for human rights and human dignity. The Council of Europe initiated a process for elaborating bioethical issues relevant to the European Convention on Human Rights of 1950, leading ultimately to the European Convention on Human Rights and Biomedicine of 1997, known as the Oviedo Convention. Article 13 of the Oviedo Convention codified in law a categorical prohibition in the name of human dignity against introducing genetic changes in the human germline. The Oviedo Convention has been ratified by 29 countries (Council of Europe 2017a).

Revolutionary ambitions

The advent of CRISPR reawakened the dormant debate about HGGE. The novelty of CRISPR elicited a sense of urgency about longstanding ethical and governance questions in areas ranging from human genetic engineering to agricultural biotechnology to engineering species and ecosystems. These discussions are patterned by a narrative of technological revolution in which the pace of technological change races ahead of society's ability to adapt and react. When *Science Magazine* declared CRISPR the 2015 breakthrough of the year, it was because the "molecular marvel" had broken through social as well as biological barriers. Its most "spectacular achievements" included "the first deliberate editing of the DNA of human embryos," which "roiled the science policy world" (Travis 2015). A *Nature* profile of the technology applied the epithet of "disruptor," noting that the rapid expansion of CRISPR-driven science "leaves little time for addressing the ethical and safety concerns" of such experiments (Ledford 2015a).

Of course, visions of a world transformed by biotechnology emerged long before "CRISPR the disruptor." Notions of the biomedical, industrial, and economic potential of biotechnology are increasingly expansive (e.g. Sharp et al. 2011), informing scientific research priorities, policy discourses, and imaginaries of social progress (NRC 2009). Such vanguard visions (Hilgartner 2015), and the forms of policy action that they promote, simultaneously reflect and re-inscribe imaginaries of relationships between technoscience and political order, including the capacities of institutions of governance to contend with emerging science and technology.

Given its perceived revolutionary potential, CRISPR was quickly assimilated into such visions. Vague narratives of biotechnological potential suddenly began to feel more concrete and immediate. *Wired* reported that "the power to quickly and easily alter DNA,…could eliminate disease… solve world hunger…[and] provide unlimited clean energy" (Maxman 2015). Only slightly less hyperbolic possibilities were predicted in elite scientific circles. As a result, CRISPR rapidly came to be seen as driving a technological revolution with far reaching political and moral consequences. "For better or for worse," *Science Magazine* declared, "we all now live in CRISPR's world" (Travis 2015).

Governing CRISPR's world

Much of the deliberation around genome editing takes for granted that "CRISPR's world" is inevitable, and that society must recalibrate its norms and rules in response. This applies not only to expectations about less controversial transformations of agriculture and industry, but of human reproduction. In the words of one pioneering figure in genome editing, human "germline editing is going to happen, and to think otherwise is naïve" (quoted in Molteni 2017). The notion that particular technological eventualities are inevitable regardless of what society wants or permits are commonplace in U.S. discourse around biotechnology (cf. Braverman 2018). Similarly common are observations around how the new powers of genome editing reveal "gaps" in regulatory regimes and social norms and law as lagging behind science. In announcing the February 2017 release of a U.S. National Academies report on human genome editing, Academies president Marsha McNutt observed that "as is always the case, the speed at which the science is advancing outpaces society's ability to grasp its implications" (NASEM 2017b).

The narrative that society and its institutions of governance necessarily lag behind science – the law lag – is a well-worn trope in U.S. discourse around science and technology, notwithstanding the observation that legal and technoscientific orders are coproduced (Jasanoff 2007). Yet, regardless of whether it holds water as a social theory, the lag narrative shapes notions of how emerging science and technology should be governed, and affects how responsibilities are imagined and allocated between scientific and democratic authorities (Hurlbut 2017). The implications of this narrative for approaches to deliberation and governance of biotechnologies are particularly significant in the case of germline genetic engineering.

The most contentious focus of ethical debate about human genome editing has been over whether it is, in principle, permissible to make heritable modifications to the human genome. The leading voices in these debates have drawn particular attention to the need to address ethical and governance challenges in an international context. The Council of Europe has affirmed the relevance of the Oviedo Convention's prohibition of HGGE and suggested that the Convention's principles should inform international deliberation (Council of Europe 2017b). The U.S. National Academies have also offered principles to guide international dialogue. On the basis of these principles, the Academies have determined that human germline genome editing (HGGE) is permissible under certain circumstances. They have furthermore determined that existing oversight and regulatory mechanisms in place in the United States are adequate to (a) oversee "basic research" on human genome editing and (b) regulate clinical applications, although they call for augmenting existing mechanisms with efforts to invite "public engagement" (NASEM 2017a).

Behind this explicit disagreement over the appropriateness of HGGE and the principles that should guide deliberation are a number of tacit differences in approach that have significant implications for governance, particularly with respect to the ways governance questions are framed and the forms of authority – scientific or political – that are privileged in addressing them. Below, I highlight three linked dynamics: (1) the ways that risks are framed such that public ethical deliberation about them is deemed to be warranted (or unwarranted) and relevant (or irrelevant) to governance; (2) the ways a priori demarcations are drawn between "basic research" and technological applications; and (3) notions of whether public norms, including where codified in law, precede and inform (and should inform) scientific and technological trajectories, or whether they are inevitably laggard and reactive to developments in science and technology such that significant advances in technical capacity require complete reappraisal of those norms. In the remainder of this chapter, I explore these dynamics by examining the current HGGE debate.

Engineering human futures

Although over the past decade there has been much work using genome editing techniques in embryonic stem cells with implications for HGGE, there was virtually no public debate until researchers in China attempted to edit human embryos directly. In early 2015, a group of prominent scientists and bioethicists responded to the Chinese research by calling for a moratorium on clinical applications of human genome editing and a deliberative process modeled on Asilomar (Baltimore et al. 2015). This led to the International Summit on Human Gene Editing, which was convened in December, 2015 in Washington, D.C. by the U.S. National Academies, the U.K. Royal Society, and the Chinese Academy of Sciences. The Summit organizers characterized the situation as urgent: now that the technology is upon us, the time has come for society to react.

Nobel Laureate and former Caltech president David Baltimore opened the Summit by explaining this sudden urgency. Although genetic engineering techniques have been around for decades, they were imperfect, and applying them to human beings was therefore "initially unthinkable." But with technological advances "the unthinkable has become conceivable." Thus, "now we must face the questions that arise: how, if at all, do we as a society want to use this capability?" (Baltimore 2015).

This framing rehearses a familiar sequence. As basic research progresses, new technologies emerge with social and ethical implications that society is forced to confront: from science to technology to society. This expresses the familiar linear model of innovation. Although a wide body of scholarship has discredited the linear model, one of its most significant impacts has been in shaping approaches to governance. In this sense, the linear model of innovation is a model of scientific responsibility and governance: scientists' choices determine technological trajectories and corollary social impacts. It reflects an imaginary not merely of technological emergence, but of *governable* technological emergence. In this imaginary, scientific authorities enjoy a unique and powerful role, both epistemic and ethical, in shaping the future: they, and only they, are in a position to know what is possible, what risks are realistic, and thus what technological futures warrant ethical evaluation. Society and its political institutions, by contrast, cannot anticipate or shape, but only react to technological developments once science has produced them – even as society is made solely responsible for addressing the "social and ethical consequences" of scientific and technological change. Put simply, the linear model's stages – from basic science, to applied technology, to society – shape allocations of authority and responsibility: between science that declares what futures will and will not come to pass and thus do or do not warrant worry, and society which is invited to contend with social, political and ethical dimensions of technology only after a technological future has been marked as plausible or inevitable by scientific authorities. Thus, the linear model's account of technological change re-inscribes demarcations between, technical and ethical, knowledge and norms, science and society, shaping notions of who can (and, therefore, should) govern the technological future. These demarcations are reified through actual practices of delegation and deference in governance.

This is evident in the slippage between Baltimore's language of "we as a society" and what he said next: "*We* are taking on a heavy responsibility *for* our society, because we understand that we could be on the cusp of a new era of human history…The actions that we take now will guide us into the future… we are initiating a process of taking responsibility for technology with far reaching implications" (Baltimore 2015).

In claiming the role of "taking responsibility" *for* society, Baltimore alluded to the 1975 Asilomar meeting on recombinant DNA – a foundational moment in the history of molecular biology and a touchstone for the idea of scientific responsibility: "In 1975 as today we believed it was prudent to consider the implications of a remarkable achievement in science. And then as now we recognized we had a responsibility to include a broad community in our discussions" (ibid.). Academies President Marsha McNutt has likewise invoked Asilomar as initiating the Academies' "long track record of providing leadership" thereby "ensur[ing] that breakthroughs continue and…that scientific and medical breakthroughs like genome editing benefit all of mankind" (NASEM 2017b).

Yet, the claims about "include[ing] a broad community" notwithstanding, one of the most consequential dimensions of Asilomar was what was *not* discussed and who was *not* included. Baltimore also opened the 1975 meeting, declaring social, ethical, and security implications out of bounds. How rDNA might be used beyond the academic laboratory, and what those uses might mean for wider society, was off the table. Instead, discussion focused narrowly on questions of technical risk assessment. That framing was used to justify not only the recommendations that came out of the meeting, but the lack of public participation in formulating them.

Yet Asilomar is seen as exemplary of responsible science – a science that takes responsibility for society by evaluating what is technologically possible, containing risks, and rendering judgment about whether and when a technology exceeds capacities of scientific self-regulation and warrants public attention (Hurlbut 2015).

Participatory technocracy

The power of that approach is particularly evident in the discussions of human genome editing. It has been recognized for half a century that genetically engineering future children touches upon fundamental questions of human rights, dignity and integrity that cannot be reduced to questions of technical risk. Partly in recognition of the hard-to-resolve ethical questions associated with HGGE, there has been a longstanding, internationally shared assumption in mainstream scientific and bioethical thinking that HGGE is off limits. This is reflected in the RAC's policy of not entertaining proposals for germline focused experiments, in the Oviedo Convention, in declarations of the UNESCO International Bioethics Committee, and in numerous other authoritative declarations. In U.S. discourse, the prohibition has sometimes been quietly qualified as provisional – it is ethically unacceptable at present. This provisionality leaves open the question of whether it could ever be acceptable in principle, while the cleaving to the linear model simultaneously discourages deliberation about what positive criteria would have to be met in order to reverse the prohibition: so long as the technology "isn't there yet," there is no warrant for raising speculative ethical questions about it. It is too early in the linear sequence from science to technology to society. Yet with CRISPR, the technology appears to have suddenly arrived, leading certain prominent authorities to declare that the prohibition was never more than a reflection of expert judgment that HGGE was technically infeasible or too risky at the moment. Because this technical judgment no longer holds, they suggest, neither does the prohibition. Thus, in one swift and apparently incremental move, the U.S. National Academies abandoned the prohibition, allowing them, in the words of one headline, to "endorse designer babies" (Regalado 2017). It declared a longstanding ethical limit untenable and unreasonable precisely because it has become technologically feasible to transgress it. And the declaration that it is no longer tenable is grounded in the same authority as the prior declaration that it was, in Baltimore's word, unthinkable – and therefore not worth deliberating about.

Behind the National Academies' judgment is a post-hoc interpretation of the longstanding prohibition as an expression of scientific responsibility in the Asilomar tradition: the prohibition was grounded in a judgment about technical risk. Because the technical means for low-risk HGGE did not yet exist, it would have been irresponsible to attempt it. As such, the prohibition is, on this account, entirely a function of the state of the technology: it ought not be understood as an a priori ethical judgment because there was not yet a technology for which the ethical implications could be assessed. The corollary is the notion that it is unnecessary (and, thus, premature) to ask broader ethical questions until the scientific community has judged that what was "unthinkable has become conceivable." Thus a norm that may have included technical considerations about safety but was by no means limited to these considerations, is, in the present controversy, being reframed as reflecting no consensus beyond technical judgments about safety.

This is illustrated by a statement of the NASEM committee co-chair Richard Hynes. "There has been a line drawn by many that says…you should refrain. That was mostly because there was no way of considering how to do that at all….so nobody was arguing that it should be done" (NASEM 2017b). Thus a complete reversal is figured as nothing more than an incremental technological step forward. Statements like Hynes' invoke a presumption that takes the form of a tacit norm: scenarios that are not deemed by scientific experts to be plausible and immediate ought not be objects of ethical concern or deliberation; scenarios that are plausible and immediate are no longer in the future and thus cannot be proscribed on ethical grounds. The corollary is that any ethical proscriptions that are made before a technological scenario is deemed plausible are subject to revision if and when that scenario becomes plausible. They must be revised through reaction to the specific technology.

This imposes a hierarchy of justifications for ethical limits, and a corollary hierarchy of authority for evaluating those justifications. Technical evaluations of risk are treated as more definitive and more compelling than other sorts of normative judgments, thereby trumping the need for (and inhibiting the articulation of) reasons other than risk for setting limits. A statement on HGGE produced by the Hinxton Group, a self-organized group of scientific experts, illustrates this.

The Hinxton statement calls for a moratorium on HGGE by privileging expert assessments of safety over other ethical frames: the technology "is not sufficiently developed to consider human genome editing for clinical reproductive purposes at this time" (The Hinxton Group 2015). Moral judgments that might emerge within or be expressed through democratic debate are treated as subsidiary to scientifically authorized judgments about risk. "A distinction should be made between objections that are based on technical or safety concerns and objections that reflect additional moral considerations." The statement continues: "any constraint of scientific inquiry should be derived from reasonable concerns about *demonstrable* risks of harm… . Policymakers should refrain from constraining scientific inquiry unless there is substantial justification for doing so that reaches beyond *disagreements based solely on divergent moral convictions*" (ibid. emphasis added). Thus the statement gives technical judgments about safety priority over democratic ambivalence about appropriateness. Disagreement and moral uncertainty are characterized as an inadequate warrant for democratically defined limits unless they are universal and univocal.

One important consequence of privileging technical modes of evaluation is that this creates an imperative for "basic" research to evaluate and address questions of safety – something that both the Hinxton group and the NASEM report strongly advocate. In these reports, and in much of the discourse emanating from scientific circles more broadly, the notion that the human germline is inviolable has been eclipsed by the inviolability of basic research. Even in cautious Germany, the German National Academy of Sciences Leopoldina has called for largely unrestricted basic research (Leopoldina 2017). Yet in the discourse that dominates debate in the United States, the justification for the inviolability of basic research involves more than merely a doubling down on scientific autonomy. It is also grounded in the notion that there is warrant for public deliberation about ethical questions if and only if there is adequate technical knowledge to declare that safety risk is no longer what stands in the way of application. At the same time, those forms of basic research that produce technical advances that overcome risks to safety are treated as both necessary and intrinsically virtuous.

This stands in significant contrast to the position adopted by the European Group on Ethics and New Technologies (EGE), which advises the European Commission. The EGE has questioned "whether germline genome editing technology *research* should be suspended…given the profound potential consequences of this research for humanity" (EGE 2016, emphasis added). The Summit organizers, the NASEM report, the Hinxton group, Leopoldina and others have taken for granted the distinction between "basic research" and "clinical applications," and thus the notion that restrictions or a moratorium on the latter are perfectly compatible with essentially unrestricted "basic" research. The EGE, by contrast, questions whether "such a clear-cut distinction can be made between basic and translational research," noting that some members "also call for a moratorium on any basic research involving human germline gene modification until the regulatory framework is adjusted to the new possibilities" (ibid.).

Conclusion

In a September 2015 Council of Europe hearing, Jean-Yves Le Déaut, the General Rapporteur on Science and Technology Assessment, made a comment that captures the political uncertainty at the core of the international controversy over human genome editing. "How can we legislate on issues which are in constant flux? Laws…may become biodegradable…Because we're talking about a moving target… with each of these changes in techniques, are we going to keep changing the law?" ("Manufacturing a New Human Species" 2017).

I have argued that the approaches to governing genome editing that have dominated U.S. discourse and U.S.-based efforts to shape international governance reflect a longstanding presumption that with ever-advancing science and technology, law and the norms it codifies are "biodegradable." This presumption is embedded in well-worn approaches to governing so-called "emerging" science and technology, approaches that treat scientific and technological emergence as linear, autonomous and inevitable, and position democratic governance as necessarily reactive. As discussions of human genetic engineering of the mid twentieth century demonstrate, the ethical stakes were recognized long before any particular technical means for doing it had been made to emerge. Yet by positioning emerging science and technology as upstream of its social "implications" – a linear, sequential relation that is imposed even in the very use of the term "emerging" – the products of such moral deliberation are marked either as premature or as biodegradable. In the controversy over genome editing, these culturally situated approaches are colonizing international space. Both the Summit and the NASEM study were undertaken in the name of the global human community, and both called for international coordination and harmonization.

The conflicts around HGGE, for instance, between the NASEM and EGE positions, are not new. In important respects, they are parallel to prior debates around biotechnology, for instance around risk and precaution in GMO regulation (Jasanoff 2005). Yet with GMOs, moral dimensions beyond public health and safety were easily drawn to the fore. The opposite holds for human genetic engineering. That the present controversy is nevertheless slipping into such well-worn frames is indicative of the powerful structuring effect that culturally situated imaginaries of the necessary and right relationships between scientific and democratic institutions exert upon the politics of public moral sense-making.

Yet it is worth remembering that "we all now live in CRISPR's world" only insofar as we imagine ourselves as powerless against an inexorably advancing technological tide and act accordingly. If we refuse this imagination, the construct at the center of controversy shifts from CRISPR/cas9 to the "we" who is "taking on a heavy responsibility *for* our society," and thus to the politics of exclusion – of knowledges, publics and moral imaginations – through which that *we* is constituted.

References

Baltimore, David. 2015. "Human Gene Editing Summit: Context for the Summit." Interntational Summit on Human Gene Editing. Keck Building, National Academy of Sciences, Washington, D.C., December 1. https://vimeo.com/album/3703972/video/149179797. Accessed July 2, 2016.

Baltimore, B. D., Paul Berg, Michael Botchan, Dana Carroll, R. Alta Charo, George Church, Jacob E. Corn, et al. 2015. "A Prudent Path Forward for Genomic Engineering and Germline Gene Modification." *Science* 348(6230): 36–38.

Braverman, Irus. 2018. *Gene Editing, Law, and the Environment: Life Beyond the Human*. New York: Routledge.

Buchanan, Allen E., Daniel Wikler, Dan W. Brock, and Norman Daniels. 2001. *From Chance to Choice: Genetics and Justice*. Cambridge, U.K.: Cambridge University Press.

Cook-Deegan, Robert. 1994. "Germ-Line Gene Therapy: Keep the Window Open a Crack." *Politics and the Life Sciences* 13(2): 217–220.

Council of Europe. 2017. "Chart of Signatures and Ratifications of Treaty 164." *Treaty Office*. www.coe. int/web/conventions/full-list. Accessed June 12, 2017.

Council of Europe. 2017. "Statement on Genome Editing Technologies." *Bioethics*. www.coe.int/en/ web/bioethics/news/-/asset_publisher/DcE9CvEiHMnp/content/gene-editing. Accessed February 17, 2017.

European Group on Ethics in Science and New Technologies (EGE). "Statement on Genome editing." European Commission, Brussels, January 11, 2016. https://ec.europa.eu/research/ege/pdf/gene_edit ing_ege_statement.pdf#view=fit&pagemode=none.

Evans, John H. 2002. *Playing God?: Human Genetic Engineering and the Rationalization of Public Bioethical Debate*. Chicago, IL.: University Of Chicago Press.

Hilgartner, Stephen. 2015. "Vanguards, Visions and the Synthetic Biology Revolution." In *Science and Democracy: Making Knowledge and Making Power in the Biosciences and Beyond*, edited by Stephen Hilgartner, Clark Miller, and Rob Hagendijk. New York: Routledge.

Hurlbut, J. Benjamin. 2015. "Remembering the Future: Science, Law, and the Legacy of Asilomar." In *Dreamscapes of Modernity: Sociotechnical Imaginaries and the Fabrication of Power*, edited by Sheila Jasanoff and Sang-Hyun Kim, 126–151. Chicago: University Of Chicago Press.

Hurlbut, J. Benjamin. 2017. *Experiments in Democracy: Human Embryo Research and the Politics of Bioethics*. New York: Columbia University Press.

Jasanoff, Sheila. 2005. *Designs on Nature: Science and Democracy in Europe and the United States*. Princeton, N.J.: Princeton University Press.

Jasanoff, Sheila. 2007. "Making Order: Law and Science in Action." In ed. Edward J. Hackett, Olga Amsterdamska, Michael Lynch, and Judy Wajcman, 3rd ed., 761–786. Cambridge: MIT Press.

Leopoldina, Nationale Akademie der Wissenschaften. "Ethische Und Rechtliche Beurteilung Des Genome Editing in Der Forschung an Humanen Zellen," March, 2017.

Ledford, Heidi. 2015a. "CRISPR, the Disruptor." *Nature* 522(7554): 20–24.

Ledford, Heidi. 2015b. "Biohackers Gear up for Genome Editing." *Nature* 524(7566): 398–399.

"Manufacturing a New Human Species." 2017. *Bioethics*. Accessed February 17. www.coe.int/en/web/bioethics/news/-/asset_publisher/DcE9CvEiHMnp/content/manufacturing-a-new-human-species.

Maxman, Amy. 2015. "The Genesis Engine." *WIRED*, August.

National Academies of Science, Engineering and Medicine (NASEM). 2017a. *Gene Drives on the Horizon: Advancing Science, Navigating Uncertainty, and Aligning Research with Public Values*. Washington, D.C.: The National Academies Press.

National Academies of Science, Engineering and Medicine (NASEM). *Public Report Release Event for Human Genome Editing: Science, Ethics, and Governance*. February 14, 2017b. Washington, D.C.

National Institutes of Health. "Recombinant DNA Research; Proposed Actions Under Guidelines." *Federal Register* 50, no. 160 (August 19, 1985): 33464.

National Research Council. 2009. A New Biology for the 21st Century. Washington, D.C.: The National Academies Press.

Regalado, Antonio. 2017. "U.S. Panel Endorses Babies Gene-Edited with CRISPR." *MIT Technology Review*. February 14.

Molteni, Megan. 2017. "China Used Crispr to Fight Cancer in a Real, Live Human." *WIRED*. November 18, 2016.

Sharp, P. A., C. L. Cooney, M. A. Kastner, J. Lees, R. Sasisekharan, M. B. Yaffe, S. N. Bhatia, et al. 2011. "The Third Revolution: The Convergence of the Life Sciences, Physical Sciences, and Engineering." Massachusetts Institute of Technology.

Sinsheimer, Robert L. 1969. "The Prospect For Designed Genetic Change." *American Scientist* 57(1): 134–142.

The Hinxton Group. 2015. "Statement on Genome Editing Technologies and Human Germline Genetic Modification."

Travis, John. 2015. "Making the Cut." *Science* 350(6267): 1456–1457.

United States, President's Commission for the Study of Ethical Problems in Medicine and Biomedical and Behavioral Research. 1982. *Splicing Life: A Report on the Social and Ethical Issues of Genetic Engineering with Human Beings*. Washington, D.C.: The Commission.

Wolstenholme, Gordon. 1963. *Man and His Future*. Boston: Little, Brown & Co.

Part 4
Diversity and justice

22

Introduction

Sahra Gibbon and Barbara Prainsack

Issues of diversity and justice featured in the 2009 edition of this *Handbook*, prominently marked by a concern with how questions of individual choice and autonomy delineated a space between eugenics and the 'new genetics' as a form of social control, even as the ability to access such 'choice' was differentiated (Rothman 2009). Such concerns have by no means dissipated in the years since the previous edition. They have diversified and become reframed as the scale of medical interventions – particularly related to reproductive genetic technologies – have increased, and the geographic scope and reach of other forms of genomic and related research have extended across a global terrain. As a result, long-standing attention in social science with eugenics, disability, enhancement and 'race' continue to generate critical engagement but in ways that are more attentive to the complexities and paradoxes of developments in genomic science and medicine. As the chapters in this section attest, this includes the ways that the still mostly elusive promise of personalised or precision medicine, which continues to drive the field of genomics, is dependent on the participation of wider groups of people defined often as themselves 'underserved' populations or communities (Prainsack 2017). It also means being attentive to the way that 'lifestyle' and other non-molecular data are becoming a valued resource – sometimes more than genetic information – and to discourses around genetic discrimination and insurance that are nonetheless reliant on a reductive understanding of genes. Of particular note is how a long-standing social science interest in how the autonomous individual as the distributed locus of responsibility in genomics is now conjoined by calls to attend equally to the role of the state in regulating research and access to genetic medicine, as well as the growing influence of corporate business in the management of biological and medical data. At the same time the increasing number of scholars and activists who call upon the principle that people have a right to 'enjoy the benefits of scientific progress and its applications', as enshrined in Article 15 of the International Covenant on Economic, Social and Cultural Rights (United Nations 1966) begs the question of what roles states should play in ensuring that people can enjoy these benefits. How far should the duty of states go in bringing the benefits of scientific progress to fruition, and where should state actors focus their attention to reach this goal? Should they actively support, or even mandate, open science, and the 'opening up' of databases? Should they invest more public funds into the improvement of social determinants of health alongside investment in genomic research? Should they focus on supporting marginalised and underserved populations and address open and implicit bias? Or should they gear up their role not only as a regulator of medicine, science and research, but also as a funder?

Challenges related to exclusion have increased in prominence and urgency at the interface of genomics, health and society. A continuing issue exists regarding the overrepresentation of people with European ancestry in genomic research. According to a study carried out in 2016, 80 per cent of the roughly 35 million samples used in genome-wide association studies – a type of study that can help to identify disease-causing genomic markers – came from people of European descent. Taken together, samples from people of African and Latin American descent, and from those belonging to native or indigenous peoples, made up less than 4 per cent of all samples analysed (Popejoy and Fullerton 2016). This means that research in this area uses samples from a small minority of the world's population, with relatively similar genetic ancestry. A number of methodological and normative challenges complicate this situation further. When we call for 'adequate' representation of all people in genomic research, we are making two tacit assumptions. The first is that we assume that there are differences between people that are clustered and sufficiently measurable for us to speak of groups. The second is that these differences are relevant to understand health and disease. Both assumptions are problematic in connection with many group labels in use today. They are particularly problematic with regard to race, as the treatment of races as groups relevant in genomic research can be understood to support the idea that race is a genetic category (Duster 2015, Kahn 2014, M'Charek 2013). Part of an answer to this problem lies in differentiating between DNA and biology here: although race is not genetic, it has a biological dimension. People who suffer racial discrimination, for example, who live in deprived neighbourhoods, and who are exposed to high levels of stress and financial worries are more likely to be of poor health than others (Marmot 2015). In this sense, race is a bio-social category. If people who are particularly strongly exposed to discrimination, deprivation, or violence are missing from genetic and genomic research, their clinical and environmental data is absent as well. Yet as earlier work has demonstrated (Reardon 2005) wider efforts at inclusion of diverse populations and communities predicated on 'humanitarian' visions of justice and equity in pursuit of 'genomics for the world' (Bustamente et al. 2011) are not straightforward either. Instead they actively inform the way that genomics is situated as a globalised resource and/or benefit.

One of the related challenges regarding inclusion and exclusion at the interface of genomics and society emerges from the increasing overlaps between research and care (ref. Cambrosio, this volume). Concepts such as Precision Medicine, or 'Learning Healthcare' with data and information infrastructures built for both clinical and research use call for the integration of disease taxonomies and data use for clinical and research purposes (NRC 2011; Faden et al. 2013; ref. Prainsack, this volume). Moreover, the ever-growing emphasis on big data epistemologies makes the mining of clinical data to obtain insights for patient care a tempting and promising opportunity (see Leonelli and Tempini, this volume). In countries without universal healthcare, and in contexts where there are formal or practical (such as open or implicit bias; e.g. Matthew 2015) access barriers to healthcare, this means that the population biases in genomics research are increased by the biases in clinical datasets: the underserved are not represented in clinical records and thus become, literally, 'missing bodies' (Casper & Moore 2009). At the same time for those who are 'included' the slippage between research and care in developing and emerging country contexts brings its own ethical complexities (Traore et al. 2015). These are shaped by histories of colonialism, public health (or the lack of it) and the use of genetic data as an 'extracted resource' that can be commodified (see Fullwiley & Gibbon, this volume).

To be clear, exclusion is not always problematic. Excluding our neighbours from access to our medicine cabinet is not problematic if they themselves have access to the medicine that they need. Exclusion is problematic when it is unjust, and it is dangerous when it deprives people of what they need to satisfy their fundamental needs and interests. While funders and policy makers can require that research projects and healthcare systems pay special considerations to the needs of underserved populations, it is very important that researchers, clinicians and all other actors in healthcare critically reflect on the dynamics and consequences of exclusion.

In recent years, social scientists and bioethicsts alike – although sometimes with different conceptual tools – have argued for greater attention to relationality and solidarity in genomic medicine and in healthcare more broadly. While the notion of personhood as relationally constituted has a long-standing history of engagement in anthropology (cf Strathern 1998) concepts such as biosociality and biological citizenship have been central in drawing critical attention to how collective identification and identities are unevenly shaped, partially emergent, and are sometimes resistant to or themselves fracture at the interface with genomic knowledge and technology (Gibbon and Novas 2008, Gibbon et al. 2010, Rose and Novas 2005, Lock 2008). At the same time, social science attention has turned to examine how questions of rights and social justice are imbricated in the shifting collectivities that emerge in relation to wider fields of science and medicine, particularly in low- and middle-income countries. As recent research illuminates (Do Valle and Gibbon 2015) concepts such as 'biolegitimacy' (Fassin 2009) or 'therapeautic citizenship' (Nguyen 2010) shed light on how science, medicine, ethics, humanitarianism and structural inequalities are often collectively imbricated and mutually reshaped. In bioethics specifically, Bruce Jennings recently diagnosed a 'relational turn' that turns away from an atomistic understanding of individual autonomy and instead conceptualises people as relational and connected beings (Jennings 2016; Mackenzie & Stoljar 2000; Baylis et al. 2008). Recent work on solidarity does not merely allude to solidarity as something that is worth achieving but seeks to use solidarity as a generative principle that guides the design of institutions and policies (Prainsack & Buyx 2017; Illingworth and Parmet 2015). Moreover, a solidarity-based perspective calls for a critical reflection of the processes of 'othering' that lead to the creation of in-groups and out-groups in the first place (Dean 1996).

The contributions to this section drill deeper into some of these issues covered in this short overview. Jackie Leach Scully's contribution discusses the ways in which mechanisms of othering in the domain of 'disability' have become more dispersed and subtle, yet no less powerful, with the turn from genetics to genomics – understood as a shift from the study of individual genes and their function to an integrated view on the activity and regulation of genes, and the role of environmental factors. The focus of ethical and political discussions has moved from discussions about what kind of embryos or fetuses could or should be aborted to selecting the healthiest embryos pre-implantation. These decisions continue to operate on the basis of often unexamined cultural norms about health and functionality. It remains to be seen whether the use of genome editing will continue, as Scully puts it, to 'decrease tolerance for strange or unusual bodies, increased rigidity of aesthetic norms, [and to increase] the growth of social stratification' (Scully, this volume). This notion of enhancement in contemporary genomics is the focus of Giulia Cavaliere and Silvia Camporesi's chapter. Like Scully, rather than portraying genome editing as something radically new in terms of their ethical consequences, these authors see it as part of a history of human enhancement where individual 'choices' and practices always reflect and in turn refine collective norms. Self-enhancement, in this way, is never merely an individual practice but always also a social one. Cavaliere and Camporesi also draw attention to the question of affordability and access as an ethical aspect of contemporary genomic medicine. Regardless of where we stand in the naturalness debate, advanced genomic technologies are problematic in the sense that they are accessible only to certain groups of people. We are therefore on the way towards a 'two-tier society of enhanced and un-enhanced' which exacerbates already present inequalities (Cavaliere and Camporesi, this volume). In this respect, even the most novel genomic technologies continue a wider social trend.

Two contributors to this section directly engage with what might be seen as the 'collective turn' in genomics, albeit from different perspectives and with varying consequences. Ine van Hoyweghen examines how a 'politics of solidarity' has emerged in policy approaches and legal definitions pertaining to genomics and the insurance industry to reposition the issue of genetic discrimination as potentially affecting a 'we' rather than just a minority. Yet, she suggests this is a rendering of solidarity which relies on out-dated reductive and deterministic understanding of genes and genetic exceptionalism that serves to mask and blunt the inherent politics of the insurance market. Engaging the shifting parameters, rhetoric and politics of participation necessary to fulfil the promise of personal medicine Barbara Prainsack illuminates the diverse 'power asymmetries' at stake in state and corporate initiatives that seek to widen the parameters of inclusion in genomic research. Notably, however, she shows how this is no longer just about the collection of genomic data, which is neither sufficient nor necessarily the most valuable data in achieving the goal of personal medicine, but how a much wider range of 'lifestyle' data is now essential to this aim, raising new questions about surveillance, control and responsibility.

Sandra Soo-Jin Lee's chapter tackles the intricate difficulties in conceptualising 'race' in genomic medicine. As noted above, although health disparities correspond with racial groups in some contexts, using race as a proxy for either phenotypic or genotypic characteristics is problematic scientifically and politically. The question of how we adequately consider (or not) race as a factor that plays a role in people's lives and health, without reifying it as something that is 'in our genes', thus remains a big challenge at the interface of genomics and health. Critically reflecting the categories that we use when curating cohort studies is a good place to start. As Lee argues, 'acknowledging that biobanking is the result of social and political processes is essential to an integrated approach to understanding how genomics emerges as a solution to some of the most enduring biomedical challenges, such as eliminating health inequities' (Lee, this volume). In conclusion Sandra Lee points out that beyond attending to the 'pot pourri' of language around race in genomics we also need, as she puts it, to examine 'what African chromosomes mean in the twenty-first century'. An answer of sorts is provided in the reflection on the genealogies and contemporary histories of genomics in Brazil and Africa examined in the chapter by Duana Fullwiley and Sahra Gibbon. They illuminate how there are both differences and parallels in the way that populations in these national contexts and regions have been and continue to be conceived as a resource for genomic research and medicine informed by colonial and postcolonial histories, as well as structural inequalities. At the same time questions of diversity, inclusion and social justice are central to how low-income and developing economy countries are becoming involved in configuring contemporary genomics as a 'pathway' to global health. As the authors point out, however, this does not create a solution to the ethical complexities of consent within the research/care hybrids that are constituted by the expanded global scope of genomic research, but raises new challenges and concerns.

References

Baylis, Françoise, Nuala P. Kenny, and Susan Sherwin. 2008. A Relational Account of Public Health Care Ethics. *Public Health Ethics* 1(3): 196–209.

Bustamante, Carlos D., Francisco M. De La Vaga, and Esteban G. Burchard. 2011. Genomics for the World. *Nature* 475: 163–165.

Casper, Monica J., and Lisa J. Moore. 2009. *Missing Bodies: The Politics of Visibility*. New York: New York University Press.

Dean, Jodi. 1996. *Solidarity of Strangers: Feminism After Identity Politics*. Berkeley: University of California Press.

Do Valle, Guilherme and Sahra Gibbon. 2015. Introduction to Health/Illness, Biosocialities and Culture. *Vibrant* 12(1): 68–74.

Duster, Troy. 2015. A Post-Genomic Surprise: The Molecular Reinscription of Race in Science, Law, and Medicine. *British Journal of Sociology* 66(1): 1–27.

Faden, Ruth R., Nancy E. Kass, Steven N. Goodman, Peter Pronovost, Sean Tunis, and Tom L. Beauchamp. 2013. An ethics framework for a learning health care system: a departure from traditional research ethics and clinical ethics. Hastings Center Report 43 (s1): S16–S27.

Fassin, Didier. 2009. Another Politics of Life is Possible. *Theory, Culture and Society* 26(5): 44–60.

Gibbon, Sahra, and Carlos Novas. 2008. *Biosocialities, Genetics and the Social Sciences*. London: Routledge.

Gibbon, Sahra, Eirini Kampriani, and Andrea zur Nieden. 2010. BRCA patients in Cuba, Greece and Germany; comparative perspectives on public health, the state and the partial reproduction of 'neo-liberal' subjecs. *Biosocieties* 5(4): 440–466.

Illingworth, Patricia, and Wendy E. Parmet. The right to health: why it should apply to immigrants. *Public Health Ethics* 8, no. 2(2015): 148–161.

Jennings, Bruce. 2016. Reconceptualizing Autonomy: A Relational Turn in Bioethics. *Hastings Center Report*. Online first: DOI: doi:10.1002/hast.544.

Kahn, Jonathan. 2014. *Race in a Bottle: The Story of Bidil and Racialized Medicine in a Post-Genomics Age*. New York: Columbia University Press.

Lock, Margaret. 2008. Biosociality and susceptibility genes: a cautionary tale. In: Sahra Gibbon and Carlos Novas (eds). *Biosocialities,Genetics and the Social Sciences*. London: Routledge: 56–79.

M'Charek, Amade. 2013. Beyond Fact or Fiction: On the Materiality of Race in Practice. *Cultural Anthropology* 28 (3): 420–442.

Mackenzie, Catriona, and Natalie Stoljar, eds. 2000. *Relational Autonomy: Feminist Perspectives on Autonomy, Agency, and the Social Self*. Oxford: Oxford University Press.

Marmot, Michael. 2015. *The Health Gap: The Challenge of an Unequal World*. London: Bloomsbury.

Matthew, Dayna Bowen. 2015. *Just Medicine: A Cure for Racial Inequality in American Health Care*. New York: NYU Press.

[U.S.] National Research Council of the Academies [NRC]. 2011. *Toward Precision Medicine: Building a Knowledge Network for Biomedical Research and a New Taxonomy of Disease*. Washington, DC: NAS.

Nguyen, Vinh-Kim. 2010. *The Republic of Therapy: Triage and Sovereignty in West Africa's Time of AIDS*. Durham: Duke University Press.

Popejoy, Alice B., and Stephanie M. Fullerton. 2016. Genomics Is Failing on Diversity. *Nature* 538(7624): 161.

Prainsack, Barbara. 2017. *Personalized Medicine: Empowered Patients in the 21st Century?* New York: New York University Press.

Prainsack, Barbara, and Alena Buyx. 2017. *Solidarity in Biomedicine and Beyond*. Cambridge, UK: Cambridge University Press.

Reardon, Jenny. 2005. *Race to the Finish: Identity and Governance in an Age of Genomics*. Princeton: Princeton University Press.

Rose, Nikolas, and Carlos Novas. 2005. Biological Citizenship. In: Aihwa Ong and Stephen J. Collier (eds). *Global Assemblages: technology, politics and ethics as anthropological problems*. Oxford: Blackwell. 439–463.

Rothman, Barbara Katz. 2009. Introduction to Diversity and Justice. In *Handbook of Genetics and Society: Mapping the new genomic era*. London: Routledge: 401–444.

Strathern. Marilyn. 1998. *The Gender of the Gift: Problems with Women and Problems with Society in Melanesia*. Berkeley: University of California Press.

Traore, K. *et al.* 2015. Understanding of genomic research in developing countries. *BMC Medical Ethics* (16).

United Nations. 1966. International Convenant on Economic, Social and Cultural Rights. www.ohchr.org/EN/ProfessionalInterest/Pages/CESCR.aspx. Accessed May 10, 2016. www.ohchr.org.

23

Disability and the challenge of genomics

Jackie Leach Scully

Introduction

The relationship between genetic science and disability is a longstanding, close, complex and often fraught one. Genetics has given a scientific basis to the observation that some disabling conditions run in families, unravelling the molecular and cellular mechanisms linking genotype (the genetic constitution) to phenotype (the embodied result). That growing understanding is a genuine advance for healthcare and biomedicine, providing new ways to prevent the suffering of disease and incapacity. But the increasing use of genomic knowledge to characterize and prevent genetically associated disabling conditions is also ethically problematic, and so is genomics' impact on cultural understandings of human identity, disability and disease. A society that adopts genomic technologies faces challenging questions not only about responsible use, but also about the conceptualization of normality and abnormality.

These concerns apply to both prenatal and postnatal genomics. Postnatal applications currently focus on better diagnosis of genetic disabilities and understanding of the mechanisms behind them, rather than treatment, since for the most part genomics has not yet provided much in the way of genuinely therapeutic interventions. The most contentious areas, however, are undoubtedly in the prenatal uses of genomic knowledge to identify anomalous genes, and the possible extension of that into true genomic modification, or genome editing, in the future. Because of that, in this chapter I will be concentrating on the impact of prenatal genomics, but important ethical issues of postnatal uses, such as genetic discrimination (Lemke 2013; McClellan et al. 2013), geneticization (Hedgecoe 1998; Weiner et al. 2017) and so on should not be forgotten.

Disability and genomics in history

Genomic science is linked to disability through a traumatic history. In the early days of genetics, a limited understanding of the mechanisms of heredity meant that many conclusions drawn about the role of heritable factors not only in disease but also in various kinds of human behaviour were, to say the least, somewhat flawed. More troublingly, the science was entangled with early to mid-twentieth-century sociopolitical eugenic attempts to improve the population's genetic profile by eradicating 'harmful' hereditary tendencies, essentially by preventing the reproduction of people with unwanted conditions or behavioural traits that were believed to have a hereditary basis (Roll-Hansen 2010; Kevles 1985).

The twentieth-century eugenic project is primarily associated with North America and Europe, though in fact it was more globally popular. It is often treated as a purely historical tragedy of little contemporary relevance. But this toxic eugenic past remains a powerful influence on debates over the ethical use of genetic, and now genomic, knowledge – understandably, since the contemporary promises of genomic science, just like eugenics, focus on the elimination of genetically based disability and disease (Miller and Levine 2013).

An unresolved question is the extent to which contemporary prenatal genetic testing and screening is continuous with the historical practices – that is, whether today's applied genomics is really just eugenics by another name, with essentially the same fundamental goal (Thomas and Rothman 2016; Shakespeare 1998). Of course, there are some indisputable differences: today's scientific knowledge is deeper and more comprehensive; genomic diagnostic technology is faster and more accurate; and unlike today, the twentieth-century eugenic project often relied on coercive measures including, in its final terrible phase, the murder of disabled children and adults (Kittay 2016). Today's prenatal and postnatal testing is offered by healthcare providers as a means to help individuals and families exercise *reproductive choice* and thereby increase their autonomy. Admittedly, the choices involved are currently limited, because few of the genetic conditions that are routinely screened or tested for can be directly treated. But patients' autonomy is nevertheless enhanced if genomic information opens up to them the options of deciding to terminate a pregnancy, or to prepare for raising a child with a disabling genetic condition.

Some critics see a significantly greater degree of continuity between old eugenics and new genomics. They hold that since the overarching aim is the same, any difference in methods or style of provision is morally irrelevant (Iltis 2016). Bioethicists, especially feminist bioethicists and disability scholars (Milligan 2011; Seavilleklein 2009), have also argued that the rhetoric of free choice and reproductive autonomy is misleading. For the state that funds screening services, the primary rationale is not about citizens' choice but about money – to prevent as far as possible the birth of individuals who, it is claimed, will be economic burdens on society. Some scholars also contend that a more purely eugenic logic is still in operation, one that seeks to eliminate deviant forms of embodiment from the community for more ideological reasons. Through a variety of indicators (such as the routinization of screening as part of standard antenatal care) the state, society and healthcare professions inevitably communicate fairly clear ideas of which choices responsible citizens are expected to make. Cultural theorists would see this in terms of Foucauldian 'technologies of the self' – the ways in which we work on ourselves in order to fit to internalized normative discourses and expectations (Taussig et al. 2005) – which in this case may be eugnic in their consequences.

Of course, the parental choice that is offered remains highly constrained because *any* kind of regulation is likely to render certain reproductive options unavailable. One example is the notorious clause in the United Kingdom's Human Fertilization and Embryology Act of 2008 forbidding clinics (and therefore their patients/clients) from using gametes or embryos known to carry severe genetic anomalies (Emery et al. 2008; HFEA 2008). In this way both implicit and explicit limits to the human variation that society or the state are prepared to accept have already been laid down as the background against which individual reproductive decisions are made.[1]

Genetics into genomics

In the first quarter of the twenty-first century, genetics is becoming subsumed into a wider field of genomics. Where genetics is the study of individual genes and their function, genomics attempts to integrate the activity of gene clusters and of nuclear and subcellular genomes, the regulation of gene expression, the role of environmental factors and epigenetics, and so on. Alongside the impressive technical developments that these advances demand, however, genomics brings *conceptual* transformations too. These have profound societal implications, as I'll discuss later.

Combined with advances in reproductive medicine, genomics contributes its expanded predictive power to the established practices of prenatal diagnosis (PND), preimplantation genetic diagnosis (PGD) (El-Toukhy et al. 2008), and the more recent moves towards preconception screening (Henneman et al. 2016; Kihlbom 2016). Both the increasing scope of the technology and the routinization of these practices raise troubling ethical questions about their implications for disability. Some of these questions are only too familiar. Since any kind of prenatal screening or diagnosis is performed so that parents have the option of ending the pregnancy, the *ethical meaning* to individuals making the decision and to societies endorsing the technology depends heavily on convictions about the morality of abortion. For those who think that abortion is a morally permissible but nonetheless serious act, termination for foetal abnormality has to be justified by the seriousness of the condition it avoids. If the phenotypic variation cannot be classed as severely impairing, then for many people termination is ruled out.

The introduction of preimplantation genetic diagnosis (PGD) to healthcare services in the United Kingdom in 1990 opened up a different moral landscape (even though its use in practice is still limited: Franklin and Roberts 2006). Many people consider the disposal of *in vitro* embryos to be a less morally serious act than termination, because it involves the discarding of a very early blastocyst rather than destruction of a later-stage foetus (Cameron and Williamson 2003). This ranking works if the major ethical difficulty in the process is the moral status of the foetus/embryo, and/or the psychological impact of abortion versus non-implantation of embryos on the women concerned.

The disability critique of genomic testing and screening

Over the past several decades, disability scholars and activists have developed a distinctive critique of genetic and genomic identification, testing and screening, and diagnosis. It contains multiple arguments, rooted in the field's challenge to the mainstream understandings of disability. Although global disability scholarship offers more than one alternative model, all share the view that the traditional biomedical approach to disability is inadequate. The biomedical model sees disability as the result of an individual's deviation(s) from some kind of bodily norm, and since the nineteenth century the norms of embodiment have increasingly been defined through quantifiable biomedical parameters of human form and function; today, this means as genomic data. In this view, disability *is* what happens when the genome is mutated from whatever has been agreed as the canonical DNA sequences producing the normative body.

From the 1970s onwards, disability scholarship has proposed alternative ways of thinking about disability. Most of these see disability not as a purely biological phenomenon but as a situation that results from the (mis)match between the body and its material and social environment. Probably the best known of these views, the strong social model developed within British disability studies also attempts to separate out the embodied *physical or mental anomaly* (the impairment), and the *social response* (or lack of response) to the impairment. According to the strong social model it is the failure of the response that in reality causes disablement, for example through the lack of environmental adaptations (Barnes 2012; Terzi 2004). The extreme version of this logic implies that impaired bodies would not in fact experience significant disability if societal responses to anomaly were better. Few disability scholars would go quite that far, but most would agree that a non-normative genome and its resulting phenotype can only be a partial explanation for disability: genetically associated disability is the result of social, cultural, political and economic factors interacting with an individual anomalous genome in ways that can be more or less disabling.

Working from this more nuanced conceptualization, the disability critique of genomic screening and testing has two main strands: (i) that these practices are unjustly *discriminatory* towards disabled people, and (ii) that they *express a negative message* about the value of disabled people (Parens and Asch 2000).

Prenatal screening and testing are unjustly discriminatory

Any kind of prenatal screening for disability, whether genetic or not, is based on predictions about the lives that people with particular kinds of body will go on to lead. With genetic or genomic screening, these predictions are based solely on genomic information; indeed, in the case of PGD or preconception screening, this is before a body in any meaningful sense actually exists. Disability scholars argue that this is discriminatory because it targets particular pieces of information that are associated with disability (and may be called 'disabled' or 'faulty' genes) but that in isolation cannot predict the future phenotype with certainty. Moreover, if disability is not a purely biological product but the result of anomalous bodies interacting with social and environmental features, then changes in society or the environment could equally well alter those interactions, such that disability is *not* the outcome. This is an additional reason why genomic knowledge cannot predict future quality of life. Relying on genome information, the argument goes, discriminates unjustly against disabled people because predictions about future lives are too often based on prejudices and stereotypes about disability rather than empirical evidence. Genomic logic also means the impairment (and by extension, the genomic informa-tion) obliterates any other significant but *genetically unknowable* characteristics of the embryo or foetus.

The expressivist objection

The philosopher Adrienne Asch introduced the idea that prenatal screening against disability 'sends out a message' that disabled lives are of lesser value than nondisabled ones (Parens and Asch 2000). Received by disabled people, the message erodes their self-esteem and capacity for self-determination; among the general public it fosters a more hostile and intolerant climate for disability.

Opinion is divided about the general plausibility of the so-called expressivist objection. Critics are sceptical that it is possible to infer a clear message from individual cases, since the context in which any personal reproductive choice is made is generally unknown, and unknowable, to outsiders (Murphy 2011; Boardman 2014). The expressivist objection becomes much more plausible when applied to healthcare policy. As Iltis notes, there is an important 'difference between individual decisions to use PNS, PND, PGD … and the systematic deployment of social resources to ensure their routine use' (Iltis 2016: 338). Regulations per-mitting interventions to prevent the birth of people with impairments could very easily be interpreted as the societal endorsement of these measures; apparent endorsement then reinforces an inhospitable cultural climate and produces a society increasingly less willing to provide the supports and adaptations that disabled lives often need. Any negative effect will also be com-pounded by what I call *justificatory expressivism*: if the termination of foetuses or discarding of embryos are considered morally serious acts, then the fact that they are permitted *at all* may send a message that the grounds for using them (i.e. the predicted non-normative embodiment) *must* present an equivalently serious problem.

So the expressivist objection may be a compelling one, at least when applied to policy. Nevertheless, so far there has been no systematic empirical investigation of whether the expressivist process works as its proponents say it does (Klein 2011). Any effects will clearly be complex and subtle, and devising appropriate ways to measure incremental changes in attitude would be challenging. The effects of the 'message sent out' will also differ between, for example, families with experiential knowledge of a condition and the general population with no previous exposure to it. Indeed, Boardman found that in thinking about reproductive choices the 'possibility of having another affected family member could not be disentangled from [families'] experiences and knowledge', when the 'worst case scenario' is not a 'depersonalized notion of disability' but a member of the family (Boardman 2014: 21–22).

A further complication is that the meaning of any genomic technology, how people 'read' it, will depend heavily on past and present circumstances: '[a] sincere desire to cure through genetic advances comes across as something more nefarious when considered in the context of disability history' (Miller and Levine 2013: 98), while present day genomic interventions would be read very differently in a sociopolitical context that normalizes the presence of disabled people versus one that is generally hostile towards disabled people's needs.

Diversity, contingency and the ethics of control

Beyond the arguments of discrimination and expressivism, some disability scholars also contend that using genomics to select against or to eradicate disability from the germline is problematic, because disability is – paradoxically – an important human good. Even mainstream bioethicists would agree that genomic diversity per se has important biological, and sometimes socio-cultural, advantages (Sparrow 2015). The biological argument is that genomic diversity constitutes a reserve of variation that enables the organism to adapt to changing environmental conditions when necessary (Savulescu and Sparrow 2013; Gyngell 2012; Powell 2012). The social argument that genomic diversity is beneficial even when it leads to disability is less easily made; and whether or not people find it convincing depends largely on whether they agree that the wider the range of 'ways of being' a society can sustain, the better, either directly (because diversity just *is* a good), or indirectly because diversity encourages valued collective responses such as inclusiveness, tolerance or solidarity. Elsewhere, I have made a case for the benefit of variant embodiment in providing an epistemic resource that improves the quality of our ethical judgements (Scully 2008), while Rosemarie Garland-Thomson has advocated 'conserving disability' on similar grounds (Garland-Thomson 2012).

The concern about loss of diversity is directed at populations, not individuals, and this highlights the potential conflict between what is best for the person and what benefits society as a whole. For instance, if diversity is an important biological and social good, a case could be made for imposing it in some situations even at the cost of decreased individual welfare (if too many people were using genomic interventions, resulting in reduced diversity, for instance) (Sparrow 2015). But actively imposing genomic diversity would be a move that many people, including disability scholars and activists, would find ethically troubling.

Procreative beneficence

The majority of ethicists take a cautious or neutral stance towards the role of genomics in disability, concluding that interventions to reduce or eliminate disability are morally permissible but not required. However, others are decidedly more positive. Working from the general obligation to promote wellbeing, Savulescu and Kahane have proposed a *principle of procreative beneficence* that underpins an actual parental *obligation* to use whatever means available to benefit the health of future children. More controversially they have gone on to suggest that parents are in fact morally obligated to produce the child '…whose life can be expected…to go best' (Savulescu and Kahane 2009: 274; Savulescu 2001). So parents have a duty to protect the child from the suffering and disadvantage of disability, and also to preserve its 'right to an open future' (Feinberg 1992), by doing their best to ensure their child has the embodiment that secures the widest range of options for its future life (Malek and Daar 2012). The characterization of the 'best possible' child also extends as far as enhancing interventions to improve its life possibilities.

Many objections have been raised against the whole idea of trying to produce the best possible health or best possible child by intervening into the genome. First, there is the problem of the current state of knowledge, or rather lack of it, of the genome and how it works. Most of the common health conditions and other traits we might want to modify are the result of interactions between thousands of genetic, epigenetic and environmental factors, while the pathway from genotype to phenotype involves molecular processes that are nonlinear, dynamic and poorly understood. This raises a perfectly reasonable concern about how feasible it is to use genomic interventions to produce a desired *phenotypic outcome* (as distinct from producing a desired sequence change). Scepticism over whether we can really know with confidence what is the 'best possible' health, or best possible child, is also hard to dismiss. And overlying this is an even broader unease, that attempting to control genomic processes shows not only a lack of respect for the evolutionary dynamics that have generated the present-day human genome, but also a hubristic failure to recognize that the resulting contingency of life (and death) is part of the human condition (Parens 1995; Sandel 2007).

What's special about genomics?

As I discussed earlier, controversy over genomics' appropriate role in disability is nothing new. The ethical arguments have been rehearsed for years in the context of 'old' genetics. So what, if anything, is really novel about 'new' genomics?

It is clear there is a significant difference in terms of scale. Because genomic science is interested in the workings of whole genomes in their cellular and organismic context, the understanding that it gives is, we assume, more detailed and complete than old-style genetics alone could offer. Better understanding makes it more likely that practical applications of the science will be effective, safe and controllable. Recent major advances in genomics were possible because of equally important technical improvements in fast-throughput sequencing, bioinformatics and so on, and the combination of knowledge with technical capacity increases the likelihood that genomic technologies will become commercially viable and incorporated into routine healthcare.

Genomic science has also driven the rapid development of *gene editing*, probably the outstanding technical advance of the past two to three years. The CRISPR-Cas9 and related systems (Wang et al. 2016) make it possible to insert modified DNA sequences in the genome accurately, effectively and efficiently. These developments mean that for the first time, the routine modification of the human genome no longer looks like a distant fantasy.

Irrespective of the novelty of the techniques, however, the key question is whether genomics brings anything new in terms of impact on and ethical consequences for disabled people in real life. One distinctive feature of genomic science is that it holds out the promise of interventions into the genome that are genuinely therapeutic, or even transformative, rather than eliminatory in the way that PND and even PGD have been, and that are (according to some authors at least) ethically preferable (see Cavaliere and Camporesi, this volume). It's because of this that societies adopting technologies of genomic manipulation face unprecedented social and ethical questions. Removing a disabling genetic anomaly from a person (curing) rather than preventing that person being born are very different acts, and the difference transforms the ethical evaluation of prenatal intervention. As discussed earlier, the moral weight of termination or embryo disposal in selective interventions has often outweighed the harmful impact of the disability. But if gene editing (by whatever technique eventually proves successful) allows direct changes to be made to the genomes of gametes, embryos or perhaps early foetal tissues, then there is no need to factor the destruction of the foetus or embryo into the ethical equation. There is probably no need to worry about balancing severity either, since the same gene editing techniques could in principle be applied whatever the severity of the anomaly. Here, individuals, communities and societies are confronted (possibly for the first time in human history) with this question: once we can simply stop genetically based disability[2] from occurring, is it right to do so?

Societal consequences

Advances in genomic science clearly have the *potential* (although so far not the reality) to transform practical interventions in disability. If genomic interventions become easy, effective and cheap enough to be widely available, the result could be an observable decrease in the prevalence of embodied disability. The long-term societal consequences of such a shift – such as decreased tolerance for strange or unusual bodies, increased rigidity of aesthetic norms, the growth of social stratification – have been envisaged in both philosophical discussion and science fiction, but no one really knows what the gradual disappearance of genetically influenced disability might mean for the way we live together.

Aside from tangible consequences, genomics is an increasingly influential mediator of contemporary understandings of the body and bodily normality, as the framing concepts and language of genomic science are incorporated into the everyday discourse of disability and normality. In line with past medicalization and later geneticization, a similar trajectory through 'genomicization' predicts that a number of areas of life, including disability, will become understood as being under the control of the genome or explicable in terms of its actions.

Perhaps the most important questions relate to policy making and regulation. In the UK genomic interventions are, and will continue to be, highly regulated. How do we as a society move towards making acceptable policy decisions? Who do we authorize to make them, who contributes to the debates, what weighting do we give to the input of healthcare professionals, the general public, and disabled people and their families? How do we reach a consensus on whether some disabilities should be eliminated, and who is to decide when consensus has been found? A further, and often ignored, point is whether disabled people can be sure of equality of access to genomic services in the future, especially when coupled to reproductive technologies, and especially if disabled people choose to use these services in unexpected ways, as in the rare but highly publicized cases of people expressing preferences for children who share their parents' impairment (Mundy 2002; Savulescu 2002).

In the north/western world, these debates and deliberations are being played out within a context of several decades of disability activism and scholarship that have generated more nuanced, and increasingly widely accepted, social-relational understandings of disability. To think of disability as something produced at the interface between the variant body and its material and attitudinal environment is, in my view, a rich and powerful perspective that opens up creative options for different kinds of intervention (or, indeed, nonintervention). In principle, it is also flexible enough to accommodate the insights of genomic science as part of the puzzle of disability. However, the pace at which genomic discoveries are made, and the often overhyped way in which they are reported, contribute to the dominance of a dangerously oversimplified narrative of the genome as the foundational explanation for all human life and behaviour, including disability. As genomics becomes practically and culturally embedded, it will be vital to ensure that it continues to dialogue with other ways of thinking about disability, variation and normality.

Notes

1 In this context it is worth remembering that the enormous economic importance and power of the contemporary bioeconomy is very different from the limited scope of mid-twentieth-century eugenic public health campaigns – perhaps a less reassuring difference between old eugenics and new genomics.
2 Genomic interventions would not affect the majority of disabilities, which are not directly linked to genome sequences. However, as genomics identify multiple genes and genetic interactions involved in complicated traits, it could be that more forms of disability – late-onset conditions, or perhaps unusual vulnerabilities to certain pathologies – become understood as having a genetic component and therefore as 'editable'.

References

Barnes, C. Understanding the social model of disability. In: *Routledge Handbook of Disability Studies*, Watson, N., Roulstone, A. and Thomas, C. (eds.), pp. 12–29. Abingdon: Routledge, 2012.

Boardman, F.K. The expressivist objection to prenatal testing: The experiences of families living with genetic disease. *Social Science and Medicine*, 2014; 107: 18–25.

Cameron, C., Williamson, R. Is there an ethical difference between preimplantation genetic diagnosis and abortion? *Journal of Medical Ethics*, 2003; 29: 90–92.

El-Toukhy, T., Williams, C., Braude, P., The ethics of PGD. *The Obstetrician & Gynaecologist*, 2008; 10: 49–54.

Emery, S., Blankmeyer Burke, T., Middleton, A., Belk, R., Turner, G. Clause 14(4)(9) of embryo bill should be amended or deleted. *British Medical Journal*, 2008; 336: 976.

Feinberg, J. The child's right to an open future. In: *Freedom and Fulfillment*, Feinberg, J. (ed.), pp. 76–97. Princeton, NJ: Princeton University Press, 1992.

Franklin, S., Roberts, C. *Born and Made: An Ethnography of Preimplantation Genetic Diagnosis*. Princeton: Princeton University Press, 2006.

Garland-Thomson, R. The case for conserving disability. *Bioethical Inquiry*, 2012; 9: 339–355.

Gyngell, C. Enhancing the species: genetic engineering technologies and human persistence. *Philosophy and Technology*, 2012; 25: 495–512.

Hedgecoe, A. Geneticization, medicalisation and polemics. *Medicine, Healthcare and Philosophy*, 1998; 1: 235–243.

Henneman, L. *et al.*, Responsible implementation of expanded carrier screening. *European Journal of Human Genetics*, 2016; 24: e1–e12.

Human Fertilisation and Embryology Act 2008. The full text can be found at www.legislation.gov.uk/ukp ga/2008/22/contents.

Iltis, A.S. Prenatal screening and prenatal diagnosis: contemporary practices in light of the past. *Journal of Medical Ethics*, 2016; 42: 334–339.

Kevles, D. *In the Name of Eugenics: Genetics and the Uses of Human Heredity*. New York: Knopf, 1985.

Kihlbom, U. Ethical issues in preconception genetic carrier screening. *Upsala Journal of Medical Sciences*, 2016; 121(4): 295–298.

Kittay, E. Deadly medicine: project T4, mental disability, and racism. *Res Philosophica*, 2016; 93: 715–741.

Klein DA. Medical disparagement of the disability experience: empirical evidence for the 'expressivist objection'. *American Journal of Bioethics Primary Research*, 2011; 2: 8–20.

Lemke, T. *Perspectives on Genetic Discrimination*. New York: Routledge, 2013.

Malek, J., Daar, J. The case for a parental duty to use preimplantation genetic diagnosis for medical benefit. *American Journal of Bioethics*, 2012; 12: 3–11.

McClellan, K.A., Avard, D., Simard, J., Knoppers, B. Personalized medicine and access to healthcare: potential for inequitable access? *European Journal of Human Genetics*, 2013; 21: 143–147; doi:10.1038/ejhg.2012.149; published online 11 July 2012.

Miller, P.S., Levine, R.L. Avoiding genetic genocide: understanding good intentions and eugenics in the complex dialogue between the medical and disability communities. *Genetics in Medicine*, 2013; 15: 95–102.

Milligan, E. *The Ethics of Consent and Choice in Prenatal Screening*. Newcastle: Cambridge Scholars, 2011.

Mundy, L. A World of Their Own. *Washington Post*, 31 March 2002. www.washingtonpost.com/archive/lifestyle/magazine/2002/03/31/a-world-of-their-own/abba2bbf-af01-4b55-912c-85aa46e98c6b/?utm_term=.fde5c449c472.

Murphy, T. When choosing the traits of children is hurtful to others. *Journal of Medical Ethics*, 2011; 37: 105–108.

Parens, E. The goodness of fragility: On the prospect of genetic technologies aimed at the enhancement of human capacities. *Kennedy Institute of Ethics Journal*, 1995; 5(2): 141–153.

Parens, E., Asch, A. *Prenatal Testing and Disability Rights*. Washington; Georgetown University Press, 2000.

Powell, R. The evolutionary biological implications of human genetic engineering. *Journal of Medicine and Philosophy*, 2012; 37: 204–225.

Roll-Hansen, N. Eugenics and the science of genetics. In: *The Oxford Handbook of the History of Eugenics*, Bashford, L. and Levine, P. (eds.), pp. 80–87. Oxford: Oxford University Press, 2010.

Sandel, M.J. *The Case Against Perfection: Ethics in an Age of Genetic Engineering*. Cambridge, MA: Harvard University Press, 2007.

Savulescu, J. Procreative beneficence: why we should select the best children. *Bioethics*, 2001; 15: 413–426.

Savulescu, J. Deaf lesbians, 'designer disability' and the future of medicine. *British Medical Journal*, 2002; 325: 771–773.

Savulescu, J., Kahane, G. The moral obligation to create children with the best chance of the best life. *Bioethics*, 2009; 23: 274–290.

Savulescu, J., Sparrow, R. Making better babies: pros and cons. *Monash Bioethics Review*, 2013; 31: 36–59.

Scully, J.L. *Disability Bioethics: Moral Bodies, Moral Difference*. Lanham, MD: Rowman & Littlefield, 2008.

Seavilleklein, V. Challenging the rhetoric of choice in prenatal screening. *Bioethics*, 2009; 23: 68–77.

Shakespeare, T. Choices and rights: eugenics, genetics and disability equality. *Disability and Society*, 1998; 13: 665–681.

Sparrow, R. Imposing genetic diversity. *American Journal of Bioethics*, 2015; 15: 2–10.

Taussig, K.-S., Rapp, R., Heath, D. Flexible eugenics: Technologies of the self in an age of genetics. In: *Anthropologies of Modernity: Foucault, Governmentality, and Life Politics*, Inda, J.K. (ed.). Oxford: Blackwell, 2005.

Terzi, L. The social model of disability: a philosophical critique. *Journal of Applied Philosophy*, 2004; 21: 141–157.

Thomas, G.M., Rothman, B.K. Keeping the backdoor to eugenics ajar? Disability and the future of prenatal screening. *AMA Journal of Ethics*, 2016; 18: 406–415.

Wang, H., La Russa, M., Qu, L.S. CRISPR/Cas9 in genome editing and beyond. *Annual Review of Biochemistry* 2016; 85: 227–264.

Weiner, K., Martin, P., Richards, M., Tutton, R. Have we seen the geneticisation of society? Expectations and evidence. *Sociology of Health and Illness*, 2017; doidoi:10.1111/1467–9566.12551/pdf.

24

Eugenics and enhancement in contemporary genomics

Silvia Camporesi and Giulia Cavaliere

The year 1989 marked not only the fall of the Berlin Wall, but also the birth of the first test-tube baby selected with preimplantation genetic diagnosis (PGD). PGD allows the screening of early embryos before implantation, and its first use marked the beginning of what, following geneticist Lee Silver (1997), we refer to as 'reprogenetics', or the use of genetic technologies in the context of reproduction to select what kind of children to bring into the world. In this chapter we draw on three examples – PGD, non-invasive prenatal testing (NIPT) and CRISPR genome editing technologies – to provide an overview of the discussion of reprogenetics at the intersection of enhancement and eugenics.

Socio-political contexts play an important role in policy-making. This is also the case for PGD and other reprogenetics practices, whose regulation varies greatly across countries. In the United Kingdom, for example, PGD is a service offered through the National Health System and is licensed by the Human Fertilisation and Embryology Authority (HFEA), which publishes and frequently updates a list of traits acceptable for screening out with PGD. This list includes traits considered 'severely' disabling, for example Huntington Disease (Huntington Chorea), cystic fibrosis, Tay-Sachs disease, and Muscular Dystrophy (Duchenne), as well as other less severely disabling conditions such as Down syndrome, achondroplasia (genetic dwarfism) and deafness. It also includes screening for genes that are implicated in the early onset of some conditions such as Alzheimer's disease (Types 3 and 4), increased susceptibility to breast cancer and to frontotemporal dementia (HFEA PGD List, 2016). In other countries in Europe, for example Switzerland or Poland, PGD is not allowed, whilst in Germany and Italy it is allowed only in very few cases deemed particularly severe, and only since 2011 and 2015 respectively (Biondi 2013, Gianaroli et al. 2016, Wilhelm et al. 2013).

While uses of PGD for such severe conditions have become more accepted over the years, one of the contentious areas around the use of PGD remains sex selection for 'non-medical reasons', i.e., for parents who want to choose the sex of their child independently of genetic conditions associated with the X-chromosome. In the United Kingdom, couples are not allowed to use PGD to choose the sex of their children unless it is for a 'medical reason', i.e., there are specific genetic conditions linked to the X-chromosome which would affect a male (XY) but not a female (XX) offspring, in which case selecting for females is allowed. In the United States, where there is no equivalent of a central regulatory system overseeing assisted reproduction such as the HFEA, PGD is offered by private clinics and can be reimbursed through insurance plans. The American Society for Human Genetics (ASHG) and the American Society for Reproductive Medicine (ASRM) publish non-binding recommendations on how clinics should provide services. In a recently published opinion (2015), the ASRM Ethics Committee stated that they have not reached a consensus on sex selection via PGD for 'family balancing reasons' (ASRM Ethics Committee 2015). Thus, they conclude that the practice remains controversial and that clinics that offer this service are encouraged to publish their policies regarding this practice. In recent years, the international debate on the ethics of sex selection has been influenced by reported data on growing sex-ratio imbalances in Asia (Van Balen and Inhorn 2003, Croll 2000). In particular, scholars have begun to argue for the need for contextual sex-selection policies, which take into account extant patterns of discrimination and gender bias that could influence parental decision-making (Dawson 2010) and that build on social science literature on the topic (Sleeboom-Faulkner 2007).

Another controversial area is the use of PGD to select children with specific desired traits. Oxford bioethicist Julian Savulescu, one of the most ardent advocates of human enhancement, argues that we have a moral duty to use PGD to choose children's traits and to screen out disorders, including to screen out some relatively minor conditions such as asthma. According to his Principle of Procreative Beneficence (PPB) (Savulescu 2001, Savulescu & Kahane 2009), 'we should use that [genetic] information to select embryos that have the best chance of the best life, other things being equal'. Sociologist Richard Twine has noted how the PPB is problematic because it is dislocated from the social: decisions regarding the use of genetic technologies in reproductive are hardly 'private', as Savulescu claims, in at least two senses – they require assistance (it is called assisted reproduction) by a national health system or a private company, and they are socially shaped. Twine writes: 'presumptions are made over the content of "best life" and "best child" with insufficient attention to how these are socially and historically mediated' (Twine 2005, p. 292). The eugenics implications of such a principle have not been left overlooked. For one, Australian bioethicist Robert Sparrow argues that Savulescu's PPB, if followed through, would bring into the world white, tall, straight males – features that provide children with a competitive advantage in the society in which they find themselves (Sparrow 2011). He refers to this idea as the 'repugnant conclusion' and adds: 'I don't think we're as far from the history of the bad old eugenics as many bioethicists would like to think' (Sparrow 2016).

Although discussions of PGD and eugenics often revolve around selecting traits that could offer some kind of positional advantage in life (e.g. selecting for enhanced intelligence, looks, or resistance to diseases), in some cases PGD has also been used to select for traits that some would consider disabilities, such as deafness (Baruch 2007, Baruch et al. 2008). Australian legal scholar Isabel Karpin has referred to the use of PGD for the selection of such traits as 'negative enhancement'. The label 'negative', however, seems to run counter the intentions of parents who choose to select for deafness as a valued trait (Camporesi 2010). In the United States, selection for such traits has been allowed, on the basis of the exercise of rights of a minority to carry on their bloodline (Sanghavi 2006) with which state authorities should not interfere (Robertson 2003). Feminist scholar Jackie Leach Scully has criticised how the case of 'choosing deafness' has 'become something of a staple of bioethical teaching and debate', pointing out how this case has been discussed without any attention whatsoever to the empirical realities that, on the contrary, 'demand careful examination' (Scully 2017, p. 211).

Sociologists Sarah Franklin and Celia Roberts' innovative fieldwork on PGD in the UK (2006) has provided the first empirical (ethnographic) data on why and how parents choose PGD. Their work has shed light onto the experiential world of couples undergoing PGD treatment and shown that one of the main reasons for resorting to PGD is the desire to eschew the emotional distress of repeated terminations, which can be avoided by selecting 'healthy' embryos. Franklin and Roberts also showed how assisted reproduction through IVF and PGD often becomes a means to re-establish traditional family norms, and meet societal expectations of forming a biological family (Franklin and Roberts 2006). Building on Franklin's work, anthropologist Charis Thompson (2005) has analysed the role of IVF and PGD in the US as biotechnological innovations 'to make parents'. Resonating with some of the findings of Franklin and Roberts' study, Thomson described the ways in which assisted reproduction clinics become normative and normalising of the ways in which couples, especially women, become 'good' fertility patients, and good parents. In other words how their bodies become 'disciplined' by the technology. The comparative work carried out by sociologist Yael Hashiloni-Dolev in Israel and Germany shows how social and religious factors can lead to dramatically different policies regarding PDG: liberal in the first case, very restrictive in the second (Hashiloni-Dolev, 2007).

Parallels between PGD and eugenics have been drawn since the very introduction of PGD in assisted reproduction (e.g. Duster 2003, King 1999, Testart 1995). The groundbreaking book by sociologist Troy Duster in 2003 was one of the very first works to connect essentialist thinking in genetics with a re-legitimisation of old mythologies typical of eugenics thinking. A comprehensive analysis of the literature on reproductive genetic technologies such as PGD and eugenics was conducted by one of the authors (Cavaliere 2018). The results show that while some scholars understand the relationship between the history of eugenics and these technologies in terms of continuity (Garver & Garver 1991, Garland-Thomson 2012, Habermas 2003, Koch 2010, MacKellar and Bechtel 2014), others do not and emphasise the discontinuity between past and present (Agar 2008, Selgelid 2000, Suter 2007, Douglas & Devolder 2013, Glover 2006, Savulescu and Kahane 2009). While the latter use the term 'eugenics' in a pejorative sense, the former have attempted to rescue the word from its negative connotation by adding the adjective 'liberal' or the prefix 'neo' or 'new'. The difference with past eugenics, they argue, is that the parents, not the State, are the *loci* of control for reproduction. Parental autonomy becomes an extension of the individual right to 'reproductive freedom' or 'procreative liberty' to make one's own choices regarding reproduction (Buchanan et al. 2001, Robertson 2003). Other scholars such as American political scientist Bruce Jennings (2000), American legal scholar and bioethicist Sonia Suter (2007) and Australian feminist philosopher Catherine Mills (2011; 2015) have discussed in their work why the discourse of 'choice' is not appropriate in the context of reproductive genetic technologies. In particular Mills (2011) argues that the discourse of 'choice' creates a false dichotomy between a duty to give children the best possible chance in life, and the duty of accepting them as they are. It is a false dichotomy because the existence itself of the genetic technologies shapes what kind of decisions are available to prospective parents, and because social conditions mediate the use of genetic technologies in a classic instance of co-production of science and society (Jasanoff 2004). Ultimately, as argued by historian of medicine Ilana Löwy (2014), it is very difficult to disentangle the intentions of the individuals when using reproductive genetic technologies, from the societal conditions that make those decisions possible.

Although medical genetics as a discipline has strived to distance itself from a history of gruesome eugenics by being non-directive in the context of reproduction (Biesecker 2001, Broberg and Roll-Hansen 1996), fears of eugenics remain central in both public and scholarly debates around reprogenetics. Scholars debate the definition of 'eugenics', which practices should be labelled 'eugenic' and whether labelling a practice as eugenic warrants its moral condemnation. However, all definitions share an understanding of eugenics as 'the attempt to improve the human gene pool' (Wilkinson 2010). Contrary to conventional wisdom, eugenics programmes were not limited to Nazi Germany; they were carried out in France, Italy, South America, the United States and Scandinavia. In addition to their geographic ubiquity, eugenics programmes, ideas and policies carried forward from the end of the nineteenth century well into the twentieth century, up to the 1970s in certain countries such as Sweden (Bashford 2010, Broberg and Roll-Hansen 1996, Lombardo 2008). It is a fear of a slippery slope to 'bad' eugenics that underlies restrictive regulations in assisted reproduction and screening technologies in countries such as Germany (Brown 2004). In Italy, it is the influence of the Catholic church and the associated belief that human life starts at conception that have had a major influence on the drafting of strict legislation against 'eugenics' programmes (Fenton 2006). Despite the effort of many scholars to provide a nuanced account of a multifaceted movement (Bashford 2010, Koch 2004, Lombardo 2008, Paul 1984, 1992), a common reference point for discussions of reproductive genetic technologies remains Nazi Eugenics.

Critics of enhancement, or as they are sometimes referred to by those in favour of enhancement and proponents of transhumanist agendas, 'bioconservatives', approach the discussion of genetic technologies in reproduction from the point of view of what it means to be a good parent. In his well-known 2004 essay 'The Case Against Perfection', American political philosopher Michael Sandel argues that we should 'appreciate children as gifts' and accept them as they are, and that doing so is a hall-mark of the virtuous parent. Sandel is not completely opposed to reproductive genetic technologies, but refers to the therapy/enhancement distinction to argue that these technologies should be used only for therapeutic purposes. According to this interpretation, a therapeutic intervention aims at restoring health (or, following Boorse [1977], species-typical normal functioning) and hence is ethically sound, whereas an enhancing intervention is aimed at going beyond this allegedly 'normal' status and, hence, is ethically problematic. However, from an ontological perspective, as scholars such as Kingma (2007), Scully (2014), Scully and Rehmann-Sutter (2001), Miriam Eilers et al. (2014), Mills (2011, 2015), and Christina Schües (2014) have pointed out, no definition of health can ever be value-free: what counts as normal and as healthy is context-dependent and value-laden, and, consequently, so is the concept of enhancement. It is in this sense that the normal is always inescapably normative. Daniels (2000) adopts a limited defence of the therapy/enhancement distinction, which seems to be a plausible middle ground for an initial demarcation of what counts as therapy (which, according to Daniels, should be prioritised), and what counts as enhancement (which should be handled with more caution). Bioliberals such as Savulescu and Harris argue instead that the therapy/enhancement distinction is not only ontologically on fragile grounds, but also irrelevant from a moral point of view. They argue that both therapy and enhancement should be granted in order to promote individuals' well-being, regardless of statistically imposed measurements and evaluations.

While PGD allows us to genetically test human embryos and to implant subsequently only those free from the prospective parents' genetic mutation, CRISPR genome editing technologies, instead, allow the targeted editing of human embryos in order to 'correct' specific mutations and possibly modify other traits before implantation. This technology is a RNA-guided tool composed of two parts: CRISPR (clustered regularly interspaced short palindromic repeat) and CRISPR-associated proteins (CAS). The RNA tool (CRISPR) functions as a guide for the CRISPR associated proteins to target and cleave specific sites in the genome (Dance 2015). The publication in April 2015 by a group of Chinese scientists of their experiments using CRISPR on human embryos (the first such reported case) ignited the controversy around CRISPR genome editing and eugenics (Cyranoski and Reardon 2015). Public responses in the United Kingdom and in the United States on the ethical standing of doing research with CRISPR on human embryos have diverged (Camporesi and Cavaliere 2016). While in the United States the announcement of the Chinese scientists raised calls for moratoria from the scientific communities (Baltimore et al. 2015), in the United Kingdom reactions have been generally more positive with scientists such as Robin Lovell-Badge and bioethicists positioning themselves against a moratorium. Among the voices arguing in favour of a moral obligation to allow CRISPR genome editing on human embryos we can find 'bioliberals' such as Julian Savulescu (Savulescu et al. 2015), and John Harris (Harris 2016). Bioliberals view the individuals' right to enhance as an extension of the sovereignty on one's own body (Agar 2008, Glover 2006, Harris 2005, Robertson 2003).

Genome editing could potentially become a tool for enhancement because it allows parents to express preferences on the genetic endowment of their future offspring. As a result, the 'ultimate' comparison with eugenics is drawn, as one could imagine that if CRISPR genome editing were used by a large enough number of people, it would change the composition of the human species, or lead to two different kinds of sub-species within *H. Sapiens*: the genetically engineered, and the 'natural'. This scenario is opposed by bioconservatives who fear that the modification of the human genome will eventually lead to a world of post-human creatures (Fukuyama 2003) and to the loss of typically human qualities (Sandel 2004), whereas it is welcomed by bioliberals such as Harris and Savulescu who do not grant any moral value to the 'natural' status (whatever that means) of our genome. A two-tier society of enhanced and un-enhanced, however, would lead to the exacerbation of existing societal inequalities between the haves and have-nots. There are reasons for a cautious approach to the governance of reprogenetics that takes into account such disparities.

American sociologist and bioethicist Erik Parens adopts an intermediate view between bioliberals and bioconservatives. In his 'Made to Order' essay discussing CRISPR genome editing technologies (2015), Parens points out the intrinsic tension between parental duty to nurture and cultivate their children's talents, and parental duty to accept their children as they are. Parens adopts a genetic exceptionalism position where he argues that there is a morally relevant difference between intervening on existing children with genetic technologies (or other more traditional means of intervening in education and childrearing), and intervening on human embryos with genome editing technologies to 'tailor' children.

While applications of CRISPR genome editing to human embryos are still speculative, another genetic technology called non-invasive prenatal testing (NIPT) has recently entered the moral economy of reproduction. NIPT is based on the analysis of cell-free DNA fragments (cfDNA) that originate from both the mother and foetus and can be found in the maternal blood circulation. It is used during the first trimester of pregnancy to make a decision about whether or not to carry forward the pregnancy. The NIPT test requires taking a small maternal blood sample at about 8–10 weeks of pregnancy; cfDNA in the maternal blood is then analysed to screen for any chromosomal abnormality in the foetus. NIPT allows for highly accurate detection rates (over % 99.5) for the three most common trisomy conditions present at birth (Down syndrome, Edwards syndrome and Patau syndrome). In 2016, the UK Government recommended that NIPT should be implemented through the NHS as part of prenatal screening. Disability activists immediately raised the concern that the introduction of a screening test apparently free of 'risks' for the mother (if we understand 'risk' only in a narrow sense as risk of miscarriage) could increase the number of terminations for children with Down syndrome, and that in the future the technology coupled with whole-genome-sequencing could lead to the eradication of a plethora of different traits. Tom Shakespeare (2006, 2013) author of 'Disability Rights and Wrongs' is now chairing a working group of the Nuffield Council on Bioethics on NIPT (Nuffield Council NIPT 2017). A report from this work, while having no binding power, will be used as evidence in the Parliament in the discussion of the type of conditions for which NIPT should be used.

Tensions arise from the use of genetic technologies in the context of reproduction, which are reminiscent of dilemmas that have been the crux of parenting and childrearing for as long as humanity has been reflecting upon childhood. What has changed are the resources available to parents – for example, the availability of genetic technologies not only to screen which children to bring into the world (through PGD or NIPT), but also to intervene directly on human embryos (through CRISPR genome editing). Genetic technologies offer us new ways of shaping future generations, both at the level of identifying among existing children those who will have a better potential for a certain path and at the level of editing the human embryo and shaping what kind of people to bring into the world. Some authors (including the authors of this chapter) reject the underlying genetic exceptionalism (the view that our genes raise some special claim of ethical concern) that motivates this thinking. In doing so we are in agreement with Lewens (2015) who argues that genetics means to impact future generations should be put on the same level as other, more traditional, ways of impacting future generations. Unfortunately, as pointed out by American evolutionary biologist and historian of science Stephen Jay Gould (1996), biological determinism (of which genetics is only the latest form of biological measurement, after the measurement of skulls or of 'intelligence' through IQ tests) comes and goes in cycles. It also appears to rise in popularity in times of political retrenchment, when the latest advancements in science are used as a means to explain social differences.

References

Agar, N. (2008). *Liberal eugenics: In defence of human enhancement*. Blackwell Publishing.

American Society of Reproductive Medicine (ASRM) Ethics Committee. (2015). Sex selection and preimplantation genetic diagnosis. *Fertility and Sterility*, 82, 245–248.

Baltimore, D., Berg, P., Botchan, M., Carroll, D. *et al.* (2015). A prudent path forward for genomic engineering and germline gene modification. *Science*, 348(6230), 36–38.

Baruch, S. (2007). Preimplantation genetic diagnosis and parental preferences: beyond deadly disease. *Houston Journal of Health Law & Policy*, 8, 245.

Baruch, S., Kaufman, D., & Hudson, K. L. (2008). Genetic testing of embryos: practices and perspectives of US in vitro fertilization clinics. *Fertility and Sterility*, 89(5), 1053–1058.

Bashford, A. (2010). Where Did Eugenics Go? In Bashford, Alison and Levine, Philippa, eds. *The Oxford Handbook of the History of Eugenics*, New York: Oxford University Press; 539–558.

Biesecker, B. B. (2001). Goals of genetic counseling. *Clinical Genetics*, 60(5), 323–330.

Biondi, S. (2013). Access to medical-assisted reproduction and PGD in Italian law: A deadly blow to an illiberal statute? Commentary to the European Court on Human Rights's decision Costa and Pavan v Italy (ECtHR, 28 August 2012, App. 54270/2010). *Medical Law Review, 21*(3), 474–486.

Boorse, C. (1977). Health as a theoretical concept. *Philosophy of Science*, 44(4), 542–573.

Broberg, G., & Roll-Hansen, N. (1996). *Eugenics and the welfare state*. East Lansing: Michigan University Press.

Brown, E. (2004). The dilemmas of German bioethics. *The New Atlantis*, (5), 37–53.

Buchanan, A., Brock, D. W., Daniels, N., & Wikler, D. (2001). *From chance to choice: Genetics and justice*. Cambridge: Cambridge University Press.

Camporesi, S., & Cavaliere, G. (2016). Emerging ethical perspectives in the clustered regularly interspaced short palindromic repeats genome-editing debate. *Personalized Medicine*, 13(6), 575–586.

Camporesi, S. (2010). Choosing deafness with preimplantation genetic diagnosis: an ethical way to carry on a cultural bloodline? *Cambridge Quarterly of Healthcare Ethics*, 19(01), 86–96.

Cavaliere, G. (2018). Clearing the shadow: The 'eugenics' argument in debates on assisted reproductive technologies and practices. Under review for *Monash Bioethics Review*.

Croll, E. (2000). *Endangered daughters: Discrimination and development in Asia*. Psychology Press.

Cyranoski, D. & Reardon, S. (2015). Chinese scientists genetically modify human embryos *Nature News*, 22 April 2015. www.nature.com/news/chinese-scientists-genetically-modify-human-embryos-1.17378 (accessed February 21, 2018).

Dance, A. (2015). Core Concept: CRISPR gene editing. *Proceedings of the National Academy of Sciences*, 112 (20), 6245–6246.

Daniels, N. (2000). Normal functioning and the treatment-enhancement distinction. *Cambridge Quarterly of Healthcare Ethics*, 9(03), 309–322.

Dawson, A. (2010). The future of bioethics: three dogmas and a cup of hemlock. *Bioethics, 24*(5), 218–225.

Douglas, T., & Devolder, K. (2013). Procreative altruism: Beyond individualism in reproductive selection. *Journal of Medicine and Philosophy*, 38(4), 400–419.

Duster, T. (2003). *Backdoor to eugenics*. New York: Routledge.

Eilers, M., Grüber, K. & Rehmann-Sutter, C. eds. (2014). *The human enhancement debate and disability: new bodies for a better life*. Hampshire: Palgrave Macmillan.

Epstein, C. J. (2003). Is modern genetics the new eugenics? *Genetics in Medicine*, 5(6), 469–475

Ethics Committee of the American Society for Reproductive Medicine (ASRM). (2015). Use of reproductive technology for sex selection for nonmedical reasons. *Fertility and Sterility*, 103(6), 1418–1422.

Fenton, R. A. (2006). Catholic doctrine versus women's rights: the new Italian law on assisted reproduction. *Medical Law Review*, 14(1), 73–107.

Franklin, S., & Roberts, C. (2006). *Born and made: An ethnography of preimplantation genetic diagnosis*. Princeton: Princeton University Press.

Fukuyama, F. (2003). *Our posthuman future: Consequences of the biotechnology revolution*. New York: Farrar, Straus and Giroux.

Garland-Thomson, R. (2012). The case for conserving disability. *Journal of Bioethical Inquiry*, 9(3), 339–355.

Garver, K. L., & Garver, B. (1991). Eugenics: past, present, and the future. *American Journal of Human Genetics*, 49(5), 1109.

Gianaroli, L., Ferraretti, A. P., Magli, M. C., & Sgargi, S. (2016). Current regulatory arrangements for assisted conception treatment in European countries. *European Journal of Obstetrics & Gynecology and Reproductive Biology*, 207, 211–213

Glover, J. (2006). *Choosing children: Genes, disability, and design*. OUP: Oxford.

Gould, S. J. (1996). *The mismeasure of man* (1981). *Revised Edition*. New York: Norton & Company.

Habermas, J. (2003). *The future of human nature*. Cambridge: Polity Press.

Harris, J. (2016). Germline Modification and the Burden of Human Existence. *Cambridge Quarterly of Healthcare Ethics*, 25(01), 6–18.

Harris, J (2005). Reproductive liberty, disease and disability. *Reproductive Biomedicine Online*, 10, 13–16.

Hashiloni-Dolev, Y. (2007). *A life (un) worthy of living: Reproductive genetics in Israel and Germany* (Vol. 34). Dordrecht: Springer (Science & Business Media).

HFEA PGD List (2016), https://www.hfea.gov.uk/pgd-conditions (accessed February 21, 2018).

Jasanoff, S. (Ed.). (2004). *States of knowledge: the co-production of science and the social order*. London: Routledge.

Jennings, B. (2000). Technology and the genetic imaginary: prenatal testing and the construction of disability. In: Parens, Erik; Asch, Adrienne, eds. *Prenatal Testing and Disability Rights*. Washington, DC: Georgetown University Press; 124–144.

King, D.S. (1999). Preimplantation genetic diagnosis and the 'new' eugenics. *Journal of Medical Ethics*, 25(2): 176–182.

Kingma, E. (2007). What is it to be healthy? *Analysis* 67(2): 128–133.

Koch, L. (2004). The meaning of eugenics: Reflections on the government of genetic knowledge in the past and the present. *Science in Context*, 17(03), 315–331.

Koch, T. (2010). Enhancing who? Enhancing what? Ethics, bioethics, and transhumanism. *Journal of Medicine and Philosophy*, 35(6), 685–699.

Lewens, T. (2015). *The biological foundations of bioethics*. Oxford: Oxford University Press.

Lombardo, P. A. (2008). *Three generations, no imbeciles: Eugenics, the Supreme Court, and Buck v. Bell*. Baltimore: John Hopkins University Press.

Löwy, I. (2014). Prenatal diagnosis: The irresistible rise of the 'visible fetus'. *Studies in History and Philosophy of Science Part C: Studies in History and Philosophy of Biological and Biomedical Sciences*, 47: 290–299.

MacKellar, C. & Bechtel, C. (Eds.) (2014). *The ethics of the new eugenics*. New York: Berghahn.

Mills, C. (2015). Liberal eugenics, human enhancement and the concept of the normal. In Meacham, D. (ed.), *Medicine and Society, New Perspectives in Continental Philosophy*, pp. 179–194. Dordrecht: Springer.

Mills, C. (2011). *Futures of reproduction: Bioethics and biopolitics*. Dordrecht: Springer.

Nuffield Council on Bioethics. (2017). Non-Invasive prenatal testing: ethical issues. London: Nuffield Council. Available at: http://nuffieldbioethics.org/wp-content/uploads/NIPT-ethical-issues-full-report.pdf (accessed February 21, 2018).

Parens, E. (2015). Made to order: Can parents be trusted with gene editing technology? *AEON*, https://aeon.co/essays/can-parents-be-trusted-with-gene-editing-technology (accessed January 19, 2018).

Paul, D. B. (1992). Eugenic Anxieties, Social Realities, and Political Choices. *Social Research*, 663–683.

Paul, D. (1984). Eugenics and the Left. *Journal of the History of Ideas*, 45(4), 567–590.

Robertson, J. A. (2003). Procreative liberty in the era of genomics. *American Journal of Law & Medicine*, 29(2003), 439–487.

Sandel, M.J. (2004). The case against perfection. *The Atlantic Monthly*, 293(3): 51–62.

Sanghavi, D.M. (2006). Wanting babies like themselves, some parents choose genetic defects. *The New York Times*, December 5th, 2006. www.nytimes.com/2006/12/05/health/05essa.html (accessed February 21, 2018).

Savulescu, J. (2001). Procreative beneficence: why we should select the best children. *Bioethics*, 15(5-6), 413–426.

Savulescu, J., Pugh, J., Douglas, T. and Gyngell, C. (2015). The moral imperative to continue gene editing research on human embryos. *Protein Cell*, 6(7), 476.

Savulescu, J., & Kahane, G. (2009). The moral obligation to create children with the best chance of the best life. *Bioethics*, 23(5), 274–290.

Schües, C. (2014). Improving Deficiencies? Historical, Anthropological, and Ethical Aspects of the Human Condition. In Eilers, M., Grüber, K. & Rehmann-Sutter, C. (eds), *The human enhancement debate and disability: new bodies for a better life*. Hampshire: Palgrave Macmillan; pp. 38–63.

Scully, J. L. (2017). Feminist Empirical Bioethics. In Ives, J., Dunn, M. and Cribb, A. (eds), *Empirical Bioethics: Theoretical and Practical Perspectives*. Cambridge: Cambridge University Press; pp. 195–221.

Scully, J. L. (2014). On Unfamiliar Moral Territory: About Variant Embodiment, Enhancement and Normativity. In Eilers, M., Grüber, K. & Rehmann-Sutter, C. (eds), *The human enhancement debate and disability: new bodies for a better life*. Hampshire: Palgrave Macmillan; pp. 23–37.

Scully, J. L., & C. Rehmann-Sutter (2001): When Norms Normalize: The Case of Genetic "Enhancement". *Human Gene Therapy*, 12(1): 87–95.

Selgelid, M.J. (2000). Neugenics? *Monash Bioethics Review*, 19(4), 9–33.

Shakespeare, T. (2013). *Disability rights and wrongs revisited*. London/New York: Routledge.

Shakespeare, T. (2006). *Disability rights and wrongs*. London/New York: Routledge.

Silver, L. M. (1997). *Remaking eden*. New York: Avon Books.

Sleeboom-Faulkner, M. (2007). Social-science perspectives on bioethics: predictive genetic testing (PGT) in Asia. *Journal of Bioethical Inquiry*, 4(3), 197–206

Sparrow, R. (2016). Diversity, Disability and Eugenics, an interview with Rob Sparrow. www.bioedge.org/bioethics/diversity-disability-and-eugenics-an-interview-with-rob-sparrow/11951 (accessed February 21, 2018).

Sparrow, R. (2011). A Not-So-New Eugenics. *Hastings Center Report*, 41(1): 32–42.

Suter, S. (2007). A Brave New World of Designer Babies? *Berkeley Technology Law Journal*, 22(2): 897–969.

Testart, J. (1995). The new eugenics and medicalized reproduction. *Cambridge Quarterly of Healthcare Ethics*, 4(3): 304–312.

Thompson, C. (2005). *Making parents: The ontological choreography of reproductive technologies*. Cambridge, MA: MIT Press.

Twine, R. (2005). Constructing critical bioethics by deconstructing culture/nature dualism. *Medicine, Health Care and Philosophy*, 8(3), 285–295.

Van Balen, F., & Inhorn, M. C. (2003). Son preference, sex selection, and the "new" new reproductive technologies. *International Journal of Health Services*, 33(2), 235–252.

Wilhelm, M., Dahl, E., Alexander, H., Brähler, E., & Stöbel-Richter, Y. (2013). Ethical attitudes of German specialists in reproductive medicine and legal regulation of preimplantation sex selection in Germany. *PloS One*, 8(2), e56390.

Wilkinson, S. (2010). *Choosing tomorrow's children: the ethics of selective reproduction*. Oxford: Oxford University Press.

25

Genomics and insurance

The lock-in effects of a politics of genetic solidarity

Ine Van Hoyweghen

Introduction

When the use of genetic technologies was fostered by the Human Genome Project in the 1990s, one of the most contentious debates on its 'social impact' was to be found in the field of insurance. Should insurers be able to use genetic testing as a means to select out insurance candidates? In 1992, Belgium was one of the first countries in the world to explicitly prohibit insurers from using genetic information (LVO, 1992, Art. 5). As the oral history goes, the wording of Article 5 was scribbled on a beer card by a Belgian policy maker in a Brussels café, after a parliamentary debate where the fear of a 'genetic underclass' was discussed. The topic of genetic discrimination in insurance has since been the focus of public and expert debates worldwide and has resulted in the establishment of Genetic Non-Discrimination Acts (GNDAs) in several countries (Quinn et al., 2014). Almost three decades after the first enactments of GNDAs it seems appropriate to reflect on the origins, rise and consequences of these regulatory approaches in Europe and beyond.

How did the issue of genetic discrimination gain such a momentum? GNDAs are one way of governing the use of genetic information in insurance markets. It is not self-evident that regulators respond to societal concerns by issuing legislation, however. What are the mechanisms that have driven lawmakers to issue GNDAs so swiftly? And what are their consequences? To understand the rise and popularity of GNDAs, it is necessary to analyse the drivers and mechanisms that helped to transform the issue from a problem affecting a small group ('the genetically at risk') into a major public issue, leading to the establishment of regulatory action on a broad scale. This chapter will focus on the capacity of the issue of genetic discrimination to summon as a 'we', the kind of issues that a group of citizens can recognize as shared and thus consider worthy of collective concern and action (Callon, 2007; Stengers, 2014). Indeed, a politics of solidarity has arisen around the notion of 'we are all genetic', where an expansive group is concerned with 'our' genes in insurance. To unravel these dynamics of the problematization of genetic discrimination, the chapter builds on work on the social mobilization of disease (Epstein, 2016) following a pragmatist approach of an issue-based politics of solidarity (Callon, 2007; Rabeharisoa et al., 2012; Moreira, 2012; Fennel, 2016).[1] Having described the drivers and mechanisms resulting in the stabilization of the issue into a politics of genetic solidarity, the chapter documents some important lock-in effects of this framing. It argues that the politics of genetic solidarity has been a productive move for the insurance industry in safeguarding the insurance principles of actuarial risk discrimination.[2] In other words, this framing has enabled the insurance industry to further maintain the status quo of using medical information for its risk selection practices. In doing so, the chapter tells a cautionary tale of the societal translation of contemporary developments in the life sciences. Given all the buzz on postgenomics, the chapter gives a sobering account of the state of the art of 'inhabiting postgenomic worlds' in documenting the problematization of genes in our 'modern' Western social policy worlds and institutions.

'We are all genetic!' – a social policy rally to genetic non-discrimination regulations

To understand the origins and rise of genetic non-discrimination regulations, it is necessary to get more insight into the drivers and mechanisms that fostered the rise of these state regulations. In the European context, discourses about fears of genetic discrimination have been wedded to narratives of a genetic determinism, histories of eugenics, and a social policy insistence upon social inclusion (see Cavaliere & Camporesi in this volume). First, underlying the fear of genetic discrimination has been a deterministic vision of genes as ultimate causal constituents framing the debate in a hyperbole (Beck & Niewohner, 2009). The location of a fundamental difference between humans, seemingly inscribed in the biology of their genes, also evoked shared memories of eugenics as well as raised important privacy concerns that further fuelled fears of genetic discrimination. Lastly, and most importantly, fears of genetic discrimination in Europe were linked to the deep-rooted understanding of insurance as an instrument of solidarity, and not one of discrimination. The issue of genetic testing was considered as *the* exemplary of the decline of an ethos of solidarity and the advent of a 'new social question', expressing wider concerns of social exclusion against the background of recent privatizations of welfare states (Van Hoyweghen, 2010). Within this framing, genetic discrimination became defined as an issue requiring *special* treatment in the public policy arena. This is how the issue got aligned within a paradigm of 'genetic exceptionalism', where genetic information is believed to differ from other medical information, and therefore should be treated in a special way.

Which actors were involved in turning genetic discrimination from a problem originally affecting a small group into such a major public issue? To some extent, genetics-specific patient groups have been active in mobilizing this view of 'genetic discrimination' as special and therefore requiring specific regulatory action. Genetics-specific interest groups, such as the Dutch Genetic Alliance (VSOP) and the Genetic Interest Group (GIG) in the UK have raised awareness of the dangers of genetic discrimination through public media, political campaigns, and lobbying. Genes have served as a symbol for their specific concerns while demanding compensation for their special status in insurance through protection of genetic information from insurance use. Importantly, these genetics-specific patient groups have enabled alliances with other – originally non-affected – publics articulating concern about genetic discrimination as an exemplary case for wider concerns of social inclusion. Further, the issue of genetic discrimination has aligned with supranational institutions and human rights organizations, having developed in important normative documents on genetic non-discrimination. In the EU, the Council of Europe's 'Oviedo Convention of Human Rights and Biomedicine' (CoE, 1997) clearly set the tone by prohibiting any form of discrimination on the grounds of one's genetic heritage and by restricting the use of genetic tests to health purposes or scientific research. This generalization of the concern of 'the genetically at risk persons' by national and supranational policy makers has been very effective in making genetic discrimination a social policy and human rights issue that needed specific regulative attention. However, the issue of genetic discrimination not only stimulated the concerns and compassion of the broader public for those considered 'genetically at risk' but also facilitated the translation of genetic discrimination into an issue that affects 'all of us'. In referring to the idea that 'we all have genes' we are faced with the prospect that genetic discrimination could happen to each and every one of us. As the director of the Genetic Alliance UK stated, 'The fact that we are all susceptible to genetic diseases is a strong factor for solidarity'. This framing of 'genes affecting all of us' has further mobilized the public in advocating for genetic non-discrimination regulations. In this framing, 'we all,' as carriers of genes, have become concerned groups in the issue of genetic discrimination. Genes have functioned as important operators of solidarity, in linking the 'genetically at risk persons' with humankind.

This work of collective integration that genes have enabled in linking people together has made possible a *common* desirability for genetics-*specific* legislative action. This makes apparent how these policy changes were brought about not by genetics-specific patient movements pure and simple, but rather by a *hybrid coalition* linking genetic patients with social policy makers, human rights organizations, administrators and multiple publics. The alignment of these different groups and shared concerns ('genetics' + 'discrimination') have pushed regulators to take swift action. Over the years, genetic non-discrimination regulations have become successful in the sense that this policy approach has spread over Europe, as there started to be a trend to legislate. With these regulations as the gold standard, national state regulators have been aligning with the paradigm of genetic exceptionalism by installing and cementing the new and separate category of 'genetic' into insurance and non-discrimination regulations. In this way, the issue of genetic discrimination has been made an *exception* in the politics of insurance markets by installing unique regulatory protection for genetics in insurance, as an exception to the general principle of actuarial discrimination in insurance.

'We are all genetic' and its social consequences

The above section has demonstrated how the issue of genetic discrimination has resulted into a politics of solidarity around genes. By establishing genetic information as a collectivizing rather than an individualizing force, a *genetic commonhood* is created that expresses a universalist body politics. Preventing genetic discrimination, in turn, becomes a matter of solidarity. This politics of genetic solidarity has been very productive in producing genetic non-discrimination regulations in many countries in the world. Enshrined in a human rights logic as well as a social policy logic of social inclusion, it reflects a normative standpoint that genes are to be protected from discrimination and as such deserve an *exclusionary* and *exceptional* protective status in insurance. But does this *genetic commonhood* really include everybody? It is tempting to argue that there is something universalizing about genes because it is a property that *all* humans contain. The common denominator has been the idea of 'genetic discrimination' that all human beings may face and thus worthy of collective concern (as a common cause) and state action. The capacity of genes to link those 'genetically at risk' with humankind ('we are all affected') creates a significant basis for universalism, where genes produce a common humanity that is grounded in a biological similarity. The common humanity subsists here in an inclusive identity of the genome, a 'genetic commons' or genetic citizenship, framed in terms of a shared biological identity (as 'carriers of genes').[3] This politics of genetic solidarity combining a biological universalism and genetic exceptionalism comes with important social consequences. First, it results in a politics of *exception* based on a biological and economic determinism. The politics of genetic solidarity *flattens* difference by a biological universalism and essentialism ('biological determinism'), while it still underpins the economic principle of actuarial discrimination in insurance ('economic determinism'). The exceptionalism is framed into a genetic determinist approach which forms the base of the 'shared identity'. This biological determinism could however give way to *re-naturalizing* difference (Reardon, 2013; Duster, 2015), which can lead to renewed forms of discrimination (Calhoun, 1995). Second, the exceptionalist status of genes in insurance and its underpinning biological determinism have resulted in the amplification of individual responsibility for lifestyle risks. In the wake of genetic non-discrimination laws, lifestyle risks have been targeted in insurance practice through a gradual shift of the insurers' focus to genetic risk *carriers* versus lifestyle risk *takers*, accentuating a 'control vs. no-control' logic (Van Hoyweghen, 2007). Ensuring legal protections for the 'genetically at risk persons' has ensued increased ratings for so-called deemed 'non-genetic', lifestyle and behavioural risks, such as smoking, fitness, and obesity. The 'shared concerns' for genetic discrimination have resulted in an increasing individual responsibility to lifestyle risks. In other words, there is no 'we' when we discuss lifestyle risks in insurance. There are only *individual* BMIs, fitness steps, and smoking habits to act more and less responsible upon. Responsibility is localized here at the individual level while insurance policies target the actions of *individual* behaviour that are said to be 'in control'. The politics of genetic solidarity thus comes with the effect of emphasizing a logic of control/no-control, articulating lifestyle agency as an *isolated, individual* event where no 'common cause' can be detected.

The lock-in effects of a politics of genetic solidarity

Having described the potential social consequences of genetic discrimination regulations, we see how the generalization of the issue of genetic discrimination into a politics of genetic solidarity seems to come with its own side-effects. Moreover, it has created important lock-in effects that have prevented the issue from further social proliferation.

With the progress in genomics knowledge, including a shift from *genetics* framed in terms of single-gene and complex disorders ('genetic diseases'), towards *genomics* according to methods of large-scale analysis, and *postgenomics* (e.g., epigenetics, showing the complex interaction between genotype, phenotype, lifestyles and other social environments), it has become increasingly difficult to differentiate between 'the genetic' – that which has been 'given' to us – and 'the non-genetic' – that which has been 'made' – for example, lifestyles and behaviours. The rise of the complexity paradigm may conflict with the reductionist legal definitions of the 'genetic' in genetic non-discrimination regulations. This blurring of the boundaries between genetic and non-genetic information in postgenomic approaches may create a new storyline in the insurance policy debate where 'genetic data' is now equated with 'medical data'. In this way, the exceptionalist rights-based logics of genes may be expanded to a rights-based logic for protection from discrimination for 'all disease', further challenging the insurance industry's fundamental 'right to underwrite' (Van Hoyweghen, 2010).

These developments have, however, not materialized. On the contrary. The 10-years-long drafting of the Council of Europe Recommendation on 'the processing of personal health-related data for insurance purposes, including data resulting from genetic tests' (CM/Rec (2016) 8) is a case in point. In 2007 the Council of Europe (CoE) asked the Group of Specialists within the Committee of Bioethics (DH-BIO) for the development of a legal instrument with respect to the use of results from genetic tests in the insurance context (CoE, 2012). Whereas the CoE request initially focused on the topic of 'genetic testing' in insurance, in 2010 the topic was expanded to consider 'medical examinations providing predictive health information other than genetic testing'. In other words, the topic of 'genetic testing' was expanded now to the notion of 'predictive testing', resulting in fierce discussion in the Group on the notions of 'predictivity', 'signs', 'symptoms' and 'lifestyle information' for insurance discrimination. In 2012, the Group of Specialists considered that, before starting the proper elaboration of a legal instrument, they would prepare a *Green Paper* (CoE, 2012) through generating comments by a broad stake-holders' consultation round. In 2016, the legal instrument finally was issued. The notion of 'predictivity' is not present here anymore, and the Recommendation clearly marks out (again) a difference between 'genetic data' and other 'health related personal data' (including lifestyle). It includes a strict ban for member states on genetic tests for insurance purposes, while for other health-related personal data, it asks for strict safeguards (CoE, 2016). In other words, the exceptionalist paradigm for genetic discrimination in insurance has been sustained and further enforced in European policymaking. Genes have still kept their special stature including their protectionary measures. This seems to be the '*bon ton*' of European state makers' public reasoning into the field.

Keeping the status quo – stabilization and the productive forces of a politics of genetic solidarity

How can we account for this stabilization of the social dynamics of genes in insurance? In order to do so, we need to pause and reflect again on the specific drivers and mechanisms of the issue. What is striking has been the swift uptake of the issue into *institutionalized* social policy-making and regulatory pathways. That is, the main drivers in this process have not been patient orga-nizations, but national policy- and law-makers, stimulated by social policy idioms of a 'genetic underclass'. This has resulted in the legal inscription of the novel category of 'genetic' into regulatory classifications and regimes. This successful uptake of a politics of genetic solidarity into the conventional policy infrastructures is, however, also a recipe for its very own failure. From a dynamic 'issue-politics', the issue of genetic discrimination has been crystallized into the stabilized regulatory category of 'genetic' through a biological universalism. At the same time, as a politics of *exception*, it has been aligned into the conventional framing of insurance markets, resulting in a de-politicization of the issue. The stabilized politics of genetic solidarity, with its combined framing of a biological universalism and genetic exceptionalism, has prevented the issue from further *proliferating* difference.

As Rabeharisoa et al. (2012) argue, for the social dynamics of an issue to proliferate, it is important to keep the problematization of issues *dynamic*. As we see in our case, the politics of genetic solidarity has turned 'the genetic' into the stabilized frameworks of the insurance workings. The politics of genetic solidarity has failed to generalize the issue into a *generalization of the exceptionalism* by not turning the issue into a 'shared concern' for general insurance discrimination. Instead, policymakers have chosen to inscribe 'the genetic' into the stabilized institutional frameworks, and to develop their actions *without changing these very frameworks*. The lesson learned here is that the swift inscription of particular issues into *institutionalized* social and legal categories comes with the important risk of *stabilizing* the issue, that way preventing it from a (further) politicization of economic markets. As a consequence, the exceptionalist politics of a genetic solidarity blunts any serious criticism on the politics of insurance market discrimination. While at first sight 'genetic solidarity' may be seen as an operator of more inclusive rights to insurance markets, it turns out that this politics of solidarity is actually productive of enduring a politics of repetition and keeping the status quo in insurance.

In *capturing* the genetic discrimination issue within the conventional social infrastructures with its routine legal and policy classification pathways, we can detect a more generative problem space underpinning these social institutions. The latter are based on a particular 'institutionalized solidarity' historically built on a specific conception of nature *versus* nurture (N/N split), where nature has received the most authority (Daston, 2002; Keller, 2010). This N/N problem-space results in a *specific* manner of attributing agency, responsibility and solidarity to circumscribed actors and motives (Hendrickx et al., 2017). Within this frame, biology as 'naturalized' gets authoritative power in attributing solidarity, resulting in a 'control vs. no-control' logic. Indeed, the very categories of 'lifestyle' and 'behaviour', as omnipresent in contemporary health policies as areas of individual intervention (as against 'inborn nature') only make sense within the N/N problem-frame. The policymakers' drive to govern the genetics issue in insurance has been underpinned by this 'naturalized' public reasoning in terms of 'control vs. no-control'. As Jasanoff (2012) argues, public reasoning is a strong force in 'naturalizing', in becoming a 'natural' thing, such as in this case, to see solidarity distributed as a nature vs. nurture (N/N) split. This demonstrates the power, still very active, of nature-nurture within contemporary social policy institutes and its forms of organized solidarity. This 'naturalized' public reasoning is also omnipresent in insurance and has contributed to keeping the status quo in insurance by 'recruiting' the genetics issue into the institutionalized categories of a solidarity 'as we know it'.

Moreover, this emphasis to a control vs. no-control logic by the genetics issue has in fact been very productive for the insurance industry. Insurers have changed their tactics considerably in recent years, moving from a defensive strategy to the genetics issue towards actually *deploying* the genetics issue to enforce their insurance principles in their current strategies to develop new digital health markets for the 'personalisation of risk'. In line with the control/no-control logic enforced by the genetics issue, the industry has gone 'InsurTech'[4] with the development of new markets focusing on wearable devices and digital health technologies for 'underwriting' in insurance. In other words, the genetic exceptionalist paradigm of GNDAs has not only enabled the insurance industry to depoliticize the issue, this politics of repetition seems to allow for a novel 'disruptive' business-development space where an amplified behaviour-based 'control'-logic is at work (Meyers and Van Hoyweghen, 2017). Genes have not only ensured the status quo of insurance workings but have fuelled legitimization to enforce the principles of actuarial discrimination by differentiating on the basis of the individual's behaviour (which is assumed to be controlled).

Conclusion

In their capacity of flattening or proliferating difference, genes bear the potential of being a catalyst for opening up the political in insurance markets. However, as this chapter has argued, this social dynamics of the problematization of genes has been stabilized in insurance markets through a politics of genetic solidarity. The chapter has documented how genetic discrimination became an issue of shared concern, framed in terms of a 'genetic solidarity' where all of us have genes and as such need to be protected from insurance discrimination. This has resulted in the enactment of genetic non-discrimination acts in many countries worldwide. The very sources for this success story are, however, the same recipe for its shortcomings. The swift inscription of the issue into the *institutionalized* categories of social policy through a politics of genetic solidarity comes with the important risk of *stabilizing* the issue, refraining the social dynamics of genes from further proliferation. By framing dilemmas of genetics and insurance as an *exception*, the issue has been depoliticized and even results in the amplification of the logic of discrimination in insurance markets. In this way, the chapter highlighted the lock-in pathways of appropriating genes in our 'naturalized' biopolitical reasoning in Western social institutions (in terms of a nature-nurture split). It demonstrates the strong 'viscosity' (Ewald, 2012) of the classificatory ordering mechanisms of insurance institutes. The 'genetic' category has been crystallized here within the biopolitical problem-space of an institutionalized solidarity 'as we know it'.

This politics of genetic solidarity is a politics of repetition in keeping the status quo of insurance discrimination and comes with important social effects. It is based on a biological determinism where difference is *flattened* through a biological universalism ('we all have genes') with an essentialist notion of a 'genetic commonhood' as a (biological) property that we all have 'in common'. As a consequence, this politics of genetic solidarity *re-naturalizes* difference, while stressing a 'control vs. no-control' logic which results in the amplification of the individual responsibility for lifestyle risks. There is no 'we' or common cause for lifestyle risks in Western social policy public reasoning. There are only *individual* lifestyle 'risk takers' who act more and less responsibly. On a wider front, the chapter illustrates the frailty of building common causes in contemporary societies (Stengers, 2014). The politics of genetic solidarity has failed to materialize into a politics that works on the *proliferation* of difference. Within this latter framing, genes are not somebody's property ('genetic commons'), but are entities one can 'common around' (Stengers, 2014; Amin and Howell, 2016). This commoning around genes is a pragmatist challenge, where common causes are not based on a shared identity (e.g., biological universalism through *flattening* difference) but on an ad-hoc communing of biosocial sorts, *proliferating* biosocial difference.

The advent of epigenetics and postgenomics has led social scientists to hypothesize on such a reconfiguration of the politics of difference. When the boundary between nature and nurture becomes blurred, this may open up the politics of difference in domains where health and markets are related. However, this chapter has offered a sobering account of the potential of genes in proliferating difference in our societies. When we are in the realms of Western social policy institutes, there are only *individualized* to-be-controlled risks.[5] This demonstrates not only the tenacity of the institutionalized solidarity of the N/N split but above all that this tenacity is exactly there, *because our social institutions and policies keep it well alive* (Hendrickx et al., 2017). Genes have not realized their potential of proliferating *in the wild*, but have been put back by Western policy and state makers. In the social policy realms of our 'modern' insurance institutions then, we can wonder whether 'we have ever been genetic' at all.

Notes

1 This approach takes central the drivers and mechanisms of rendering a singular event into a major public problem. In the field of the social mobilization of disease, Rabeharisoa and others (2012) for example studied the singularization/generalization dynamics of 'rare' diseases, identifying the patterns of bridging of collective action frames and identities that can lead to connections across differences and broader mobilization while building collective causes through proximities between singular situations ('in common'). On a broader level, it matches a general movement which simultaneously affects political and economic life: the connection between the life sciences and 'the proliferation of the social' (Strathern, 1999) mobilizing new concerned groups in society to politicize and reconfigure economic markets (Callon, 2007).

2 Private insurers portray discrimination as a neutral concept and defend that discrimination can be considered fair, when it is based on actuarial principles ('actuarial risk discrimination'), where policyholders are charged a premium in accordance with the risk they bring into an insured pool during the process of risk selection, or 'underwriting'.

3 This notion of a biological universalism is also enshrined in supranational human rights legislation, where 'genetics' has received a unique protective status. For example, the UN Universal Declaration on the Human Genome and Human Rights frames a 'genetic commons' or a 'gene pool' as a stable

property, based on an idea of a 'shared identity' and its related claims to a common human heritage (UNESCO, 1997).

4 Insurtech is a portmanteau of 'insurance' and 'technology' that was inspired by the term Fintech. The belief driving insurtech companies is that the insurance industry is ripe for innovation and disruption through the use of digital technology innovations.

5 While there is increasing mentioning of epigenetics in the social policy domain, again, these accounts are attributing individual responsibility to pre-defined actors (e.g., 'soft heredity') (Lock, 2013; Meloni, 2016; Palsson and Lock, 2016), re-inscribing epigenetics into the conventional biopolitical rationality of a solidarity 'as we know it', with its specific attribution of causes, agency and responsibility.

References

Amin, A. and Howell, P. (Ed) (2016) *Releasing the Commons: Rethinking the futures of the commons.* London & New York: Routledge.

Beck, S. & Niewohner, J. (2009) 'Localising genetic testing and screening in Cyprus and Germany: contingencies, continuities, ordering effects and bio-cultural intimacy'. In: Atkinson, P., Glasner, P. & M. Lock (Eds). *Handbook of Genetics and Society: Mapping the new genomic era.* New York: Routledge.

Calhoun, C. J. (1995) *Critical Social Theory: Culture, History, and the Challenge of Difference.* Cambridge, Mass.: Wiley-Blackwell.

Callon, M. (2007) 'An essay on the growing contribution of economic markets to the proliferation of the social'. *Theory, Culture & Society*, 24, 7–8, 139–163.

Council of Europe (CoE). (1997) Convention for the protection of human rights and dignity of the human being with regard to the application of biology and medicine: Convention on human rights and biomedicine, CETS No. 164, Oviedo.

Council of Europe (CoE) Steering Committee on Bioethics (CDBI). (2012) Consultation Document on Predictivity, Genetic Testing and Insurance. (DH-BIO/INF 1).

Council of Europe (CoE) (2016) Recommendation CM/Rec(2016)8 of the Committee of Ministers to the member States on the processing of personal health-related data for insurance purposes, including data resulting from genetic tests.

Daston, L. (2002) *The Morality of Natural Orders. The Tanner Lectures on Human Values.* Princeton: Princeton University Press.

Duster, T. (2015) 'A Post-Genomic Surprise: The Molecular Re-inscription of Race in Science, Law, and Medicine,' *British Journal of Sociology*, 66, 1, 1–27.

Epstein, S. (2016) 'The politics of health mobilization in the United States: The promise and pitfalls of "disease constituencies"', *Social Science & Medicine*, 165, 246–254.

Ewald, F. (2012) *Assurance, Prevention, Prediction… dans l'univers du Big Data.* Paris: l'Institut Montparnasse.

Fennel, C. (2016) Are we all Flint? *LIMN*, 7, http://limn.it/are-we-all-flint.

Hendrickx, K., Meyers, G., Wauters, A., Van Hoyweghen, I. (2017). Biopolitesse: naar een kosmopolitiek van verantwoordelijkheid en solidariteit. *Sociologos*, 38, 4, 322–341.

Jasanoff, S. (2012) *Science and Public Reason.* New York: Routledge.

Keller, E.F. (2010) *The mirage of a space between nature and nurture.* Durham & London: Duke University Press.

Lock, M. (2013) The Lure of the Epigenome, *The Lancet*, 381, 9881, 1896–1897.

LVO (Wet van 25 juni op de Landverzekeringsovereenkomst) (1992) *Belgisch Staatsblad*, 20 augustus 1992, 18283–18333.

Meloni, M. (2016) *Political Biology: Science and Social Values in Human Heredity from Eugenics to Epigenetics.* London: Palgrave Macmillan Press.

Meyers, G. & Van Hoyweghen, I. (2017) Enacting Actuarial Fairness in Insurance: From Fair Discrimination to Behaviour-based Fairness. *Science as Culture*, 2017, art.nr. http://dx.doi.org/10.1080/09505431.2017.1398223.

Moreira, T. (2012) 'Health Care Standards and the Politics of Singularities: Shifting In and Out of Context', *Science, Technology, & Human Values*, 37, 4, 307–331.

Palsson, G. & Lock, M. (2016) *Can Science Resolve the Nature/Nurture Debate?* London: Polity Press.

Quinn, G., de Paor, A., Blanck, P. (Eds). (2014) *Genetic Discrimination: Transatlantic Perspectives on the Case for a European Level Legal Response.* New York: Routledge.

Rabeharisoa, V., Callon, M., Filipe, A.M., Nunes, J., Paterson, F., Vergnaud, F. (2012) The dynamics of causes and conditions: the rareness of diseases in French and Portuguese patients' organizations' engagement in research. CSI Working Paper Series, nr 026, Paris: Mines Paris Tech.

Reardon, J. (2013) On Science and Justice, *Science, Technology & Human Values*, 38, 2, 176–200.

Stengers, I. (2014) *In Catastrophic Times: Resisting the Coming Barbarism*. Open Humanities Press.

Strathern, M. (1999) 'What is intellectual property after?' In: J. Law and J. Hassard, eds. *Actor network theory and after*. Oxford: Blackwell, 156–180.

UNESCO (1997) *Universal Declaration on the Human Genome and Human Rights*. Paris: UNESCO.

Van Hoyweghen, I. (2007) *Risks in the Making: Travels in Life Insurance and Genetics*. Amsterdam: Amsterdam University Press.

Van Hoyweghen, I. (2010) Taming the wild life of genes by law? Genes reconfiguring solidarity in private insurance'. *New Genetics and Society*, 39, 4, 431–455.

•

26

Power asymmetries, participation, and the idea of personalised medicine

Barbara Prainsack

•

From personalised to 'precision' medicine

The concept of personalised medicine gained considerable traction in the beginning of our millennium. One of the reasons for this was that the Human Genome Project – a multinational endeavour to map a full human genome starting in early 1991 – had been expected to deliver key insights into the causes for health and disease (Hilgartner 2017). It was hoped that it would also lead to the ability to match drug treatments to the genetic makeup of individual patients (Collins et al. 2003; Hedgecoe 2004; Tutton 2014). Today, genetic tests can indeed determine whether a person is genetically predisposed to producing fewer or more of the enzymes involved in the metabolism and intestinal absorption of certain drugs, for example. If a patient produces fewer of these enzymes than other people, she should be given a lower dose, because more of the drug will remain in her body for longer, putting her at a higher risk for adverse drug events. Another example of the translation of findings from the Human Genome Project into the clinic is tumour profiling: if a patient's breast cancer tumour does not express specific hormone receptors, certain types of cancer treatments would have no effect. Such patients can be given different treatments to spare them the side effects.

The use of genetic insights in the quest to develop treatments for complex disorders such as diabetes, late onset Alzheimer's disease, or bipolar disorder – conditions that affect hundreds of millions globally – has been less fruitful. One important lesson learned from the Human Genome Project has been that the DNA sequence of a person alone does not tell us very much. Apart from monogenic disorders where a single genetic factor is often sufficient in causing a disease – such as Tay Sachs, an inheritable fatal neurodegenerative condition – most health problems emerge from a complex interaction of genetic and environmental factors (see also Keller 2010; Gurwitz 2013). The latter include a person's upbringing, her lifestyle, her social and natural environment, and even her environment *in utero*. All of these factors can have an impact on how genes are regulated and expressed, and vice versa. The field of epigenetics is devoted to studying these interactions (Niewöhner 2011; Meloni 2014; Locke and Palsson 2016; Lock, in this volume, and Lloyd and Raikhel, in this volume). Also in cases where people's DNA sequence determines drug response, for example, these findings are typically not unique to specific individuals. Instead they apply to entire groups that share certain DNA markers in common. This situation contributed to the widespread critique in the early 2000s that 'personalised' medicine was a misnomer, because it was not in fact truly personal (see also Juengst et al. 2016). Moreover, people who belonged to an ethnic group in which a specific marker had been found to be particularly common were sometimes seen as if they all had the marker themselves. Practices such as these gave rise to concerns about the emergence of a new molecular racism (Nelson 2016; Duster 2015; Kahn 2013).

Personalisation obtains a broader meaning

Fuelled by the insight that DNA information alone was not very informative about health and disease, and by the greater availability of other molecular and non-molecular datasets, personalised medicine gradually started to obtain a wider meaning (NAS 2011; ESF 2012; Gamma 2013; Lupton 2013; Weber et al. 2014; Montgomery et al. 2016). Many visions of personalised medicine from the mid-2010s onwards have included the use of a broad variety of data from potentially any domain: from clinical care and medical research, but also from public archives, social media, from people's personal devices, wearable sensors, and the Internet of Things (IoT) – networks of interconnected devices that 'speak' to each other without needing human input. Some organisations have preferred the term 'precision medicine' to mark a departure from the older, DNA-focused version of personalised medicine towards an understanding that included much wider datasets, envisaged a more active role for patients in decision making, and was more explicitly patient-centred also in determining goals and measuring outcomes (NAS 2011; see also Juengst et al. 2016).

Participation, surveillance, and corporate power: challenges of and for precision medicine

Patient participation and empowerment

While the broadening of the notion of personalisation can be considered a positive development in many respects, it comes along with its own challenges. First, the strong rhetoric of participation and empowerment within personalised medicine and precision medicine initiatives tends to gloss over the differences in dynamics of participation of empowerment. Calls and measures to 'empower' patients and to strengthen patient participation are underpinned by a wide range of goals and values. It is useful to differentiate between at least four different types of patient empowerment and participation (Prainsack 2017): First *individualistic* empowerment is closely tied to the notion of individual autonomy and choice; it considers patients empowered if they can choose freely between different market options available to them (see also Mold 2010). Discourses of individualistic empowerment tend to focus on formal access and choice instead of the social and economic circumstances that make participation possible. Second, *instrumental* empowerment supports listening to the voices of patients because it helps to improve healthcare delivery, drug development, or other services and processes (e.g. Del Savio et al. 2016). Instrumental empowerment does thus not attribute intrinsic value to patient participation. This is different in the third case, of *democratic* empowerment, where widening the range of people participating in decision-making processes or other social and political practices is seen as having value in itself. Democratic empowerment also places a lot of emphasis on deliberation and the inclusion of underserved or otherwise marginalised groups. The fourth kind of empowerment, *emancipatory* empowerment, places emphasis on liberating people from oppressive hegemonies in practice and in thought, such as women's health or some global health movements, for example.

Teasing apart these different types of empowerment helps us to ask more nuanced questions about what personalised medicine means for patients (see also McGowan et al. 2017). The inclusion of ever wider ranges of information in the types of evidence analysed for the purpose of personalisation, many of which do not come from 'traditional' places such as clinical care or medical research institutions, means that patients are expected to collect and contribute these data. At the same time, patients often have no or only very little say in how these data are used (Prainsack 2017). Moreover, to minimise the risk that large sets of patient data are shared with, or otherwise accessed by, commercial companies for consumer scoring, marketing or more dubious purposes (e.g. Ebeling 2016), it has been suggested that patient data should be stored and managed by patients themselves (e.g. Murgia 2017). While this solution would give considerable control to patients, it would also mean that patients, rather than healthcare providers and other relevant organisations, would bear the primary responsibility for keeping records up to date and confidential.

Continuous health surveillance

Another challenge connected to contemporary visions and practices of precision medicine is surveillance. Traditionally, biomedicine has collected 'snapshots' of data and information about patients, namely when they have a problem and seek help. Some proponents of precision medicine believe, however, that data from in between doctor's visits may hold the key to understanding health and disease (Ausiello and Shaw 2014; Agus 2016). By recording data from people's bodies and lives also during the times when people feel well, they argue, it is possible to create a baseline of physiological functioning for each patient that can serve as a 'control sample' for times of disease. Others argue that such continuous health surveillance would not only make medicine truly personalised, but it could help to create data doubles of patients that we can use to predict health problems before they become discernible in the living person (Zazzu et al. 2013). This agenda, of course, moves genomics even farther away from centre stage (Zwart 2016): Genetic and genomic information, because they are for the most part 'stable', become background information to changes in people's health and disease. The informational gold mines are information on behaviours that vary, such as blood sugar levels, blood pressure, sleeping patterns, or gut microbes, as well as on how they interact with each other (see also Strohman 2003).

New types of corporate power in the health domain

The growing emphasis on continuous monitoring of various aspects of people's bodies and lives means increasing reliance also on consumer technology (Sharon 2016). Consumer technology companies such as *Alphabet* (the parent company of *Google)*, *Apple*, or *Fitbit* do not only produce the devices used for monitoring, but they also regularly own the data that users collect. And more often than not the algorithms that they use to analyse user data, including the calculation of activity scores on the basis of various types of different activities from running after a bus to playing tennis, for example, are treated as commercial secrets (Pasquale 2015).

The increasing availability of digital data and the advances in machine learning have made healthcare a very promising area of investment also for enterprises without a healthcare focus. Machine learning requires large datasets to 'train' software, which increases the appetite for health data on the side of technology companies even further. *Apple*, for example, provides platforms for app developers to help patients, carers, and clinicians to collect, monitor and analyse data for healthcare and fitness purposes. While *Apple* does not develop all the apps offered via their 'health and fitness ecosystem' (Apple 2016) themselves, they serve as a gatekeeper: App developers can offer their services on *Apple*'s phones and watches only if they are approved by the company. In this way, *Apple* assumes the role of a quasi-regulator deciding what instruments *Apple* users can use to monitor their food intake or sleep patterns, for example. Similarly, *Google* and *Amazon*, via their home assistant technologies that help people run their homes by switching lights on and sending e-mails for them, get access to wide ranges of information on behavioural patterns and conversations. Data from such home assistant technologies could be part of comprehensive patient monitoring too (see Tempini and Leonelli, in this volume), as they include information on when people go to bed and wake up, what programmes they watch on television, or how often they have sex.

Because of their know-how and their privileged access to deep learning experts and other leading computer scientists, healthcare providers and insurers often happily partner with consumer technology companies to develop tools for the clinic. The partnership between the National Health Services (NHS) in England and an artificial intelligence company called *DeepMind*, owned by *Google*'s parent company *Alphabet*, is one of the most prominent examples (*Google DeepMind* 2016; Shead 2016; Powles and Hodson 2017): In 2016, to help hospitals react faster to critically ill patients, *DeepMind* received access to over a million patient records without these patients' consent, and without any public debate preceding the move. Another example is the acquisition of *Merge*, an IT company with access to millions of medical images, by *IBM*, to improve their artificial intelligence capabilities (*Merge* 2015). Such access to large sets of patient data, in turn, allows companies to improve their algorithms also for other parts of their services and products (e.g. Hodson 2016).

Although there is also significant public investment in tools and infrastructures for data-driven medicine – such as the U.S. National Institutes of Health 'Big Data to Knowledge' (BG2K) initiative, or England's £4bn Digital Health strategy (Wachter et al. 2016) – some large private corporations have an advantage in that they already have established expertise on deep learning technologies – or the datasets needed for them. Deep learning technologies are particularly useful in diagnostic situations, or in situations where computers have to tell noise from signal, such as in medical imaging. As noted, deep learning systems cannot be merely written, but they need to be 'trained' by real patient data – e.g. by teaching the algorithm which spots on breast X-rays are tumours and which ones are not. This is the reason that the aforementioned artificial intelligence company *DeepMind* needed to access patient data via their partnership with NHS England, and why *IBM* acquired *Merge* to improve their *Watson* software. An example of a company that already has access to such patient data is *23andMe*, a U.S.-based genetic testing company that holds DNA samples, lifestyle and phenotypic information of over a million users, making them one of the largest DNA databases outside of the forensic database sector. This has turned them into an important player in the disease genetics research domain and has helped them to attract significant public funding (Aicardi et al. 2016). An important difference between such privately held databases and those funded and established by public actors – such as the 'All of Us' cohort study for Precision Medicine in the United States, or patient cohorts in publicly funded healthcare systems in Europe such as UK Biobank – is that the former contain a selective group of mostly wealthy, highly educated, and often predominantly white patients. This means that diagnostic tests and treatments developed on the basis of these data are tailored to these groups while they may not be suitable for others (see Lee, in this volume, and Kelly, Wyatt and Harris, in this volume). Publicly funded population-wide cohorts are more inclusive, although diversity is also a notorious problem there (e.g. Burchard et al. 2015; Oh et al. 2015).

Strategies to address power asymmetries in personalised medicine

Roughly a year after then U.S. president Obama announced over 200 million USD of funding for precision medicine in his State of the Union Address in 2015, the Chinese government announced theirs – with the equivalent of about 9 billion USD of funding (Cyranoski 2016). The European Union, with its Innovative Medicines Initiatives – a public–private partnership to accelerate the development of new treatments (imi.europa.eu) and its funding focus on personalised healthcare – as well as many other countries and regions all over the world also dedicated funding to studying how people's bodies, their lifestyles, and their environments interact to affect health and disease. The U.S. precision medicine initiative – which is now known under the name 'All of Us' – stands out in its emphasis on addressing health disparities and the needs of minorities and other understudied populations (NIH 2016). 'All of Us' participants will be asked to make their electronic health records, as well as information on their lifestyles and environmental exposures, available for research. Some of these data will be collected on smartphones and wearable devices. Participants will also have their blood taken, undergo baseline exams for vital signs, medical history, etc., and fill in questionnaires. These data will be used to study the foundations and dynamics of health and disease – with special attention to the health of minorities and people with low socio-economic status – going far beyond genomics.

In terms of data security, the main strategy of the 'All of Us' initiative in the United States is to enhance individual control over the use of personal health data. This is very much in line with the dominant global discourse, which treats better individual-level control as the most effective strategy to mitigate the power asymmetries that the collection and use of health data is embedded in. Scholars and practitioners are working towards making consent more fine-grained (Dove et al. 2012) or dynamic (Kaye et al. 2015). It is also becoming mainstream for organisations and governments to consider individual patients or research participants as having a right to access the data that are held about them in any kind of database (Lunshof et al. 2014; Wilbanks and Topol 2016).

Other scholars consider more individual control over data collection and use as an insufficient remedy of the challenges posed by data-driven personalisation (Leonelli 2016a, 2016b; Prainsack and Buyx 2017). They call for a stronger role of public and *collective* civil society actors in the health data domain. They propose better public oversight over how personal data are collected and used (Pasquale 2015), or for the creation of data commons (Evans 2016). The latter type of collective-action focused approaches are underpinned by the idea that because of the power asymmetries that the production and use of health data is embedded in, data protection is not only an individual right but also a collective and public concern (Casilli 2014). Some have argued that treating health data as an important asset for personal, commercial and health-related value creation requires that those who use data without creating public benefit need to give something back to people and communities that the data come from (Parry 2004), and that better harm mitigation strategies are needed (Prainsack & Buyx 2017).

Conclusions

At the time of the conclusion of the Human Genome Project, personalised medicine was largely understood as matching drug treatments to groups of patients sharing genetic markers in common. A decade later, the concept of personalisation has broadened to include a wide range of other data, ranging from gene expression to gut bacteria to details about physical activity. Importantly, data on physiological functioning of patients should also be included; the idea of continuous patient monitoring with the help of unobtrusive wearable or remote sensors is gaining more traction. The vision of some proponents of precision medicine is to bring these data together to create virtual data doubles of patients that can be studied and observed in order to detect problems before they become visible in the living patient. Henrik Vogt and colleagues call this a technoscientific holism 'directed at all levels of functioning, from the molecular to the social, continual throughout life and aimed at managing the whole continuum from cure to disease to optimization of health' (Vogt et al. 2016: 307).

While the goals of precision medicine, as well as its aspirations to be patient-centred and participatory, are laudable shifts away from a much narrower and genomics-focused understanding of personalisation, they change the power structures in the field in ways that benefit not only patients but also corporate players. Many of the data required for data-driven personalisation do not come from the traditional places such as the clinic or medical research institutions but from people's personal devices, from public archives, or social media. Together with the growing importance of artificial intelligence, access to deep learning expertise and large datasets becomes an important determinant of power and influence in the health domain. Large consumer technology companies have had a head start in both respects, not least because public actors contract private companies to help them digitise and personalise medicine instead of investing in publicly owned infrastructures and technology solutions. This means that personalised medicine unfolds against the backdrop of the growing power of commercial actors that shape practices and rules on the ground.

Against this backdrop it is important not to treat claims to empowerment and participation at face value but to look at the values and goals that they serve. I have argued in this chapter that there are at least four different types of empowerment at play here: *individualistic* empowerment focusing on widening consumer choice; *instrumental* empowerment that treats listening to patients as necessary to obtain other goals, such as improving healthcare; *democratic* empowerment, which treats widening and deepening of participation as having intrinsic value; and *emancipatory* empowerment with a focus on liberating people from oppressive structures or arrangements (see also Prainsack 2017).

Suggestions to address these power asymmetries range from increasing individual control over data to strengthening public oversight to harm mitigation strategies (see also Dove, in this volume). Some public and private actors are also devoting closer attention and greater resources into addressing the needs of underserved and otherwise marginalised populations, such as the 'All of Us' initiative in the United States. It remains to be seen how the changing political leadership and climate will affect initiatives aiming at widening participation. In order to avoid that the pronounced rhetoric of patient empowerment within personalised medicine initiatives obscures our view of the diverse values and goals that underpin these initiatives, it is useful to analyse what type of participation is promoted or expected from patients and whether the personalisation of their healthcare includes aspects that matter to them.

References

Agus, David B. 2016. *The Lucky Years: How to Thrive in the Brave New World of Health*. New York: Simon and Schuster.

Aicardi, Christine, Maria Damjanovicova, Lorenzo Del Savio, Federica Lucivero, Maru Mormina, Maartje Niezen, and Barbara Prainsack. 2016. "Could DTC Genome Testing Exacerbate Research Inequities?" Bioethics Forum – The Blog of the Hastings Center Report, January 20. Available at: www.theha stingscenter.org/Bioethicsforum/Post.aspx?id=7711&blogid=140 (accessed 2 January 2017).

Apple. 2016. Getting the most out of HealthKit [video]. Available at: https://developer.apple.com/ videos/play/wwdc2016/209/ (accessed 2 January 2017).

Ausiello, Dennis, and Stanley Shaw. 2014. "Quantitative Human Phenotyping: The Next Frontier in Medicine." *Transactions of the American Clinical and Climatological Association* 125: 219.

Burchard, Estaban G., Sam S. Oh, Marilyn G. Foreman, and Juan C. Celedon. 2015. Moving Toward True Inclusion of Racial/Ethnic Minorities in Federally Funded Studies: A Key Step For Achieving Respiratory Health Equality in the United States. *American Journal of Respiratory and Critical Care Medicine* 191/5:514–521.

Casilli, Antonio A. 2014. Four Theses on Mass Surveillance and Privacy Negotiation. *Medium* (26 October). Available at: https://medium.com/@AntonioCasilli/four-theses-on-digital-mass-surveillance-and-the-ne gotiation-of-privacy-7254cd3cdee6 (accessed 2 January 2017).

Collins, Francis S., Eric D. Green, Alan E. Guttmacher, Mark S. Guyer. 2003. A vision for the future of genomics research. *Nature* 422/6934: 835–847.

Cyranoski, David (2016). China's bid to be a DNA superpower. *Nature* 534: 462–463.

Del Savio, Lorenzo, Alena Buyx, and Barbara Prainsack. 2016. Opening the Black Box of Participation in Medicine and Healthcare. *ITA-manuscript* 16/01. Available at: http://epub.oeaw.ac.at/ita/ita-manuscrip t/ita_16_01.pdf (accessed 17 January 2017).

Duster, Troy. 2015. A Post-Genomic Surprise: The Molecular Reinscription of Race in Science, Law, and Medicine. *British Journal of Sociology* 66/1: 1–27.

Ebeling, Mary F.E. 2016. *Healthcare and Big Data: Digital Specters and Phantom Objects.* New York City: Palgrave.

Edward S. Dove, Susan E. Kelly, Federica Lucivero, Mavis Machirori, Sandi Dheensa, and Barbara Prainsack. 2017. Beyond individualism: Is there a place for relational autonomy in clinical practice and research? *Clinical Ethics* 12/3: 150–165.

Evans, Barbara J. 2016. Barbarians at the Gate: Consumer-Driven Health Data Commons and the Trans-formation of Citizen Science. *American Journal of Law and Medicine* 42(4): 243–265.

European Science Foundation (ESF). 2012. *Personalised Medicine for the European Citizen – towards more precise medicine for the diagnosis, treatment and prevention of disease.* Strasbourg: ESF.

Gamma, Alex. 2013. The Role of Genetic Information in Personalized Medicine. *Perspectives in Biology and Medicine* 56/4: 485–512.

Google DeepMind. 2016. DeepMind Health: Clinician-Led Technology. Available at: https://deepmind. com/health (accessed 17 January 2017).

Gurwitz, D. 2013. Expression Profiling: A Cost-effective Biomarker Discovery Tool For The Personal Genome Era. *Genome Medicine* 5/5: 41.

Hedgecoe, Adam. 2004. *The Politics of Personalised Medicine: Pharmacogenetics in the Clinic.* Cambridge: Cambridge University Press.

Hilgartner, Stephen. 2017. *Reordering Life: Knowledge and Control in the Genomics Revolution.* Cambridge, MA.: MIT Press.

Hodson, H. 2016. What DeepMind brings to Alphabet. *The Economist* (17 December). Available at: www. economist.com/news/business/21711946-ai-firms-main-value-alphabet-new-kind-algorithm-factory-wha t-deepmind-brings (accessed 2 January 2017).

Juengst, Eric, McGowan, Michelle L., Fishman, Jennifer R., and Settersten, Richard A. 2016. From 'Personalized' to 'Precision' Medicine: The Ethical and Social Implications of Rhetorical Reform in Genomic Medicine. *Hastings Center Report* 46/5, 21–33.

Kahn, Jonathan. 2013. *Race In A Bottle: The Story Of BiDil and Racialised Medicine in A Post-Genomic Age.* New York: Columbia University Press.

Kaye, Jane, Edgar A. Whitley, David Lund, Michael Morrison, Harriet Teare, and Karen Melham. 2015. Dynamic Consent: A Patient Interface For Twenty-First Century Research Networks. *European Journal of Human Genetics* 23/2: 141–146.

Keller, E. F. (2010). *The Mirage of a Space: Between Nature and Nurture.* Durham & London: Duke University Press.

Leonelli, Sabina. 2016a. *Data-Centric Biology: A Philosophical Study.* Chicago, IL.: Chicago University Press.

Leonelli, Sabina. 2016b. Locating Ethics in Data Science: Responsibility and Accountability in Global and Distributed Knowledge Production. *Philosophical Transactions of the Royal Society:* Part A. 374: 20160122. http://dx.doi.org/10.1098/rsta.2016.0122.

Locke, Margaret, and Palsson, Gisli. 2016. *Can Science Resolve the Nature/Nurture Debate?* Cambridge: Polity.

Lunshof, Jeantine E., George Church, and Barbara Prainsack. 2014. Raw Personal Data: Providing Access. *Science* 343(6169): 373–374.

Lupton, Deborah. 2013. The Digitally Engaged Patient: Self-Monitoring and Self-Care in the Digital Health Era. *Social Theory & Health* 11/3: 256–270.

McGowan, Michelle L., Choudhury, Suparna, Juengst, Eric T., Lambrix, Marcie, Settersten, Richard A. Jr., and Fishman, Jennifer R. 2017. 'Let's Pull These Technologies Out Of The Ivory Tower': The Politics, Ethos and Ironies of Participant-Driven Genomic Research. *BioSocieties* online first: doi: doi:10.1057/s41292–41017–0043–0046.

Meloni, Maurizio. 2014. How Biology Became Social, And What It Means For Social Theory. *The Sociological Review* 62/3: 593–614.

Merge. 2015. IBM closes deal to acquire Merge healthcare. 13 October. Available at: www.merge.com/News/Article.aspx?ItemID=667 [accessed 3 January 2017].

Mold, A., 2010. Patient Groups And The Construction of The Patient-Consumer in Britain: An Historical Overview. *Journal of Social Policy* 39/04: 505–521.

Montgomery, Kathryn, Chester, Jeff, and Kopp, Katharina. 2016. *Health Wearable Devices in the Big Data Era: Ensuring Privacy, Security, and Consumer Protection*. Washington, D.C.: American University. Available at: www.democraticmedia.org/sites/default/files/field/public/2016/aucdd_wearablesreport_final121516.pdf [accessed 29 December 2016].

Murgia, Madhumita. 2017. How Smartphones are Transforming Healthcare. *The Financial Times* (17 January). Available at: www.ft.com/content/1efb95ba-d852-11e6-944b-e7eb37a6aa8e?segmentid=acee4131-99c2-09d3-a635-873e61754ec6 [accessed 15 January 2017].

[US] National Academy of Sciences (NAS). 2011. *Toward Precision Medicine: Building a Knowledge Network for Biomedical Research and a New Taxonomy of Disease*. Washington, DC: NAS.

National Institutes of Health (NIH) (2016). NIH funds precision medicine research with a focus on health disparities. 28 July. www.nih.gov/news-events/news-releases/nih-funds-precision-medicine-research-focus-health-disparities [accessed 26 December 2016].

Nelson, Alondra. 2016. *The social life of DNA: Race, reparations, and reconciliation after the genome*. Boston: Beacon Press.

Niewöhner, Jörg. 2011. Epigenetics: Embedded Bodies And The Molecularisation of Biography and Milieu. *BioSocieties* 6/3: 279–298.

Oh, Sam S., Joshua Galanter, Neeta Thakur, Maria Pino-Yanes, Nicolas E. Barcelo, Marquitta J. White, Danielle M. de Bruin, Ruth M. Greenblatt, Kirsten Bibbins-Domingo, Alan H.B. Wu, Luisa N. Borrell, Chris Gunter, Neil R. Powe, and Esteban G. Burchard. 2015. Diversity in clinical and biomedical research: a promise yet to be fulfilled. *PLoS Medicine* 12/12: e1001918.

Parry, Bronwyn. 2004. *Trading The Genome: Investigating The Commodification of Bio-Information*. New York City: Columbia University Press.

Pasquale, Frank. 2015. *The Black Box Society: The Secret Algorithms That Control Money and Information*. Cambridge: Harvard University Press.

Powles, Julia, and Hal Hodson. 2017. Google DeepMind and healthcare in an age of algorithms. *Health and Technology* [online first: doi doi:10.1007/s12553–12017–0179–0171]

Prainsack, Barbara. 2015. Is personalized medicine different? (Reinscription: The Sequel). Response to Troy Duster. *British Journal of Sociology* 66/1: 28–35.

Prainsack, Barbara, and Alena Buyx. 2017. Thinking Ethical And Regulatory Frameworks in Medicine From The Perspective of Solidarity on Both Sides of The Atlantic. *Theoretical Medicine and Bioethics* 37: 489–501.

Prainsack, Barbara. 2017. *Personalized Medicine: Empowered Patients in the 21st Century?* New York City: New York University Press.

Prainsack, Barbara, and Alena Buyx. 2017. *Solidarity in Biomedicine and Beyond*. Cambridge, UK: Cambridge University Press.

Sharon, Tamar. 2016. The Googlization of Health Research: From disruptive innovation to disruptive ethics. *Personalized Medicine* 13/6: 563–574.

Shead, Sam. 2016. The NHS refused to reveal how much it's paying Google's DeepMind. *BusinessInsider* (24 December). Available at: www.businessinsider.de/nhs-deepmind-contract-foi-2016-12?r=UK&IR=T (accessed 2 January 2017).

Strohman, Richard C. (2003) Genetic Determinism as a Failing Paradigm in Biology and Medicine. *Journal of Social Work Education*, 39:2, 169–191.

Tutton, Richard. 2014. *Genomics and the Reimagining of Personalized Medicine*. Farnham: Ashgate.

Vogt, Hendrik, Bjorn Hofmann, and Linn Getz. 2016. The New Holism: P4 Systems Medicine and the Medicalization of Health and Life Itself. *Medicine, Health Care, and Philosophy* 19/2: 307–323.

Wachter, Robert, and National Advisory Group on Health Information Technology in England (2016). *Making IT work: Harnessing the power of health information technology to improve care in England*. London:

Department of Health. Available at: www.gov.uk/government/publications/using-information-techno logy-to-improve-the-nhs (accessed 2 January 2017).

Weber, Griffin M., Kenneth D. Mandl, and Isaac S. Kohane. 2014. Finding the Missing Link for Big Biomedical Data. *The Journal of the American Medical Association* 331/24: 2479–2480.

Wilbanks, John T., and Eric J. Topol. 2016. Stop the privatization of health data. *Nature* 535: 345–348.

Zazzu, Valeria, Babette Regierer, Alexander Kühn, Ralf Sudbrak, and Hans Lehrach. 2013. IT Future of Medicine: from molecular analysis to clinical diagnosis and improved treatment. *New Biotechnology* 30/4: 362–365.

Zwart, H. A. E. 2016. The obliteration of life: depersonalization and disembodiment in the terabyte era. *New Genetics and Society* 35/1: 69–89.

27

Excavating difference: race in genomic medicine

Sandra Soo-Jin Lee

Recent advances in genomic sequencing technologies have resulted in increased focus on population differences. They have also generated interdisciplinary debate over the relationship between race and genes (*See Prainsack this volume and also Fullwiley and Gibbon, this volume*). Despite the adage that "there is more genetic variation within populations than between," (Lewontin 1972, Cavalli-Sforza 2005) groups have increasingly become the unit of analysis for research related to precision medicine. Classification of populations continues to be a salient taxonomy in genetic research, not only ascribing identity to individuals, but also onto the biospecimens that are donated and stored in biobanks. Although some predicted that the "genomic revolution" spurred by the completion of the Human Genome Project would render "race" obsolete, social scientists have demonstrated how deeply entrenched racial taxonomies are in genetic science, continuing to have downstream impact on the practice of clinical medicine (Reardon and TallBear 2012, Montoya 2007, Pálsson 2009, Koenig et al. 2008, Lee 2015).

Over the last several decades, social scientists have considered how the collection and classification of genetic samples have the potential to reify the relationship between genetics and race, and recapitulate enduring assumptions about racial biology (Kahn 2014, Lee 2006, Morning 2009). This chapter discusses how scientific practices have contributed to a powerful infrastructure that maintains a focus on race as the predominant axis for differentiating human populations and for understanding the meaning of genetic variants for this concept. It will describe the techniques of "excavating" difference from the human genome and the practices of naming and sorting that entrench race as a fundamental category in genomic medicine.

In 2005, a meeting on Human Genetic Variation and Health was held at the Royal Society in London. The meeting was organized by David Goldstein, a population geneticist and currently Director of Genomic Medicine at Columbia University, along with Michael Stumpf, Professor of theoretical systems biology at Imperial College London and Nicholas Woods, Professor of Genetics at University College, London. The meeting convened a select group of scientists working in the emerging field of pharmacogenomics that seeks to determine how genetic variation impacts response to medicines. As an important arena for realizing the significant investment in the genomic sciences, the meeting was an opportunity to discuss how to move the field forward toward clinical translation. In opening the meeting, Goldstein referred to an article that had been published earlier in the month by the *Wall Street Journal* entitled, "DNA Disease Links

Aren't Always as Real as They First Appear" (Begley 2005). The focus of the article was the high rate of false positive findings in pharmacogenomic research that suggested a problem of replication. Lamenting the high-profile critique of the field's progress, Goldstein cautioned that "when our dirty laundry gets aired in such a public way, it's time to clean up our act." Goldstein's statement conveyed the urgency for new and transformative research methodologies that would address the lack of validation hampering the emerging field of pharmacogenomics. Goldstein emphasized that in order to demonstrate the efficacy of their research and progress toward identifying genetic variants with drug response, scientists would need to control the confounding effect of population differences.

Not a new problem, confounding and the need to control for population stratification has long been a challenge in human genetic research. Early work prior to the completion of the Human Genome Project demonstrated that failure to account for population stratification may lead to false statistical associations between a disease phenotype and arbitrary markers that have no causative properties (Lander and Schork 1994, Slatkin 1991). The problem occurs when the population under study is assumed to be homogeneous with respect to allele frequencies but in fact comprises subpopulations that have different allele frequencies of the gene in question. If these subpopulations vary randomly in their frequencies of specific genes, then subpopulation membership can be a confounder and an association between the candidate gene and disease may be incorrectly estimated without properly accounting for expected variation from population structure. The existence of genetic subgroups or substructure in a population may lead to spurious associations if the subgroups are not equally represented in cases and controls. For example, if one subgroup has a higher prevalence of disease, then this subgroup will likely be overrepresented among cases compared to controls. Scientists understand that this may alter their results where any allele that has a higher frequency in that subgroup by chance may appear to be falsely associated with the disease. Theoretically, if cases and controls are matched by their genetic ancestry, then the confounding due to population stratification should be eliminated (Cardon and Palmer 2003). Unfortunately, in most cases the relevant population structure may not be known.

To address this significant challenge, the search for ancestry has become a focal point in an effort to minimize spurious results in pharmacogenomics and disease association studies. These are studies that identify proposal ancestry through the use of admixture markers that showed the greatest variation between continental populations (Gravel 2012). In addition to creating larger sets of samples to sufficiently power studies, Goldstein and others at the Royal Society meeting emphasized the need for careful characterization of DNA, suggesting that in practice, it may not be possible to precisely match cases and controls by genetic ancestry based on self-report, especially in admixed populations. "The challenge is," Goldstein suggested, that "individuals may not be completely aware of their precise ancestry and we need better techniques to determine this information."

The Royal Society meeting agenda made plain a shift in focus on population differences in pharmacogenomic research and the explicit use of ancestry estimation as a research tool. Admixture mapping relies on using sets of autosomal ancestry informative markers (AIMS) that have been identified over the past several years to distinguish major ethnic groups by their genetic ancestry. Single nucleotide polymorphisms (SNPs) distributed throughout the genome that have very large differences in allele frequencies between African and European continental populations, but are not known to have functional specificity have been used to distinguish sections of the genome from each other. These SNPs have been selected to show little difference between disparate African populations and few differences in allele frequencies between different European subpopulations. The major axis of differentiation is between continents.

The term admixture refers to the result of two different "parental" populations with different allelic frequencies, or genetic profiles that combine to produce what scientists refer to as "admixed" offspring whose genomes are mixtures of the genomes represented in the parental populations. Scientists suggest that it is most advantageous to apply this approach to populations that have descended from a recent mix of two ancestral groups that have been geographically isolated for many tens of thousands of years. The prototypical example used by scientists are African Americans who have both West African and European American ancestry. The approach assumes that in close proximity to a disease causing or drug response gene there will be more evidence of specific ancestry from the population that exhibits greater risk of disease or influencing particular drug response. If one can calculate the ancestry along the genome for an admixed sample set, one can use this information to identify the gene variants of a particular disease or drug response.

At the London meeting, admixture was described in terms of a "mosaic" illustrated by three sets of chromosomes: African, European and a combination of African and European. Recombination of these sets of chromosomes through time over generations ultimately resulted in a contemporary set where each chromosome is interspliced by red and yellow components, representing a mixture of African and European ancestry. Suggesting a mosaic mixture where individual parts come together to make a discernible pattern, admixture mapping attempts to desegregate parts of the genome through time and identify sections by racial ancestry. As such, chromosomes from an admixed individual – African American – might be depicted with red sections that are considered "European" and yellow sections that are considered "African" in origin. The goal of admixture mapping is to identify where these different sections begin and end along the genome and thus, minimize the possible location of where a functional variant may reside. To create an admixture map, scientists have searched through databases containing approximately hundreds of thousands of SNPs for which frequencies have been estimated in African and European population samples.

These techniques have been described by some geneticists as disentangling the "mixing of two or more genetically differentiated populations" (Bamshad et al. 2004). The technique of admixture mapping is emblematic of the reification of race and genes as it is based on assumptions of discrete differences – sections of the genome identified as European are asserted as categorically distinct from the parts of the genome labeled African. In this way, human DNA is parsed and disassembled into components where the rubric of difference allows for only the African and the European, leaving large swaths of DNA as "translucent" or putatively raceless and not having any origin. Omi and Winant (2014) offer a framework for understanding this process of *racialization*, which they define as occurring when racial meaning is ascribed to a relationship that was previously unclassified. Sections of the genome – the combination of bases and pairs – are imbued by race through admixture mapping. In this way, the genome becomes the source of race, where the translucent bits and pieces of DNA are increasingly excavated for their underlying "color" through ever more efficient and extensive identification of genetic markers.

The utility of admixture mapping emerges as a solution to the inherent "messiness" of applying racial categories. This is a particular dilemma for scientists working with populations from the Americas and their relatively recent colonial history of human migrations. Studies of Latin America reveal how race is inextricable from the history of the nation (Wade 2014, Santos et al. 2009). The U.S. category of "Hispanic" presents challenges for genome wide association studies given that this general population is applied to many heterogeneous groups that are believed to represent broad genetic variation. Admixture mapping is used by scientists, for example, to genetically distinguish Mexicans from Puerto Ricans. Racial and ethnic difference, as one genome scientist characterized it, is "noise" that needs to be identified and then controlled in pursuit of the genes that are functionally important for particular diseases or drug response.

The problem of replication in genetic ancestry research and the challenges of controlling population stratification makes admixture mapping approaches welcome new strategies. Scientists have argued that non-replication may be the result of poorly defined phenotype, such as attributed racial and/or ethnic identity to individuals recruited into studies, which make it difficult to identify, with any precision, how underlying genetic substructure among populations, may ultimately impact study results. Although these characteristics may not have a clear causative role at the molecular level for a particular drug response, failing to account for variation within a study population of what is often interchangeably referred to as "black," "African," "African-American," "white," "Caucasian," or "European," can misconstrue findings. Scientists suggest that narrow phenotype definition may be a simple approach to clarifying the genetic architecture of study populations and may increase rates of replications (Sillanpaa and Auranen, 2004). This suggestion belies the ongoing challenge of defining what race, ancestry and difference mean in the context of designing genetic association studies. It does not explain the dogged, enduring focus on race as simultaneously the primary organizing principle and technical challenge. In our study that examined how genetic epidemiologists incorporated concepts of population differences and race into their research, one molecular biologist points out:

> When you identify a variant that's associated with the trait, you want to know: is that variant on the European section of that person's genome or the African section of the genome? That's not a trivial issue and it's certainly not a trivial thing to sort out. So when we find their variant is associated with a trait in European-Americans and then that's replicated in the African-Americans, is that the European-American showing up in the African-American? It just takes a lot more sorting out.

In this way, sorting it out begins and ends with what Amade M'charek has said are the fictions of race-making – "the fictions that help to 'clean up the mess,' the messiness of the coexistence of different facts in tension and in conflict with one another" (M'charek 2013: 441).

Bioethicists Pamela Sankar and Mildred Cho have shown that the language used by scientists in reporting results of population differences in the biomedical literature reveals the problem in its potpourri of terms that transect different and often incomparable units of analysis. They write:

> Non-equivalent use of labels is illustrated by the common juxtaposition of terms such as "white" with "African-American," where skin color and geographical location are treated as equivalent. Another example is the juxtaposition of "Asian-American" with "Mexican-American," which implies that people of Asian ancestry now living in the United States represent a level of genetic diversity that is equivalent to that of people of Mexican ancestry now living in the United States. Such examples indicate a need for more consistent attention to definition of groups and to the need to explain the rationale for their equivalence.
> *(Sankar and Cho 2002: 1337)*

An enduring challenge identified by some scientists working on problems related to population genetics is the inability to find neutral terms for genetic difference. The question of language is a hallmark issue that emerges often when scientists participate in discussions on implications of population-based genomic research for social understandings of difference.

Geneticists, sensitized to the potential controversy over identifying populations in genomic research, have suggested alternatives to the term, "race," proposing "ancestry" or "biogeographical ancestry" instead. Population differences are self-evident according to scientists who cite studies that suggest a tight relationship between patterns of genetic variation and geographical ancestry. Over the last decade there has been growing consensus among scientists that with a high degree of accuracy and reliability individuals can be consistently allocated to groups that represent broad geographical regions using a relatively modest number of multilocus genotypes. The challenge for scientists is not whether differences exist, but how to sidestep what one genome scientist in the study described as "the potential landmine of race." Scientists struggle with how to speak about human genetic variation without risking reductionism and reification of race. The landmine refers to the history of constructing the infrastructure and coalescing the basic materials for genomic research that has long used racial taxonomies.

There is danger in placing too much emphasis on linguistic solutions such as admixture mapping at the expense of defining conceptually what work race and racial categories do in the context of human genetic variation research. The efforts toward finer precision where the human genome can be parsed into sections of different continental ancestries must include serious consideration of what African chromosomes mean in the context of society in the twenty-first century and how such discourse may alter how we think about race in everyday life. Assumptions among scientists of the biological underpinnings of race contribute to the new trajectory of scientific inquiry for what anthropologist Duana Fullwiley describes as the *molecularization of race* – the process by which the concept of race becomes inextricable from genetics and genetic differences (Fullwiley 2007).

Scientists struggle to seek new language to characterize what they understand to be the highly circumscribed search for population differences in genetic ancestry in the context of their specific research questions. The challenge for scientists is maintaining the borders in which population difference is constituted and interpreted. These borders of inquiry and responsibility begin to crumble when a shift in vantage point brings into view the populations from which samples are obtained. Race, the product of prevailing social values that inform political arrangements and structures of power, is the prism through which science is realized. Social studies of genomic science can reveal how political history has determined who is sampled, whose chromosomes are painted yellow vs. red, and how much of the genome is translucent. Race emerges in the relationships that are drawn and the racial rubric determines the kinds of question asked and the possible permutations of the comparisons drawn. The metaphors we use to describe difference inform our visions of the world and make us act on the world in the way we perceive it or perhaps want to envision it or, indeed, may also prevent us from acting. It is clear that what is perceived as real or not real is real in its consequences.

Despite this, the scientist's dilemma of finding language that sufficiently brackets his or her work in controlling for population stratification is a red herring for those who are concerned with the new subjectivities in human genetic variation research that go beyond race. The debate over whether ancestry is, in reality, race in new shoes occludes from view the implications of new techniques such as admixture mapping, which does not actually address race at the population level and, in fact ignores the formation of social groups.

Social scientists have demonstrated how the collection and classification of genetic samples have the potential to conflate and render invisible contextual factors that may significantly impact the onset of disease. Created through a social process and imbued by a set of values and assumptions, critical resources for genomic medicine such as biobanks are embedded with the taxonomies that reflect how samples are named and represented historically. In this way, biobanks are artefacts that reveal how societies understand "difference" as critical biomedical facts. Analyzing the biobank as a cultural artefact demands a framework of technological politics that considers how technical systems emerge in response to technological imperatives (Koenig 1988).

Science, Technology and Society (STS) scholars have demonstrated in detail how technical systems are created by the conditions of modern politics. Winner (1980) emphasizes that although not everything can be reduced to the interplay between social forces, technical systems matter and reveal that political phenomenon are integral to scientific practice and knowledge. The biobank in its organization of human DNA is infused with the politics of representation and informs the sociopolitical agenda that infuses scientific priorities and interpretation of data.

Acknowledging that biobanking is the result of social and political processes is essential to an integrated approach to understanding how genomics emerges as a solution to some of the most enduring biomedical challenges, such as eliminating health inequities (*cross-refs to other chapters addressing this same goal*). Many have cautioned against an exclusive focus on genetics at the expense of seriously considering the impact of social context on disease and illness. Scientists must acknowledge and account for the prism of race and the taxonomies of racial kinds that are used to characterize difference to create viable and impactful models that integrate the lived experience of race. Only by excavating the full range of social and environmental factors that contribute to disease and that interact with the human genome can we intervene on persistent inequity and health disparities.

It is difficult to imagine where social justice may emerge from the ongoing search for difference in genomics that has now come of age in the era of precision medicine. For in the turn toward statistical race (Lee 2009), the individual genome, rather than populations, becomes the terrain for race making. Prompted by a need for replication and for controlling population stratification, scientists pursue ancestry informative markers that parse the human chromosomes with continental origins with ever greater precision. These markers, chosen for their ability to distill difference between and not within continental groups, recapitulate the racial rubric that assumes original pure parental populations of Africans, Europeans, Asians and Native Americans. Scientists map back the intrinsic hybridity of contemporary genomes with greater precision, re-inscribing difference into smaller, distinctive parts. However, statistical race remains at the level of DNA, with little understanding of how identification of the parts will translate into clinical applications that will address the political economy of health. Merely a tool to eradicate cumbersome noise, statistical race in the context of genomic medicine does little to disrupt the durable categories of differences built by excavating the genome for race.

References

Begley, Sharon (2005) DNA-disease links aren't always as real as they first appear. *The Wall Street Journal.* Jan. 14.

Bamshad, M., Wooding, S., Salisbury, B. A., Stephens, J. C. et al. (2004) Deconstructing the relationship between genetics and race. *Nat Rev Genet*, http://isites.harvard.edu/fs/docs/icb.topic185351.files/RACEgen.pdf.

Cardon, L. R. and Palmer, L. J. (2003) Population stratification and spurious allelic association. *Lancet*, 361 (9357): 598–604.

Cavalli-Sforza, L. L. (2005) The Human Genome Diversity Project: past, present and future. *Nat Rev Genet*, 6(4): 333–340.

Fullwiley, Duana (2007) The molecularization of race: Institutionalizing human difference in pharmaco-genetics practice. *Science as Culture*, 16, 1.

Gravel, S. (2012) Population genetics models of local ancestry. *Genetics*, 191: 607–619.

Kahn, Jonathan. (2014) *Race in the Bottle: The Story of BiDil and Racialized Medicine in a Post-Genomic Age.* New York: Columbia University Press.

Kleinbaum, D. G., Kupper, L. L. and Morgenstern, H. (1982) *Epidemiologic Research: Principles and Quantitative Methods.* New York: Van Nostrand Reinhold.

Koenig, Barbara, Sandra Soo-Jin Lee and Sarah Richardson (2008) *Revisiting Race in a Genomic Age.* Studies in Medical Anthropology Series. New Brunswick: Rutgers University Press.

Koenig, B. (1988) The Technological Imperative in Medical Practice: The Social Creation of a "Routine" Treatment. In *Biomedicine Examined*. M. Lock and D. Gordon, eds. Pp. 465–496. New York: Kluwer Academic Publishers.

Lander, E. S. and Schork, N. J. (1994) Genetic dissection of complex traits. *Science*, 265: 2037–2048.

Lee, Sandra (2009) Defining Statistical Race and Phenotypic Race and Their Implications for Health Disparities. *Current Pharmacogenomics and Personalized Medicine*, 7, 238–242.

Lee, Sandra Soo-Jin (2015) Biobank as Political Artifact. Special Issue: "Race, Racial Inequality, and Biological Determinism in the Genetic and Genomic Era." *The Annals of the American Academy of Political and Social Science*. September, 661(1): 143–159. DOI: doi:10.1177/0002716215591141.

Lee, Sandra Soo-Jin (2006). Biobanks of a 'racial kind': mining for differences in the new genetics. *Patterns of Prejudice*, 40(4–5).

Lewontin, Richard C. (1972). The apportionment of human diversity. *Evolutionary Biology*, 6: 381–398.

M'charek, Amade (2013) Beyond Fact or Fiction: On the Materiality of Race in Practice. *Cultural Anthropology*, 28, 3: 420–442.

Montoya, Michael J. (2007) Bioethnic Conscription: Genes, Race, and Mexicana/o Ethnicity in Diabetes Research, *Cultural Anthropology* 22(1): 94–128.

Morning, Ann. (2009) Toward a Sociology of Racial Conceptualization for the 21st Century. *Social Forces* 87(3): 1167–1192.

Omi, M. and Winant, H. (2014) *Racial Formation in the United States*. New York: Routledge Press.

Pálsson, G. (2009) Biosocial relations of production. *Comparative Studies in Society and History*, 51(2): 288–313.

Reardon, Jennifer and Kim TallBear (2012) Your DNA is our history. *Curr Anthropol* 53(5): s233–s245.

Sankar, P. and Cho, M. (2002) Toward a New Vocabulary of Human Genetic Variation. *Science*, 298: 1337–1338.

Santos, Ricardo Ventura, Fry, Peter H., Monteiro, Simone, Maio, Marcos Chor, Rodrigues, Jos Carlos, Bastos-Rodrigues, Luciana, and Pena, Sergio D. J. (2009). Color, Race, and Genomic Ancestry in Brazil Dialogues between Anthropology and Genetics. *Curr Anthropol*, 787–819.

Sillanpaa, M. J. and Auranen, K. (2004) Replication in genetic studies of complex traits. *Ann Hum Genet*, 68: 646–657.

Slatkin, M. (1991) Inbreeding coefficients and coalescence times. *Genet Res*, 58: 167–175.

Wade, Peter (2014) *Mestizo Genomics: Race Mixture, Nation and Science in Latin America*. Durham: Duke University Press.

Winner, Langdon (1980) Do Artifacts Have Politics? *Daedalus*, 109(1): 121–136.

28

Genomics in emerging and developing economies

Duana Fullwiley and Sahra Gibbon

Low and middle-income countries have become a site of increasing research interest and investment with the transnational expansion and spread of genomic knowledge and technologies (Kumar 2012, Seguin et al. 2008). This reflects a dynamic terrain in which genomics is being harnessed to address a range of healthcare challenges. With initiatives such as *MalariaGEN* (Malaria Genomic Epidemiology Network 2008) this includes not only infectious diseases, but also the growing financial and social burden of common diseases as these intersect with genomic susceptibility amid larger environmental forces and economic precariousness (Fullwiley 2011, H3Africa 2013). At present genomics is increasingly situated as a pathway to global health, framed as a useful tool to understand the genetic patterning of ailments affecting all of humanity in its diversity (Bustamante et al. 2011, Popejoy and Fullerton 2016). These developments demand critical and engaged social science attention (Beaudevin and Pordie 2016, Taussig and Gibbon 2013). Central to this analysis is how countries, regions, and populations themselves are amassing 'genomic profiles', where the unique characteristics of their populations are now thought to be of exceptional value. The ways that peoples, tribes, races, and 'nation-state' groupings came to be conceived of in population genetic terms have diverse genealogies tied to how their value is perceived today in practice. These issues of genomics in developing economies link transnational research and investment to the questions of inclusion and social justice that social scientists have explored in Europe and the United States (Fullwiley 2008, Nelson 2016, Reardon 2012, Lee and Prainsack *this volume*), albeit in new ways.

Some social science literature is emerging on how notions of inclusion and population specificity are unfolding outside Euro-American contexts (Benjamin 2009, Fullwiley 2011, Gibbon 2015, Sleeboom-Faulkner 2010, Wade et al. 2014, Whitmarsh 2008). In this chapter we further these conversations by focusing on two regions of the world, Latin America and Africa, to show how diverse histories of colonialism as well as entrenched structural inequalities inform the limits and possibilities of genomic research and medicine. We present case studies of research on identifying disease biomarkers in cancer genetics and pharmacogenomic research related to warfarin dosage in Brazil and a wide range of technologies used in sickle cell programmes in Senegal, as well as the H3Africa initiative (which is currently funding major genomics studies in Benin, Botswana, Ethiopia, Mali, Nigeria, South Africa, Uganda, and Zimbabwe). In Senegal the most common technology to assess sickle cell markers in clinical care and research is still basic electrophoresis, while several specific pharmacogenetic and population health investigations have taken place collaboratively between scientists in France and Senegal. Researchers working in varied sites under the rubric of H3 Africa employ or plan to use 'genotyping chips used in genome-wide association studies, sequencing of all the genes in the human genome (called exome sequencing) and eventually, whole genome sequencing' (H3 Africa 2013). In late 2016 the genome sequencing company Illumina announced the development of a special sequencing chip specifically for the project, called the 'H3Africa Consortium Array'.[1] At stake in all of these instances is how questions of human diversity, global inclusion, equity and ethics are actively shaping how genomics is proceeding in these contexts.

Genomic medicine and research in Brazil: the historical legacy and contemporary relevance of population diversity

The notion of Brazil as a 'racial laboratory' has fascinated and facilitated the interests of both foreign and Brazilian scientists since the mid-nineteenth century with the themes of race and racial mixture integral to thinking about the country's historical formation (De Souza and Santos 2014). As a result Brazil has long been constituted as both a valuable site and source of genetic data. Visits and exchanges by North American geneticists from the 1930s were central to the development of national and international fields of genetic research and medicine (Sequeiros et al. 2015). As Ricardo Santos and colleagues suggest, Brazil was constituted as a 'significant site of cognition' in the production of knowledge related to population genetic variation after World War II (2014). Like other areas of the 'global south' (Anderson 2012) Brazil was both a resource and domain of investigation regarding new methods and approaches where, as part of a broader internationalisation of science, it was constituted as an 'idealised' field site for inquiry into and understanding of human biological variation. While 'miscegenation' meaning the 'mixing of different racial groups' was often seen as foundational to scientific inquiry, both 'mixture' and 'purity' could be examined in relation to the so-called 'race question' (Santos et al. 2014); themes which also have a long history in Brazil of being linked to public health (Lima 2007). In the twenty-first century the dynamics between genomics, population variation and nationhood have continued to be generative not only in Brazil but also within the wider context of Latin America in countries such as Mexico and Columbia (Wade et al. 2014). While an emphasis on 'tri-hybrid' genetic ancestries (European, Indigenous, and African) has been commonly used to parse admixture across the region, the use of stable parental population categories and in some cases an emphasis on more 'regional' homogenised specificity, point to the diverse articulations of what Wade et al. refer to as 'mestizo genomics' (2014). In Brazil particular colonial and post-colonial histories of race classification and discrimination, a national and nationalising discourse of race mixture and a contemporary focus on multiculturalism in areas of health and education shape the meaning and significance of molecularised categories of population difference (Kent et al. 2014).

The fields of genomic research and medicine relating to cancer genetics and pharmacogenomics in Brazil are informed by this historical legacy and also by the recent contemporary global medical and scientific relevance attributed to human genetic variation. The work to identify cancer-susceptibility variants or develop pharmacogenetic targets and understand the various contributions of environment and biology to disease aetiology or drug response (Suarez-Kurtz 2010, Palmero et al. 2009) is directly linked to the ability in Brazil to 'include and account for genetic heterogeneity' (Gibbon et al. 2011). Yet population mixture is also dynamically constituted in ways that bring into view unity and diversity or national uniqueness and transnational utility. In the context of Brazilian cancer genetics an emphasis on addressing rising rates of cancer through examining genetic variation in terms of 'needing to know and characterise…the particular aspects of our population' (INCA 2009) would appear to lay claims to genomic sovereignty (Benjamin 2009). At the same time this articulation of specificity creates potentials for different kind of leverage. Whilst facilitating an articulation of national need linked to certain population characteristics it can also be situated as a resource within a global and globalising discourse about the scientific or medical value and utility of identifying diverse patterns of human genetic variation.

Pursuit of these fields of genomic medicine in Brazil reflect therefore the broader international and transnational terrain in which a focus on population difference and diversity has shaped and continues to shape post-genomic research (see Lee, *this volume*). As a result, an ability in Brazil to address what is framed as extensive diversity has engendered a range of heterogeneous research collaborations, networks and consortiums. In the case of pharmacogenetics this includes participation in international multi-centric studies of warfarin dosage (Santos et al. 2015). Similarly, Brazilian cancer genetics is constituted and made possible through a range of international collaborations with researchers and leading institutes in the United States, Portugal and France, supported directly and indirectly by both global charitable and pharmaceutical research investment.[2] In both cases claims relating to the diversity of the Brazilian population engage and facilitate trans-national scientific interest.

Nevertheless, attempts to 'universalise' the data from this research raise more or less explicit concerns or critiques. Many Brazilian researchers discover that their research findings are only relevant to a wider international research community when translated into specific taxonomies and categories of difference. In the case of pharmacogenomic studies in Brazil this has led to the exclusion of samples from populations classified as 'brown' or '*Mestizo*' from multi-centre studies which are seen as too indeterminate to fit the required racial framework that is often demanded within a 'political economy' of publications (Santos et al. 2015, see also Montoya 2007). By contrast an explicit research interest in European genetic ancestry (framed in some publications in terms of a 'Caucasian haplotype') in Brazilian cancer genetic research would appear to bring about an apparent alignment between specific high profile 'global' population genetic research categories or priorities and an urgent need to understand or explain Brazilian regional disparities in cancer incidence, given variable admixture. Yet this co-exists alongside a recognition, amongst this research community, of the still 'unknown' nature of genetic variants and cancer risk in Brazil, which must be understood in terms of a continuum of national population diversity as this concerns correlations between genetic ancestry and skin colour (Gibbon 2016). In both cases we see how in variable ways articulations of difference and similarity or specificity and diversity are alternately incorporated, reframed and sometimes explicitly rejected as part of process described by Santos et al. in terms of 'racialising to deracialise' (2015).

Alongside an engagement with genetic variation, a range of other rationales and justifications also underpin the pursuit of these particular fields of genomic research in Brazil, informed by concerns with social justice and health care inequities.

Underserved communities, inequities and the judicialisation of health

The way that a notion of 'under-served' communities is articulated in Brazilian genomics, as well as in broader arenas of Brazilian public health, has generally not been orientated to the needs of specifically racialised groups. Despite an emerging multi-cultural agenda that intersects with genomic research in Brazil and across Latin America (Wade 2017) there is not currently the same delineation of such populations as there has been in the United States, where an explicit focus on the needs of 'Latina' and 'African American' communities has been fore-grounded (Fullwiley 2008, Lee this volume, Joseph 2014, Reardon 2012). While there is an emphasis in Brazil on regional specificity linked to different colonial histories of migration this does not map directly or easily onto a notion of 'under-served communities'. Instead the question of inequities is marked by wider concerns with national and cross-regional social or economic inequities in health care resources more broadly.

While there is therefore articulation and recognition of inequalities in public health by those working across a variety of genomic research fields in Brazil these do not always align with race. In the case of pharmacogenomics, explicit efforts to 'de-molecularise' race and genomics are underlined and pursued in an effort to address and ameliorate inequities in public health policy. It is argued that these would be exacerbated through a 'racialised' application of pharmacoge-nomics by imposing 'arbitrary' race categories that do not account for the heterogeneity of the Brazilian population, with potentially dangerous consequences in the case of the warfarin dosage (Santos et al. 2015). Cancer genetics by contrast is constituted by Brazilian cancer genetics researchers as a form of 'neglected' preventative public health that entails actively engaging with 'under-served' communities and families (Gibbon 2015). As Melo et al. demonstrate, the question of inequities and genetic medicine in the context of public policy debates concerning rare genetic disease is coloured by concerns about extending universal access to drugs and public health care services (2015).

Efforts to address health care inequities through the promise of genomics as preventative medicine are particularly significant in a context such as Brazil where the public health system is precarious, uneven and under resourced (Paim et al. 2011). While this association contributes to the visibility and also legitimacy of genomic research, this is an ethically complex endeavour, especially where the boundary between research and care is thin. While the inter-dependencies between terrains of care and research are a characteristic feature of 'translational' research in cancer genetics more broadly in Euro-American contexts (Cambrosio et al. this volume), these intersections also generate particular ethical dilemmas in developing country contexts (Traore et al. 2015). While some see this as a form of 'hidden innovation' in Brazil that is enabling of both national and transnational research (de Souza *unpublished thesis*), aligning clinical practice and research at the interface with public health and national or transnational networks has varied outcomes. Whilst informing as it mobilises calls for genomics as public health, it also propels patients into research in search of scarce medical care and resources with specific consequences (Petryna 2009).

High technology medicine and related interventions are now increasingly entangled with the growing judicialisation of health in Brazil (Biehl 2013, Diniz et al. 2012). The constitutional commitment to provide health care for all by the Brazilian state has led patients, families and patient organisations to pursue and strategically instrumentalise (mostly free) legal rights to medications and also other health care sources, including the use of genetic testing in diagnosis. With hundreds of thousands of judicial cases it has become a significant means in Brazil by which access and inclusion to technologies and medicines associated with genomic research and interventions, are now adjudicated; a situation made particularly evident in relation to 'rare' genetic disease (Gibbon and Aureliano forthcoming).

A recent special issue of the leading Brazilian anthropology journal *Vibrant*, examining ques-tions of identity in the context of novel developments in the life sciences highlights the need to consider how questions of social vulnerability, inequality and legitimacy are entangled with novel fields of genomics (Do Valle and Gibbon 2015). With the likely deepening and further entrenching of social differences at a time of economic and political crisis, such issues will continue to be a feature of the terrain across which genomic research and medicine in

Brazil will necessarily have to traverse. This raises questions about how individualised ethical frameworks for consent which have been developed for pursuing genomic research and medicine outside of the region are translated and made relevant in contexts such as Brazil. This is particularly true when questions of inequity, human rights and social justice have and continue to be more immediate concerns (Penchezadeh 2015) and where there are different histories of both bioethics and social medicine (Goldim and Gibbon 2015). While novel articulations of bioethics are emerging in North America and Europe in response to the shifting parameters of research regarding data sharing and public participation (see Prainsack this volume), there are also particular concerns in pursuing 'broad consent' in the creation of biorepositories in developing country contexts, as explored below. It remains to be seen to what extent diverse histories and experiences, such as those outlined in Brazil, can shape evolving discussion of ethics in an era of global and globalising genomic research.

Uses of ethnicity and tribe in Africa for science

In contrast to the country of Brazil, European colonists often saw African populations throughout the continent as fixed in time and rooted to specific spatial geographies where ethnicity was naturalised as race (see Pales and Linhard 1952; cf Tapper 1999, chapter 3). Ideas of group mixing, although acknowledged, were at times written off or ignored for scientific and mapping projects that consisted of surveys of ABO, MN, and RH blood groups along with sickle haemoglobin for 'comparative' *raciologie* (Fullwiley 2011: 171). In the French West African colonial capital of Dakar, these serological studies conducted by the anthropological mission, the *Musée de l'homme* and the National Blood Transfusion Centre were explicitly *not* about health. Rather the objective was to use biomarkers like the sickle cell trait to establish that what had been considered cultural and linguistic ethnicity was in fact also "biological" race (Fullwiley 2011: 168; Pales and Linhard 1952: 85). In other contexts, such as colonial Zimbabwe (then Rhodesia), population-based sickle cell frequencies were used to arbitrate land and resource disputes based on imagined purity and autochthony (Tapper 1999: chapter 3). In what is now Ivory Coast, different dynamics of ethnic consolidation took place that had implications for housing, economic labour casts and, in the postcolonial era, for the humanitarian distribution and triaging of anti-retroviral HIV treatments (Nguyen 2010: 124–125).

The admittedly messy dynamics that underlie the production of singular bloc ethnicities happened in contexts where mixing and human diversity were surely realities. Nonetheless, the tendency of colonial medicine and anthropology was to highlight ossified notions of ethnicity, race, and even tribe when describing African peoples as study subjects. In both colonial and post-colonial states the trend has been to construct and to speak about relatively singular groups, while this often happens through global dynamics whereby African people are active agents in constructing or edifying ethnic identities as a result of exposure to European power (Nguyen 2010: 124–125; Matory 1999: 79–81). What is interesting is that after the wave of independence movements beginning in the 1950s, when many new nation-states emerged, a continued pattern of labelling arose to delineate populations – albeit now in terms of the nation-state. These ascriptions were often synonymous with, or superimposed onto, prior notions of ethnicity or tribe. That is: concepts of population remained tightly yoked to geography, whereby new stakes for claiming ancestry for political reasons (linked to land and market shifts) created new value for African peoples' biology (Crane 2013; Peterson 2014). At times this confluence of factors bolstered older ideas of biological and genetic sameness in new terms (Fullwiley 2011: 5–7).

New nation-states, ancestry, disease, and the value of African populations

One clear example of this development is that starting in the early 1980s geneticists discovered that three African and one Middle Eastern sickle cell populations differed in the genetic markers (haplotypes) inherited with the gene for sickle haemoglobin (hbS). French researchers obtained samples in Africa from patients in several cities that were chosen for their Pasteur Institutes and other markers of scientific promise that might afford contacts with local physicians working on the disease. With funding from the multinational pharmaceutical company Sanofi, the scientist leading the research attended conferences in the sites of interest, whereby she made arrangements to access samples that her team would analyse after transporting the biological material back to France (Fullwiley 2011: 179–185). The African haplotypes discovered and characterised by this team were named for the polities of 'Senegal', 'Benin', and 'Central African Republic, CAR'. The results that came of these initial studies shifted how the vast array of sickle cell phenotypic difference would be conceptualised the world over. Each haplotype was correlated with a general degree of severity, versus mildness.

The HbS haplotype mapping that established ethno-national ideas of 'mild' versus 'severe' sickle cell ushered in two important moments for genetics in Africa. The first was that, despite earlier cartographic studies of human variation and bloc ethnicity via sickle cell frequencies in the late colonial period, the new haplotype studies were framed in terms of genetic 'origins' and 'ancestry' now tied to new national borders. Yet even though these genetic signatures conceptually signalled notions of old human biological descent, Parisian researchers followed the logic of a 1980s Francophone-informed geopolitical map in calling the African sickle cell haplotypes by names such as 'Senegal' and 'Benin'. In this process, ethnicity often receded to the background in favour of ethnicised nation-state labels, even though patients in Nigeria might have the 'Benin' type and patients in Gambia could have the 'Senegal'. This expansive frame where nation-state, region, and later 'Africa' as a whole could be conceived of in genetic terms would persist. The second was that researchers in Senegal began to protest unfair treatment of African researchers in this and later 'collaborations' where they were not mentioned in journal publications (Fullwiley 2011: 138). Subsequent efforts to conduct such studies in Senegal were met with clear demands for fairness, calls to build infrastructure and to create training opportunities. In the 2000s sickle cell researchers in Senegal refused to merely serve as '*envoyeurs de sang*' (blood senders) who mail samples to scientists in the North (ibid.: 190).

Health politics and North–South research to meet local needs

In Senegal where the disease was thought to be the most 'mild' on the continent, sickle cell centres and eventually country programmes focused on improving patients' quality of life through minimal medical interventions in part out of scarcity. Also at work was a science narrative intimating that the exceptional 'genetics' of Senegalese people had alleviated some fraction of the disease burden. Although newborn screening programmes have been piecemeal, country specialists have shown that early intervention and follow-up care increase survival and basic indices of thriving, socially as well as biologically (Diagne et al. 2000, Diop et al. 1999). With this country success story of sickle cell made "mild" through a hybrid of genetic localisation, economic scarcity and medical making-do, the head of the national programme requested that the state finance a comprehensive care and research centre for populations outside of the capital beginning in the late 2010s. The goal was to research and track success – and to save lives with minimal intervention. After several failed promises for state funding, specialists eventually secured funds from the foundation arm of the third largest pharmaceutical company in France,

Pierre Fabre. Since 2013 the *Fondation Pierre Fabre* and specialists at the Gaston-Berger faculty of health sciences in Saint-Louis, Senegal are conducting a long-term study on neonatal drug and care regimens that they hope to eventually institutionalise at a population level. In short, private financing, humanitarian philanthropy, development market logics, and population health converge in a research-care hybrid for a national "population" group that is seen to possesses unique genetics that favour better health. Local specialists refer to mild sickle cell as a "saving grace," a biosocial boon that must be documented for research and public health efforts.

If researchers in Paris were informed by the francophone imperial map to define the sickle cell haplotypes in sub-Saharan Africa, a similar linguistic set of colonial and post-colonial relations inflect how initiatives under the rubric of the Human Health and Heredity (H3Africa) project are emerging across the continent today. H3Africa, which launched in 2010, is a US National Institutes of Health and UK Wellcome Trust joint initiative that has as its mission to support genomic research in Africa and to set up research infrastructures across the continent. Its goal is also to investigate the interplay of genomic and environmental factors that determine disease susceptibility and drug responses in African populations (H3Africa Vision 2013), which are theorised to benefit the whole world. Thus far, the majority of projects are in Anglophone countries. To date, the structural legacies of British colonialism has limited actual inclusion from researchers from countries with French, Spanish, or Portuguese as their official languages. Out of the 27 projects currently funded as part of the consortium, all but three are headed by PIs from Anglophone countries. This will surely affect how genetic variation is described down the line. At present, and since the launch, H3Africa researchers have globalised 'Africa' as an entity; a place where humans' unique genomic characteristics are keys to science.

H3Africa's organisers hopes and founding philosophies are manifold as concerns diversity, inclusion, fairness, and notions of African population biology. These were evident in pronouncements made by scientists from the start. One was that real relationships and "capacity building" would be a defining feature of the initiative. At the highly publicised press conference signalling the first round of $38 USD million in funding, Dr Bangani Myosi, Head of the Department of Medicine at the University of Cape Town, reiterated concerns held by sickle cell scientists in Senegal years prior. Denouncing what he called 'a colonial mode of doing science', he told the international audience that collaborations must be ethical in terms of halting extractive practices based in asymmetrical power relations. This would also require envisioning how to address the inability to exploit research findings in resource-poor settings that resulted in samples – and the final achievements of published science – nearly always being taken abroad (Wellcome Trust, Transcript 2010).

Myosi went on to discuss the African 'genetic origins' of common and rare diseases. Francis Collins, head of the National Institutes of Health, doubled down on this notion of African populations' ancestral utility and potential health gift to global humanity. After detailing the age and unique structures of genetic diversity of African populations, he laid out the vexing issue of trying to assess common disease risks and concluded: 'Africa will answer the question' (Wellcome Trust, Transcript 2010).

Thus the ethical treatment of African scientists would be crucial to harnessing the power of African genomes for humanity. To address both, building the necessary infrastructure would be paramount. This concerned establishing biorepositories and laboratories where samples could be stored and shared for eventual intra-African research collaborations, as well as global ones. Just how the imagined tens of thousands of African people who would be asked to join the research would understand what storing and sharing DNA and tissue samples via a biorepository would mean raises a second layer of ethical exigencies the consortium must now address. These include biobanking, export of samples for global secondary uses, and 'broad consent' to store and re-use samples in the first place. Issues of ownership and governance of DNA and biological tissues are also at stake. To date, many of the research guidelines in African countries that might be used to assess the ethics of these practices were developed in response to clinical trials, epidemiological studies, or broader health research – not genomics (De Vries et al. 2017). This is disconcerting because genomes allow for myriad research queries and analyses that go far beyond those of more traditional clinical trials or epidemiological projects. The implications of such broad potential use of genomes should not be underestimated or played down in discussions of broad consent and the facility it allows researchers to exploit genomes with today's technologies.

These issues are not specific to genomic research in Africa. Appeals to join genomic research are also currently taking place much more generally. New rules around broad consent and data sharing, detailed in revisions to The Common Rule, received public comments and underwent debate in the United States in 2016, the implications of which many study participants may not fully grasp (Skloot 2015). Issues of commercial uses, privacy, and re-identification are but a few of the concerns that discussions around true informed consent of DNA samples must address in Africa, Latin America, the United States, and elsewhere. It will be up to scientists involved in these projects to do more to think through past structural inequalities, histories of resource extraction, and issues of equity as they ethically engage and educate people whose DNA is the basic necessity that permits human genetic research in the first place.

Notes

1 In their press release Illumina promises that 'H3Africa data will inform strategies to address health inequity and will lead to health benefits in Africa and beyond'. See 'H3Africa Consortium Array Available Soon': New cost-effective chip offers unprecedented content for African genomic research. www.illumina.com/company/news-center/feature-articles/h3africa-consortium-array-available-soon-.html (Accessed May 19, 2017).
2 This includes US organisations such as the Susan Komen Foundation with programmes of sponsored research on addressing 'Latina' populations and 'ethnic diversity' in cancer genetic research.

Works cited

Anderson, W. (2012). Racial hybridity, physical anthropology, and hum biol in the colonial laboratories of the United States. *Current Anthropology* 53(5), S95–S107.

Beaudevin, C. and Pordie, L. (2016) Diversion and globalisation in biomedical technologies. *Medical Anthropology* 35(1): 1–4.

Benjamin, R. (2009) A Lab of Their Own: Genomic Sovereignty as postcolonial science policy. *Policy and Society* 28: 341–355.

Biehl, J. (2013). The Judicialization of Biopolitics: Claiming the right to pharmaceuticals in Brazilian courts. *American Ethnologist* 40(3): 419–436.

Bustamante, C. D, De la Vega, F. and Burchard, E. (2011) Genomics for the world. *Nature* 475: 163–165.

Crane, J. T. (2013) *Scrambling for Africa: AIDS, Expertise, and the Rise of American Global Health Science*. Ithaca: Cornell University Press.

De Souza, V. S. & Santos, R. V. (2014) The emergence of human population genetics and narratives about the formation of the Brazilian nation (1950–1960). *Studies in History and Philosophy of Biological and Biomedical Sciences* 47: 97–107.

De Vries, J., Nchangwi Munung, S., Matimba, A., McCurdy, S., Oukem-Boyer, O., Staunton, C., Aminu, Y., Tindana, P., and the H3Africa Consortium (2015). Addressing ethical issues in H3Africa research – the views of research ethics committee members. HUGO.

De Vries, J., Munung, S. N., Matiba, A., Mc Curdy, S. *et al.* (2017) Regulation of genomic and bio-banking research in Africa: A content analysis of ethics guidelines, policies and procedures from 22 African Countries. *BMC Medical Ethics* 18(8): 1–9.

Diagne, I., Ndiaye, O., Moreira, C., Signate-Sy, H., Camara, B., et al. (2000) Sickle cell disease in children in Dakar, Senegal. *Archives De Pediatrie* 7(1): 16–24.

Diop, S., Thiam, D., Cisse, M., Toure-Fall, A. O., Fall, K., et al. (1999) New results in clinical severity of homozygous Sickle Cell Anemia in Dakar, Senegal. *Hematology and Cell Therapy* 41(5): 217–221.

Do Valle, C. G. and Gibbon, S. (2015) Introduction. Health/illness, biosocialities and culture. *Vibrant (Virtual Brazilian Journal of Anthropology)* 12(1).

Fullwiley, D. (2011) *The Encultured Gene. Sickle Cell Health Politics and Biological Difference in West Africa.* Princeton, Princeton University Press.

Fullwiley, D. (2008) The biologistical construction of race. *Social Studies of Science* 38(5): 695–735.

Gibbon, S. (2016) Translation population difference: The use and re-use of genetic ancestry in Brazilian cancer genetics. *Medical Anthropology* 9: 1–15.

Gibbon, S. (2015) Anticipating Prevention: constituting clinical need, rights and resources in Brazilian Cancer Genetics. In *Anthropologies of Cancer in Transnational Worlds*, edited by Holly F. Mathews, Nancy Burke, and Eirini Kampariani. New York: Routledge.

Gibbon, S., Santos, R. V. and Sans, M. (2011) *Racial Identities, Genetic Ancestry and Health in South America. Argentina, Brazil, Colombia and Uruguay.* New York; Palgrave Macmillan.

Gibbon, S. and Aureliano, W. (in press) Inclusion and Exclusion in the Globalisation of Genomics; the case of rare genetic disease in Brazil. *Anthropology and Medicine.*

Goldim, J. and Gibbon, S. (2015) Between personal and relational privacy: understanding the work of informed consent in cancer genetics in Brazil. *Journal of Community Genetics* 6 (3): 287–293.

H3Africa (2013) Vision of the H3Africa Initiative. http://h3africa.org/about/vision (Accessed February 17, 2017).

INCA (2009) *Rede Nacional de Cancer Familial.* Rio de Janeiro. Ministerio de Saude.

Joseph, G. (2014) Genetics to the People. BRCA as public health and the dissemination of cancer risk knowledge. In Gibbon et al. (eds) *Breast Cancer Genetic Research and Medical Practices: Transnational Perspective in the time of BRCA.* Routledge, London.

Kent, M., Santos, R. V., Wade, P. (2014) Negotiating imagined genetic communities: Unity and diversity in Brazilian science and society. *American Anthropologist* 116(4): 736–748.

Kumar, D. (2012) *Genomics and Health in the Developing World.* Oxford, New York: Oxford University Press.

Lima, N. T. (2007) Public Health and social ideas in modern Brazil. *American Journal of Public Health* 97(7): 1168–1177.

Malaria Genomic Epidemiology Network (2008) A global network for investigating the genomic epidemiology of malaria. *Nature* 456, 732–737.

Matory, J. L. (1999) The English professors of Brazil: On the diasporic roots of the Yoruba Nation. *Comparative Studies in Society and History* 41(1): 72–103.

Melo, D. et al. (2015) Genetics in primary health care and the National Policy on Comprehensive Care for People with Rare Diseases in Brazil: opportunities and challenges for professional education. *J Community Genet* 6: 231–240.

Montoya, M. (2007) Bioethnic conscription: Genes, race and Mexicana/o ethnicity in diabetes research. *Cultural Anthropology* 22(1): 94–128.

Mello, M. and Wolf, L. (2010) The Havasupai Indian Tribe Case – Lessons for Research Involving Stored Biologic Samples. *The New England Journal of Medicine* 10(1056): 1–4.

Nelson, A. (2016) *The Social Life of DNA: Race, Reparations and Reconciliation after the Genome.* Boston: Beacon Press.

Nguyen, Vinh-Kim (2010) *The Republic of Therapy: Triage and Sovereignty in West Africa's Time of AIDS.* Durham: Duke University Press.

Paim et al. (2011) The Brazilian health system, advances and challenges. *The Lancet* 377(9779): 1778–1797.

Pales, L. and J. Linhard (1952) La Sicklémie (sickle cell trait) en Afrique Occidentale Française: Vue de Dakar. *L'Anthropologie* 56: 53–86.

Palmero, E. I., M. Caleffi, L. Schuler-faccini, F. Roth, I. Kalakun, C. Netto, G. Skoniesky, J. Giacomazzi, B. Weber, R. Giugliani, S. Camey & P. Ashton-Prolla (2009) Population prevalence of hereditary breast cancer phenotypes and implementation of a genetic cancer risk assessment program in southern Brazil. *Genetics and Molecular Biology* 32: 447–455.

Penchezadeh, V. (2015) Ethical issues in genetics and public health in Latin America with a focus on Argentina. *Journal of Community Genetics* 6: 223–230.

Peterson, K. (2014) *Speculative Markets: Drug Circuits and Derivative Life in Nigeria.* Durham: Duke University Press.

Petryna, A. (2009) *When Experiments Travel: Clinical Trials and the Global Search for Human Subjects*. Princeton, NJ: Princeton University Press.

Popejoy, A. B. & Fullerton, S. M. (2016) Genomics is failing on Diversity. *Nature* 538(7624): 161–164.

Reardon, J. (2012) The Democratic, anti-racist genome? Technoscience at the limits of Liberalism. *Science as Culture* 21(1): 25–48.

Santos, R., Oliveira da Silva, G. and Gibbon, S. (2015) Pharmacogenomics, human genetic diversity and the incorporation and rejection of color/race in Brazil. *Biosocieties* 10(1): 48–49.

Santos, R. V., Lindee, S. and Souza, V. S. (2014) Varieties of primitive: Human biological diversity in Cold War Brazil (1962–1970). *American Anthropologist* 116(4): 723–735.

Sequeiros, J., Gibbon, S. and Clarke, A. (2015) Genetics and ethics in Latin America. *Journal of Community Genetics*. 6(3): 185–187.

Seguin, B. *et al.* (2008) Genomic medicine and developing countries: creating a room of their own. *Nature Reviews Genetics* 9: 487–493.

Skloot, R. (2015) Your cells. Their research. Your permission? *The New York Times*. Dec. 31.

Sleeboom-Faulkner, M. (2010) *Frameworks of Choice: Predictive and Genetic Testing in Asia*. Amsterdam: Amsterdam University Press.

Suarez-Kurtz, G. (2010) Pharmacogenetics in the Brazilian population. *Frontiers in Pharmacology* 1: Article 118.

Tapper, M. (1999) *In the Blood: Sickle Cell Anemia and the Politics of Race*. Philadelphia: University of Pennsylvania Press.

Taussig, K. S. and Gibbon, S. (2013) Public health genomics. Anthropological interventions in the quest for molecular medicine. *Medical Anthropology Quarterly* 27(4): 471–488.

Traore, K., Bull, S. *et al.* (2015) Understandings of genomic research in developing countries: a qualitative study of the views of MalariaGEN participants in Mali. *BMC Medical Ethics* 16.

Wade, C. Lopez-Beltran, E. Restrepo and R. V. Santos (eds.) (2014) *Mestizo Genomics: Race Mixture, Nation, and Science in Latin America*. Durham, NC: Duke University Press, pp. 33–54.

Wade, P. (2017) *Degrees of Mixture, Degrees of Freedom: Genomics, Multiculturalism and Race in Latin America*. Durham, Duke University Press.

Wellcome Trust (2010) Transcript of the Human Heredity and Health in Africa Press Conference in London, 22 June: www.genome.gov/Pages/Newsroom/Webcasts/Transcript-H3-AfricaPressConference.pdf (accessed February 15, 2017).

Whitmarsh, I. (2008) *Biomedical Ambiguity. Race, Asthma and the Contested Meaning of Genetic Research in the Caribbean*. Cornell: Cornell University Press.

Part 5
Crossing boundaries

29

Introduction

Janelle Lamoreaux

Social studies of science, situated in various disciplines, have long pointed to the artificiality of dichotomies between nature and culture, life and death, male and female, and the social and the biological. Canonical works in science studies, such as Sarah Franklin and Margaret Lock's *Remaking Life and Death* (2003), Donna Haraway's *Simians, Cyborgs and Women* (1991), and Marilyn Strathern's *After Nature* (1992), show the ways in which the reification of dichotomies shapes understandings of the world, both in and beyond science and medicine. This section explores emergent sciences in which many of these commonly held epistemological and ontological boundaries are breaking down. A broad array of disciplines increasingly research scientific objects that cross boundaries through techniques that do the same.

The section highlights how the increasing understanding of genes as only a part of more complex systems is a major aspect of this shift. As genetics is reframed as epigenetics and (post-)genomics, objects that were once defended as bounded entities are now being studied with more open boundaries and flexible meanings. Genomes and an increasing number of other -omes (the proteome, the metabalome, and even the interactome) are defined more by their propensity to intermingle than to independently determine. An emphasis on entanglements has transformed, and been transformed by, techniques, methods and theories that drive both the natural and the social sciences to new understandings of health.

Social scientists have a complicated relationship with the shifting grounds of contemporary biological sciences. More meaningful quests to practice "good science" mean social scientists both argue for diminished boundaries between the social and the biological, and find themselves warning about the potential for boundary-crossing endeavors to re-entrench hierarchies (Vasiliou et al., 2016). The development of CRISPR genome editing technologies, for example, has raised bioethical fears about genetic enhancement rooted in eugenic histories as intervention at the level of the genome becomes increasingly accessible to scientists (Kirksey, 2016a). But this same technology has created an opportunity for citizen-scientists to experiment with boundaries between not only nature and culture, but also expert and lay knowledge (Kirksey, 2016b). While optimistic attention is paid to the transgressive potential of some research, many social scientists remain skeptical of the ability of (post-)genomic research to address longstanding health inequities that are distributed along gender, ethnic, and class lines. As some boundaries are crossed, others remain. For instance, social scientists argue that epigenetic research does not do away with genetic determinism but creates new kinds of reductionism, an "epigenetic determinism" (Kuzawa and Sweet, 2009) primarily focused on maternal responsibility for fetal health even when interactions between genes and environments or social and biological factors are being stressed (Landecker, 2014; Richardson et al., 2014; Waggoner and Uller, 2015 and many more).

The increasingly global nature of (post-)genomics, however, provides an opportunity to consider that what are thought of as commonly held boundaries or sticky dichotomies, for example between genes and environments or the "social" and "biological," are not universal (Lamoreaux 2016). Cosmological differences and varied sociopolitical histories shape how researchers think about the boundaries of the objects and subjects they study, even while increased international research infrastructures attempt to universalize scientific standards, practices and technologies. Today biomedical boundary crossings often occur across national borders and regional territories, time zones and climate zones, laboratories and disciplines, funding streams and government regulations, as well as bioethical standards and institutional review boards (Ong and Chen, 2010; Petryna, 2009; Sunder Rajan, 2012). Such international crossings create anticipated and unexpected forms and findings during and after research. Terms such as "global assemblage" (Ong and Collier, 2004) and "global biological" (Franklin, 2007) capture the contingent nature of such configurations. Yet what more can we learn about the frictions (Tsing, 2005) and world-makings (Zhan, 2009) that increasingly occur through "global science," including postcolonial science (Anderson and Adams, 2007; Harding, 2011)? As we see in the chapters that follow, the potential consequences of proliferating boundary crossings in contemporary genomic research vary depending on how knowledge and knowledge practices are situated (Haraway, 1991) in global and disciplinary histories and hierarchies.

The section opens with a chapter by Margaret Lock, an anthropologist who has crossed topical and methodological boundaries throughout her career. Lock's early contribution to the burgeoning field of medical anthropology, *Encounters with Aging* (Lock, 1993), used both quantitative and qualitative research to disrupt the idea of biological universalism through the condition of menopause. In this chapter, we see a continuation of Lock's early interest in what she called "local biologies" reappearing in her research on epigenetics (Lock, 2013). Lock's chapter provides a historical overview of epigenetics, charting the patchy (Tsing, 2015) development of this research area and technique in the work of C. H. Waddington. Outlining Waddington's theory of the epigenetic landscape and the central role of development in his thinking and research, Lock connects early epigenetic theories to their molecularized, post–Human Genome Project uptake. Lock then explores the postgenomic revitalization of epigenetics through one particular subfield: behavioral epigenetics. Lock ends with a note both hopeful and critical: that the "slowly consolidating impacts" of what are now being referred to as "internal and external environments" are potentially paradigm-shifting, but "as yet unsubstantiated."

A co-authored essay by Stephanie Lloyd and Eugene Raikhel furthers this critical perspective on epigenetics, again concentrating on the behavioral subfield. They argue that the potential complexity of gene–environment interaction research is contained through scale, and then elucidate this claim through a careful analysis of suicide risk studies. Time and space both complicate and inform the work of epigenetic researchers who strive to understand the causal factors of suicide. Experiences of trauma early in the life course may ignite a series of "cascading effects" in some "suicide completers." But even when epigenetic changes (studied via donated, post-mortem brains) can be linked to traumatic events, such patterns of methylation have unpredictable afterlives. With regard to space, the chapter concludes that epigenetic research focused on "environments," which one might assume include factors at the social, economic or political scale, places responsibility for health at the level of the individual, solidifying "the tendency in Western society to see illness from an internalizing perspective, with the relevant causes seen as immediate and localized in the body."

The next chapter helpfully moves away from a focus on the interpretation and consequences of boundary-crossing objects and research in the United States and Europe to a focus on stem cells in Taiwan. Jennifer Liu's contribution problematizes a universal understanding of stem cells and the sciences that study them through attention to both the local and global of international stem cell research. Discussing the therapeutic prospects, economic potential and ethical problems surrounding stem cells, Liu suggests that the local contours of these objects depend upon the histories and politics of the local contexts in which they are elicited and researched. Liu shows through her ethnographic research in a stem cell laboratory that the international, or even national, standardization of stem cells is a goal more than a reality, painting a picture of multi-species resistance to singularity and simplicity through practice.

Carrie Friese's chapter on animal models also takes us from the human to the animal and back again, discussing the key role of animals in the production of genomic knowledge both currently and historically. The chapter starts by defining how models in general and model organisms in particular have been understood in scientific pursuits, and how the relationship between animal models and the genetic sciences developed in the twentieth century. Friese discusses Drosophila, mice, bacteria, phages, C. elegans, and humans to provide a historical analysis of "the twin standards of standardization and generalization." The chapter then turns to a discussion of how practices of care influence the "diversity" of animal models in the genomic laboratory. Friese writes, "The environments of the laboratory become a crucial data point in this context." This observation furthers the contributor's main point: that animal models and the genomic sciences co-produce one another.

Amber Benezra's chapter explores microbial ecologies through natural and social scientific perspectives. Benezra describes microbiomes as socio-natural entanglements, shaped by "how and where we are born, what food we eat, who we live with and love…" The microbiomes are also thought to shape humans, making us "symbiotic assemblages," as much microbe as we are human. This is not to say that microbe–human relations are always characterized by what Donna Haraway has referred to as multi-species co-flourishing (Haraway, 2008). The focus of much microbiome research looks into the negative impacts of life events and circumstances on microbial "communities." Benezra reviews trends in these scientific studies, showing how the rise in microbial research has been made possible through the specific high-throughput sequencing technologies that study them via gnotobiotic animal models and metagenomics. She also discusses how social scientists studying microbial ecologies have attempted to expand the idea of community, environment, species, and health inequity that underlie efforts to improve human–microbe relations in locales such as the gut, mouth, and vagina. The chapter includes a discussion of the potential ways that the social science of microbial ecologies is itself crossing disciplinary boundaries, as Benezra's previous collaborative publication makes clear (Benezra et al., 2012).

The next chapter continues to explore epistemological boundaries, this time through controversy rather than collaboration. Nicole Nelson and Aaron Panofsky's chapter on Behavior Genetics charts the beginnings and subsequent iterations of this "biosocial" science, asking how shared epistemologies shape and are shaped by understandings of what constitutes biology, the social, and science. Drawing on Karin Knorr-Cetina's concept of epistemic cultures (Knorr-Cetina, 1999), which are distinguished through shared understandings of how we know what we know as well as what we should strive to know, Nelson and Panofsky portray disciplinary debates among behavior genetics as "contact zones" (Haraway, 2008), spaces of co-mingling and interchange which redefine the entities involved. They show how behavior genetics became a discipline – both an academic field, and a cautious way of being – first, through actively seeking opportunities to collaborate with "socialization" researchers, and, second, by managing and moderating the degree of claims made by those participating in the discipline. In the end, Nelson and Panofsky argue that even if heterogeneity complicates the boundaries of behavior genetics, such contact and conflict is what makes this an interesting field to research.

As with Behavior Genetics, perhaps the title of the last chapter in the section, written by Deborah Scott, Dominic Berry and Jane Calvert, should appear in scare quotes. Questioning the continued usefulness of the category of synthetic biology all together, their contribution asks what happens to an emergent field when its defining characteristic becomes widespread practice. With the rise of interventionist techniques across the biological sciences, has the category of synthetic biology not only lost pertinence as a specialized field of research, but also as a unique approach to understanding the limits of nature and culture, the biological and the social? Given such questions, Scott, Berry, and Calvert "keep the ambiguity of its constitution central" to their analysis, not focusing exclusively on science conducted under the label synthetic biology, but raising issues pertinent to new and emerging sciences and technologies more generally. These include questions of the moral economy of the biosciences, governance, and the publics of synthetic biology, as well as a reflection on natural and social science collaborations that have occurred under the synthetic biology moniker. Their helpful synthesis of collaborative efforts concludes that, "As social sciences and humanities researchers invested in making biotechnologies diverse, plural and reflexive, we may come to find our goals are best served by moving beyond synthetic biology."

Collectively considering the work of authors who study a variety of boundary-crossing (post-)genomic objects and subjects in various locales, it seems that while some boundaries are being crossed, others are reiterated or even dissolved. Certainly species boundaries experience each of these dimensions, as in the case of microbial and animal research. Likewise disciplinary boundaries are shifted, defended and made obsolete, as in the case of "behavior genetics" and "synthetic biology." Epigenetic research seems to push understandings of actions around boundaries even further – reinterpreting temporal and spatial borders, even while being constricted by them, both theoretically and physically. Reflection on this section in total allows one to ask: What kinds of boundaries are being crossed in (post-)genomics today? Who is doing the boundary crossing – natural scientists, social scientists, both? With what kind of affectual relationship – collaborative, combative, or combined? Finally, and perhaps most importantly, how might attention to boundaries crossed also elicit attention to how boundaries are (re)created?

References

Anderson, W., Adams, V., 2007. Pramoedya's chickens: Postcolonial studies of technoscience. *Handb. Sci. Technol. Stud.* 181–204.

Benezra, A., DeStefano, J., Gordon, J. I. 2012. Anthropology of microbes. *Proceedings of the National Academy of Sciences of the United States of America.* 109(17), 6378–6381. doi:10.1073/pnas.1200515109.

Franklin, S., 2007. Stem Cells R Us: Emergent Life Forms and the Global Biological, in: Ong, A., Collier, S. J. (Eds.), *Global Assemblages.* Blackwell Publishing Ltd, pp. 59–78.

Franklin, S., Lock, M. (Eds.), 2003. *Remaking Life and Death: Toward an Anthropology of the Biosciences,* 1 edition. ed. School for Advanced Research Press, Santa Fe: Oxford.

Haraway, D. J., 2008. *When Species Meet.* University of Minnesota Press, Minneapolis, MN.

Haraway, D. J., 1991. *Simians, Cyborgs, and Women: The Reinvention of Nature.* Routledge, New York.

Harding, S. (Ed.), 2011. *The Postcolonial Science and Technology Studies Reader.* Duke University Press Books, Durham.

Kirksey, E., 2016a. The CRISPR Hack: Better, Faster, Stronger. *Anthropol.* Nov. 8, 1–13. doi:10.1080/19428200.2016.1152860.

Kirksey, E 2016b, Who is Afraid of CRISPR Art?, *Somatosphere*. Retrieved May 31, 2017, from http://somatosphere.net/2016/03/who-is-afraid-of-crispr-art.html,

Knorr-Cetina, K., 1999. *Epistemic Cultures: How The Sciences Make Knowledge*. Harvard University Press, Cambridge, MA.

Kuzawa, C.W., Sweet, E., 2009. Epigenetics and the embodiment of race: Developmental origins of US racial disparities in cardiovascular health. *Am J Hum Biol*. 21, 2–15. doi:10.1002/ajhb.20822,

Lamoreaux, J., 2016. What if the Environment is a Person? Lineages of Epigenetic Science in a Toxic China. *Cultural Anthropology* 31(2), 188–214doi:10.14506/ca31.2.03

Landecker, H., 2014. Pregnancy: Study the mother's DNA as well. *Nature* 513, 172–172. doi:10.1038/513172b,

Lock, M., 2013. The Epigenome and Nature/Nurture Reunification: A Challenge for Anthropology. *Med. Anthropol*. 32, 291–308. doi:10.1080/01459740.2012.746973,

Lock, M., 1993. *Encounters With Aging Mythologies of Menopause In Japan and North America*. University of California Press, Berkeley.

Ong, A., Chen, N.N. (Eds.), 2010. *Asian Biotech: Ethics and communities of fate*. Duke University Press, Durham, NC.

Ong, A., Collier, S. J. (Eds.), 2004. *Global Assemblages: Technology, Politics, and Ethics as Anthropological Problems*, 1 edition. ed. Wiley-Blackwell, Malden, MA.

Petryna, A., 2009. *When Experiments Travel: Clinical Trials and the Global Search For Human Subjects*. Princeton University Press, Princeton.

Richardson, S. S., Daniels, C. R., Gillman, M. W., Golden, J., Kukla, R., Kuzawa, C., Rich-Edwards, J., 2014. Society: Don't blame the mothers. *Nature* 512, 131–132. doi:10.1038/512131a

Strathern, M., 1992. *After Nature: English Kinship in the Late Twentieth Century*. Cambridge University Press, Cambridge, U.K.

Sunder Rajan, K. (Ed.), 2012. *Lively Capital: Biotechnologies, Ethics, and Governance in Global Markets*. Duke University Press, Durham, NC.

Tsing, A. L., 2015. *The Mushroom at the End of the World: On the Possibility of Life in Capitalist Ruins*. Princeton University Press, Princeton.

Tsing, A. L., 2005. *Friction: An Ethnography of Global Connection*. Princeton University Press, Princeton.

Vasiliou, S. K., Diamandis, E. P., Church, G. M., Greely, H. T., Baylis, F., Thompson, C., Schmitt-Ulms, G., 2016. CRISPR-Cas9 System: Opportunities and Concerns. *Clin Chem*. 62, 1304–1311. doi:10.1373/clinchem.2016.263186

Waggoner, M. R., Uller, T., 2015. Epigenetic Determinism in Science and Society. *New Genet. Soc*. 34, 177–195. doi:10.1080/14636778.2015.1033052,

Zhan, M., 2009. *Other-Worldly: Making Chinese Medicine Through Transnational Frames*. Duke University Press, Durham, NC.

30

Epigenetics

Margaret Lock

Introduction

Epigenetics implies "over or above genetics," but its denotation differs among the sub-disciplines of epigenetics, and also changes as new discoveries come to light. Over the past decade epigenetics has rapidly expanded into an enormous field of inquiry that includes stem cell biology, cancer biology, investigations into genome instability and DNA repair. Following a general introduction, discussion in this chapter is confined to the vibrant sub-field known as behavioral epigenetics.

The polymath Conrad Hal Waddington – developmental biologist, geneticist, and paleontologist – originally coined the term in the 1940s. Waddington wrote that he was inspired by the Aristotelian word "epigenesis" when thinking about what he wanted to express, although that concept had long since been out of use by the mid-twentieth century. Waddington was familiar with classical theories of development, and wrote about the theory of "preformationism" widely held from the time of the Greeks on, in which it was believed that all the characters of an adult, mental and physical, are present in a newly fertilized egg, packed into such a small space that nothing is visible to the naked eye, and hence is not amenable to verification.

Waddington likened the dominant thinking of his day, in which genes were understood as determinants of adult phenotypes, to a "new-fangled version" of preformationism. He went on to assert that embryologists had reached quite a different conclusion on the basis of research, and elaborated a theory of "epigenesis" in which adult characteristics do not exist in a newly fertilized egg; on the contrary, they "arise gradually through a series of causal interactions between the comparatively simple elements of which the egg is initially composed" (Waddington 1957:156). Waddington's stated position was influenced by the dawning realization of several researchers of his time that embryological development must involve networks of gene interactions that form a complex integrated system, and that the bifurcated subjects of genetics and embryology should be brought closer together, even though many embryologists feared that their field might then be overtaken by genetics. Waddington, trained in both fields – he had worked in Germany with the Nobel Laureate embryologist, Hans Spemann, and in California with the geneticist Thomas Hunt Morgan – made "development" central to his arguments specifically because of its double meaning: the growth of individuals, and evolutionary change.

For Waddington, development denoted the set of conditions that enable what we now describe as "multi-potent stem cells" to become differentiated into cells with specific tissue functions – nerve cells, liver cells, and so on. He insisted that genes are responsible for guiding only "the mechanics of development," and his thinking is recognized today as a "paradigm-changing idea" (Gilbert 2001: 3). Furthermore, an appreciation of what continues to be recognized as "critical periods" in development was embedded in Waddington's approach from the outset.

The epigenetic landscape

In his book *Organisers and Genes* published in 1940, Waddington depicted "the epigenetic landscape" as a symbolic representation of developmental processes. The image is of a marble rolling down an undulating plateau in concert with other marbles, eventually coming to rest at the lowest points. The marble represents a developing egg and the gradual transformation of its cells into tissue types, the process of which is controlled by genes and their mutual interactions that modulate the manner in which the egg/ball descends the slopes, and at intersections selections are made.

Waddington introduced the concept of "canalization" to account for why organisms reliably produce similar phenotypes, even though most species exhibit considerable genetic variation and are exposed to a wide range of environmental variables during development. Part of Waddington's intention was to demonstrate that no straightforward relationship exists between a gene and its phenotypic effects and, further, should a mutation arise, its effects may well be buffered by other genes – a process he termed "genetic assimilation" that he explicitly linked to Darwinian thinking. Canalization is accepted to the present day, although debate persists about what drives this process (Siegal and Bergman 2002).

Waddington recognized that the metaphor of the epigenetic landscape had limitations (1940: 92), nevertheless it is usually assumed to be the starting point for a genealogy of epigenetics. In the preface to the first edition of his book Waddington notes that his greatest debt goes to the biochemist Joseph Needham, who was also an extraordinarily influential sinologist best known for his monumental seven volume work on the history of science in China. It is possible that the image of the epigenetic landscape was inspired to a degree by discussions with Needham who, in the 1940s, was teaching himself Chinese, and would have been well acquainted with traditional understanding of disease causation in China in which individual bodies are understood as ceaselessly striving to restore and retain equilibrium within the spheres of society, environment, and the cosmos in which they are embedded.

Waddington's work essentially languished for the first three decades of its existence but was extensively revived after the mapping of the Human Genome, to become molecularized in form. In the postgenomic era, epigenetics is thoroughly revitalized and focuses on "alternative developmental pathways, on developmental networks underlying stability and flexibility, and on the influence of environmental conditions on what happens in cells and organisms" (Jablonka

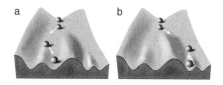

Figure 30.1 The Epigenetic Landscape as conceptualized by Conrad Waddington.

and Lamb 2005: 89) at both individual and population levels. Epigenetics has expanded into an enormous field of inquiry that includes stem cell biology, cancer biology, and investigation into genome instability, DNA repair, epigenetic epidemiology, behavioral epigenetics and more. The concept of environment, having been rendered essentially of no importance in hard line deterministic genetics is resuscitated, to take on singular importance with respect to both cell functioning and the health and illness of individuals and their families in specific contexts (Lock and Palsson 2016).The present review is largely confined to a consideration of behavioral epigenetics. Within this field, the majority of researchers are in broad agreement about the ultimate objectives of their specialty, nevertheless, a presumption that one or more teams of researchers represent the entire field would be a mistake.

Epigenetics and the reactive genome

Mapping the human genome commenced in 1990 and continued for more than a decade. In 2001, by which time it was near completion, it was clear that many surprises had come to light, some of which had been predicted by scientists prior to mapping, but that had been ignored.

The Human Genome Project (HGP) revealed that humans have approximately 20,000 genes, and not 100,000 as had been claimed. Numerous plants have many more genes than do humans, and the diminutive worm *C. elegans* has about the same number as ourselves. The size of a genome bears no relationship to its complexity, and the genome is not a template for the organism as a whole. Approximately 1.2 percent of DNA segments code for proteins and the remaining 98.8 percent was initially labeled disparagingly as "junk" (Carey 2015).

Although non-coding sections of the genome initially appeared to have no obvious function, and are frequently remnants of bacterial and viral genomes, they separate out the coding parts of the genome, thus inhibiting unwanted mutational changes during DNA transmission between generations. Furthermore, numerous non-coding DNA sequences are highly conserved, indicating that they have most likely been present in genomes for hundreds of millions of years, hence suggesting that they are crucial to both evolutionary change and the processes of life. Moreover, it is well established that the activities of non-coding RNA (ncRNA) comprise a comprehensive regulatory system that functions to create the "architecture" of organisms, without which chaos would reign. To this end, ncRNA profoundly affects the timing of processes that occur during development, including stem cell maintenance, cell proliferation, apoptosis (programmed cell death), the occurrence of cancer and other complex ailments (Mattick 2004).

These findings relate to the structure and function of the genome itself; over the past decade molecular epigeneticists have added further insights to this already complex picture. Researchers in behavioral epigenetics claimed several years ago that they had tracked the molecular links between nature and nurture, thus providing evidence that nature/nurture is entangled and not divisible (Labonté et al. 2012). This assertion was based on findings demonstrating how environmental stimuli and stressors originating both externally and internally to the body initiate trains of molecular activity that modify how DNA functions, often with lasting effects on human behavior and wellbeing including, at times, increased mental illness and suicide rates (McGowan et al. 2009).

It is of note that DNA is among the most non-reactive, chemically inert molecules in the world, with no "power to reproduce itself" (Lewontin 1992). Hence, not surprisingly, the majority of biologists, whatever their specialty, accept that cellular differentiation is governed by something akin to what Waddington described as the epigenetic landscape. In addition, robust findings make clear that, independently of developmental processes, environments both external and internal to the body bring about epigenetic alterations in the genome that are deeply involved with health and illness throughout life (Cortessis et al. 2012; Feil and Fraga 2012). In other words, the environment can "sculpt the genome and affect the phenotype throughout life" (Szyf et al. 2008: 46).

With the advent of epigenetic insights, the philosopher of science Evelyn Fox Keller notes: "The role of the genome has been turned on its head, transforming it from an executive suite of directional instructions to an exquisitely sensitive…system that enables cells to regulate gene expression in response to their immediate environment" (2014: 2425). The biologist Scott Gilbert characterizes this situation as one in which the genome is best understood as "reactive" (2003: 92). Furthermore, if genes are conceptualized as "real" entities, then they must be understood as composite rather than as unitary, somewhat analogous to "the solar system, or a forest, or a cell culture" as Barnes and Dupré put it (2008: 53). A dynamic epigenetic network with a life of its own has been exposed that creates a context-dependent reactive system in which DNA is just one part. Thus, contingency displaces determinism.

Gene regulation – above all how, and under what circumstances, genes are expressed and modulated – is at the heart of epigenetic investigation, and whole cells, rather than DNA segments, are the primary targets of inquiry. Effects of evolutionary, historical, and environmental variables on cellular activity, developmental processes, health and disease are, in theory, central to this research endeavor although, to date, this is by no means the case in most basic science investigations (Lock 2015:154).

It is of note that biologists generally agree that epigenetics does not overturn the Darwinian theory of natural selection, although epigenetic research has been characterized by some as neo-Lamarckian. Of course, the notion was long ago abandoned that use or disuse of body parts brings about evolutionary change, but epigeneticists recognize an important aspect of Lamarck's thinking with which Darwin agreed, namely that forces internal and external to the body contribute directly to the makeup of ensuing generations (the concepts of genotype and phenotype were not available to Lamarck or Darwin).

Sculpting the genome

An article published in the mid-1990s by the epigeneticist Michael Meaney is a frequently cited classic. This research made use of a model of maternal deprivation created in rats by removing young pups from their mothers shortly after birth, thus terminating maternal licking and grooming crucial to their development. The deprivation altered the expression of genes that regulate behavioral and endocrine responses to stress, and hence hippocampal synaptic development. It was found that these changes could be reversed if pups were returned in a matter of days to their mothers. Furthermore, when the birth mother was a poor nurturer, placement of her deprived pups with a surrogate mother who licks and grooms them adequately enabled the pups to flourish. Crucially, it was shown that pups or foster pups left to mature with low licking mothers not only exhibited a chronically increased stress response but also passed this behavior on to their own pups. Hence, variation in maternal behavior brings about biological pathways causing significantly different infant phenotypes that can persist into adulthood (Meaney et al. 1996).

A substantial body of work based largely on animal models, but increasingly in humans, substantiates these findings. Since the 1990s a literature has accrued showing a strong relationship between "childhood maltreatment" and negative mental health outcomes ranging from aggressive and violent behavior to suicide. Current investigations are gradually exposing the pathways that enable epigenetic processes to come about regarded as crucial mediators in the biological embedding of childhood maltreatment. The conclusion drawn from this research is that the "epigenome is responsive to developmental, physiological and environmental cues" (Lutz and Tureki 2014).

The epigenetic mechanism best researched to date is methylation, a process in which methyl groups are added to a DNA molecule. DNA methylation is found in all vertebrates, plants, and many non-vertebrates. Enzymes initiate such modifications that do not alter the DNA sequence, but simply attach a methyl group to residues of the nucleotide cytosine, thus rendering that portion of DNA inactive. Animal research has shown definitively that such modifications can be transmitted intergenerationally, and some research strongly suggests that this is also the case among humans (Pembrey et al. 2014), although certain researchers argue that this is not as yet irrevocably established. Relatively recently it has been recognized that environmental exposures originating outside the body bring about changes to the three-dimensional chromatin fiber, that in turn influence the expression of DNA (Lappé and Landecker 2015; Szyf et al., 2008) (see Figure. 30.2).

Chromatin, the "stuff" of chromosomes, the task of which is to compact DNA, is composed of DNA, RNA, proteins, among which are histones, and yet other molecules. Small amounts of DNA are wound around eight histones to form a beadlike structure – the nucleosome – from which parts of the histone molecules protrude. Strings of nucleosomes are in turn twisted up to become chromatin fiber, compacted as a tight package. Chromatin is no longer thought of as an inert scaffold, as was formerly the case, but is conceptualized as a dynamic body, profoundly influenced by environmental input, of which methylated DNA is just one part. Moreover, the "epigenetic marks" of chromatin may be transmitted from generation to generation. Hence, chromatin carries experience forward over lifetimes and, at times, to future generations. Epigenetic mechanisms have been shown to play an important role in learning and memory suggesting how the trauma of colonization, slavery, war, and displacement may be transmitted through time (Rumbaugh and Miller 2011).

The idea of an epigenome – a distinct layer over or enveloping the genome – as the majority of involved scientists think of it, is a misnomer. The genome and epigenome are not independent but rather a flexible, comingled entity, orchestrated by shape-shifting chromatin that may bring about hereditable changes. In addition, strips of DNA can be altered, often during

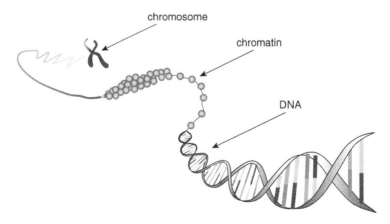

Figure 30.2 The Structure of Chromatin.

replication, some of which changes result in mutations that may or may not be hereditable. Epigenetic mechanisms other than methylation exist, including histone acetylation and deacetylation, that also regulate gene expression, but they have not yet been researched to the same extent, as has methylation.

The genome cannot be conceptualized as timeless – a fixed base upon which a plastic, malleable epigenome performs certain activities. Genes are "catalysts" rather than "codes" for development (Meloni 2014), and it is the structure of information rather than information itself that is transmitted (Lappé and Landecker 2015). DNA is not changed *directly* by environmental exposures. Rather, whole genomes respond ceaselessly to a wide range of environments, exposures, and experiences and chromatin mediates such responses that, in turn, modulate DNA expression.

It is of note that the methylation processes described above are manifest in several time scales – evolution; trans-generational inheritance; individual lifetimes; life-course transitions (including infancy, adolescence, menopause, and old age) and seasonal change modifications. The effects of these passages of time are miniaturized in individual bodies making them researchable at the molecular level (Landecker and Panofsky 2013).

The embodiment of trauma

In 2011 Moshe Szyf titled a presentation he gave at a Montréal gathering: "DNA methylation: A molecular link between nurture and nature." At the time this talk was given, evidence for such a link had accrued primarily from animal research and from one human study based on a sample of 25 individuals who had suffered severe abuse as children and later committed suicide. At autopsy, the donated brains of these individuals showed a significantly different pattern of DNA methylation than did those of a control group of 16 "normal" individuals. A second control group of 20 individuals who had committed suicide but where no known abuse had taken place was also included in the study. The findings are presumed to substantiate a mechanism whereby nature and nurture meld as one. In this particular case, childhood adversity is associated with sustained modifications in DNA methylation across the genome, among which are epigenetic alterations in hippocampal neurons that may well interfere with processes of neuroplasticity (Labonté et al. 2012 see also Lloyd and Raikhel this volume).

The researchers acknowledged that the sample was small, and that the study cannot be validated. The absence of a control group that experienced early life abuse and did not die by suicide is another shortcoming. Furthermore, the abuse that the subjects experienced was exceptionally severe (Tureki, personal communication). Szyf and colleagues readily agree that understanding of these processes is rudimentary. Recent work has established that DNA-methylation-induced tissue changes are not limited to brain tissue alone, but can also be detected in white blood T-cells. This is good news for researchers who will be able to make use of blood samples procured from living subjects for some of their research projects, rather than be limited to the use of donated brains.

Minaturized environments

It has been repeatedly demonstrated that prenatal exposure to maternal stress, anxiety, and depression has lasting effects on infant development linked to the appearance of psychopathology later in life. A review of 176 articles, based on both animal and, to a lesser extent, findings from human research, notes: "the *in utero* environment is regulated by placental function and there is emerging evidence that the placenta is highly susceptible to maternal distress and is a target of epigenetic dysregulation" (Monk et al. 2012: 1361), added to which a large body of research suggests that postnatal maternal care can induce further disruptions. Such findings are based largely on correlations; although researchers are beginning to map segments of the pathways whereby environmentally induced epigenetic marks are apparently associated with behavioral outcomes pre- and post-natally. Antenatal depression and anxiety are understood as major contributors to a toxic placental environment. In this type of research the environment is effectively scaled down to molecular activity inside a single organ of the body – the uterus and its fetal contents. The poverty and often violent living conditions of many mothers-to-be may well be virtually ignored by researchers and their attention be confined almost exclusively to the pregnant belly and its contents (Singh 2012).

One of the most cited studies in which the fetal environment has become an object of study is in connection with pregnant women affected by the so-called Dutch Hunger Winter in the 1940s; research has continued over two ensuing generations. Thirty thousand people died from starvation as a result of a food embargo imposed by the Germans in World War II that brought about a complete breakdown of local food supplies, adding to the misery of an already harsh winter. Birth records collected since that time have shown that children born of women who were pregnant during the famine not only had low birth weights but also exhibited a disproportionally high range of developmental and adult disorders later in life including diabetes, coronary heart disease, breast and other cancers. The second generation, although prosperous and well nourished, produced low-birth-weight children who inherited similar epigenetically transmitted health problems (Heijmans et al. 2008). Furthermore, exposure to severe food deprivation during the first trimester of pregnancy showed a substantial increase in hospitalized schizophrenia for women once adult, but not for men.

Extensive research has also shown that deprivation associated with social isolation, notably in orphanages (Nelson et al. 2013), and that brought about by an impoverished life, result in epigenetic changes in the brain that reduce the chances of affected individuals being able to cope effectively with daily life while increasing the risk of mental illness (Hamzelou 2016). Space does not permit documentation of the extensive damage caused to brain functioning by toxic chemicals throughout life. In the epoch of the Anthropocene every one of us is exposed daily to toxic insults that can bring about epigenetic changes, but the frequency of these insults is highly stratified by income, and they are unequally distributed across space.

It is important to keep in mind that behavioral epigenetics is a very young science, and that numerous findings are as yet unsubstantiated. Nevertheless it is possible that a paradigm shift is slowly being consolidated in which environments external and internal to the body are recognized as initiating and sustaining human life, development, well being and malaise over the life course.

References

Barnes, B. and J. Dupré (2008) *Genomes and What to Make of Them*. Chicago and London: University of Chicago Press.

Carey, N. (2015) *Junk DNA: A Journey Through the Dark Matter of the Genome*. New York: Columbia University Press.

Cortessis, V. K., D. C. Thomas, A. J. Levine, C. V. Breton, T. M. Mack *et al.* (2012) 'Environmental epigenetics: prospects for studying epigenetic mediation of exposure-response relationships', *Human Genetics* 131: 1565–1589.

Feil, R. and M. F. Fraga (2012) 'Epigenetics and the environment: emerging patterns and implications', *Nature Reviews/Genetics* 13: 97–109.

Gilbert, S. F. (2001) 'Ecological Developmental Biology: Development Biology Meets the Real World', *Developmental Biology* 233: 1–12.

Gilbert, S. F. (2003) 'The Reactive Genome', In G.B. Müller and S.A. Newman eds. *Origination of Organismal Form: Beyond the Gene in Developmental and Evolutionary Biology*. Cambridge Mass: MIT Press. Pp. 87–101.

Hamzelou, J. (2016) 'Neuroscientist sees kids suffer brain damage due to dire poverty', www.newscientist.com/article/mg23130931-200.

Heijmans, B.T. *et al.* (2008) Persistent epigenetic differences associated with prenatal exposure to famine in humans. *Proceedings of the National Academy of Sciences* 105: 17046–17049.

Jablonka, E. and M. J. Lamb (2005) *Evolution in Four Dimensions: Genetic, Epigenetic, Behavioral, and Symbolic Variation in the History of Life*. Cambridge, Mass.: MIT Press.

Keller, E. F. (2014) 'From gene action to reactive genomes', *The Journal of Physiology* 592: 2423–2429.

Labonté, B., Suderman, M., Maussion, G., Navaro, L., Yerko, V., Mahar, I., Bureau, A. (2012) 'Genome-wide epigenetic regulation by early-life trauma', *Archives of General Psychiatry* 69: 722–731.

Landecker, H. and A. Panofsky (2013) 'From Social Structure to Gene Regulation, and Back: A Critical Introduction to Environmental Epigenetics for Sociology', *Annual Review of Sociology* 39: 333–357.

Lappé, M. and H. Landecker (2015) 'How the genome got a lifespan', *New Genetics and Society* 34(2): 152–176.

Lewontin, R. (1992) The dream of the human genome. *New York Review of Books*, May 28th.

Lock, M. (2015) 'Comprehending the Body in the Era of the Epigenome', *Current Anthropology* 56: 151–177.

Lock, M. (2017) 'Recovering the Body' *Annual Reviews of Anthropology* 46: 2–14.

Lock, M. and G. Palsson (2016) *Can science resolve the nature/nurture debate?* Cambridge: Polity Press.

Lutz, P. E., and G. Tureki (2014) 'DNA methylation and childhood mal-treatment: from animal models to human studies', *Neuroscience* 264: 142–156.

Meaney, M. J., J. Diorio, F. D. Widdowson, J. Laplante, P. Caldji, *et al.* (1996) 'Early environmental regulation of forebrain glucocorticoid receptor gene expression: implications for adrenocortical responses to stress', *Development Neurosciences* 18: 49–72.

Mattick, J. S. (2004) The Hidden Genetic Program of Complex Organisms', *Scientific American* 291: 60–67.

McGowan, P. O., A. Sasaki, A. C. D'Alessio *et al.* (2009) 'Epigenetic regulation of the glucocorticoid receptor in human brain associated with childhood abuse', *Nature Neuroscience* 12: 342–348.

Meloni, M. (2014) 'How Biology Became Social, and What It Means for Social Theory,' *The Sociological Review*. Published online, 26 March; DOI: doi:10.1111/1467–954X.12151.

Monk, C., J. Spicer and F. A. Champagne (2012) 'Linking prenatal maternal adversity to developmental outcomes in infants: the role of epigenetic pathways', *Developmental Psychopathology* 24: 1361–1376.

Nelson, C. A., N. A. Fox and C. H. Zeanah (2013) *Romania's abandoned children: deprivation, brain development, and the struggle for recovery*. Cambridge, MA: Harvard University Press.

Pembrey M. Saffery, R., Bygren, L. O. (2014) 'Human transgenerational reponses to early-life experience: potential impact on development, health and biomedical research', *Journal of Medical Genetics* 51: 563–572.

Rumbaugh, G. and C. A. Miller (2011) 'Epigenetic Changes in the Brain: Measuring Global Histone Modifications', *Methods in Molecular Biology* 670: 263–274.

Siegal, M. L. and Bergman A. 'Waddington's canalization revisited: Developmental stability and evolution', *Proceedings of the National Academy of Sciences*, 99: 10528–10532.

Singh, I. (2012) 'Human development, nature and nurture: working beyond the divide', *Biosocieties* 7: 308–321.

Szyf, M., P. McGowan and M. J. Meaney (2008) 'The Social Environment and the Epigenome', *Environmental and Molecular Mutagenesis* 49: 46–60.

Waddington, C. H. (1940) *Organizers and Genes*. Cambridge: Cambridge University Press.

Waddington, C. H. (1957) *The Strategy of the Genes: A Discussion of Some Aspects of Theoretical Biology*, London: George Allen & Unwin.

31

Environmental epigenetics and suicide risk at a molecular scale

Stephanie Lloyd and Eugene Raikhel

Environmental epigenetics is a rapidly growing field interested in identifying the molecular traces of life experiences within the body. While many epigenetic changes within the body are considered to be fleeting, others are conceptualized as relatively stable and are thought to be associated with long-term health implications. It is the latter case we are interested in; in particular, we are concerned with models of suicide risk that epigenetics researchers are attempting to construct by linking the experience of early childhood abuse with specific methylation[1] patterns. These methylation patterns, identified in people's brains, are considered to confer increased risk of suicide.

In our ongoing consideration of environmental epigenetics research focused on suicide risk, we have been struck by the degree to which arguments that hinge on certain assumptions about scale – and how these arguments document the interactions of genes and environments in specific ways – have been central both to the research in epigenetics we have been tracking, and to attempts in the human sciences to argue for biosocial research. In this chapter, we draw upon anthropological literature on risk and recent work in linguistic anthropology which emphasize scaling as a semiotic process in order to think through some of the conceptual questions which have emerged from calls for a renewed biosocial or biocultural agenda in the human sciences.

As Summerson Carr and Michael Lempert have recently argued, scholars often "ontologize scalar perspectives, rather than ask how they were forged and so focused. Indeed, it is all too easy to proceed with our analyses as if the oft-critiqued but still convenient tiers of macro, meso, and micro were the ready-made platforms for social practice, as if social life simply unfolded in more or less intimate, proximate, local, grounded, or contained situations" (Carr & Lempert 2016: 8), with profound consequences for our means of conceiving risk. Instead, Carr and Lempert argue for a "pragmatics of scale" – an attention to "the social circumstances, dynamics, and consequences of scale-making as social practice and project" (2016: 9).

Our central argument in this chapter is that while environmental epigenetics research has the potential to profoundly complicate the relationship between "the biological" and "the social," and between "genes" and "environment," at least some of this potential for complexity is contained through researchers' use of scale. In other words, we suggest that arguments which assume certain scales as given, attempt to domesticate and contain an otherwise unruly complexity.

Risk, genes, and environments

Attention to risk is not new for anthropologists. Mary Douglas described it as a key cultural construct of our time (1986). As such it has been of great interest to medical anthropologists who have drawn attention to the impact of probabilistic reasoning about risk, the embeddedness of risk in social hierarchies and structural inequalities, the impacts of risk labelling, alongside the scientific neutrality associated with the term (Ho 2003; Lock 1998; Nguyen & Peschard 2003; Obermeyer 2000; Rhodes et al. 2005). Since the 1990s, social scientists have been particularly attentive to the concepts of genetic risk, genetic reductionism, and genetic determinism (Conrad 1999; Franklin & Roberts 2006; Heath et al. 2008; Lock et al. 2006; Rapp 1999).

Genetic determinism, epitomized by a "gene for" logic, in which single genes are considered capable of explaining certain of our traits and predicting complex phenotypes (Billings et al. 1992; Conrad 1999), has been widely criticized by biological and social scientists (Cunningham-Burley & Kerr 1999). Beyond specialists interested in studying or identifying these genes, sociologist Peter Conrad has noted that the one-gene-one-disease model of genetic determinism predominates popular discourse about genetics, despite the fact that single-gene disorders are relatively rare (Conrad 1999). Early literature on genetic determinism raised concerns about privacy and discrimination (Draper 1991; Greely 1992; Rothstein, 1997), although genetic reductionism was perhaps a more appropriate term for these discriminatory practices. Concerns about genetic determinism have been particularly pronounced in the area of behavioral genetics, which must include not only the environment in risk assessments, but also changing classification systems (Billings et al. 1992; Conrad 1999). Indeed, this literature has paid significant attention to the difficulties associated with negotiating the probabilities and uncertainties of medically relevant genetic information (Lock 1998).

As scholars became increasingly interested in identifying the role of the environment in "genetic conditions," social scientists transitioned from discussions of genetic determinism to genetic reductionism. Genetic reductionism refers to an interpretation of "the gene" as the material entity that is the single underlying *true cause* of expressed bodily traits or "human nature" (Sloan, 2000, p. 17, emphasis added). For example, in her pioneering analysis of "geneticization," sociologist Abby Lippman described the process by which "priority is given to searching for variations in DNA sequences that differentiate people from each other and to attributing some hereditary basis to most disorders, behaviors and physiological variations" (Lippman 1991: 13), potentially resulting in the reduction of complex traits to an essentialized biology grounded in small genetic differences. In effect, while the role of factors other than specific genes is recognized, causal narratives identify genetics as the essential cause of a person's condition. To the extent that this dominant conceptual model was increasingly understood to reveal and explain health, disease, normality and abnormality, as well as to affect the availability of intellectual and financial resources available to address health problems, Lippman argued that genetic models had the potential to profoundly affect our values and attitudes, as well as to contribute to a lack of attention to the environmental context of disease and risk.[2] In parallel, a number of social science researchers focused on studies of occupational and environment-based models of risk, drawing attention to the political processes through which specific social groups are placed "at-risk" by virtue of their political–economic position (Lupton 1995). This explicitly critical analysis looks for the production of risk in political and economic structures and processes, that is, those processes exterior to the body.

Sara Shostak has observed that each of these definitions of risk – whether focused on genetic profiles (interior/proximal) or environmental factors (exterior/distal) – "highlights and/or obscures a different set of locations, whether inward (bodily), outward (environmental), or interactive" (Shostak 2003: 2332). Though different "locations" of risk were not perceived as independent of one another, they were generally separated for analytic purposes. Early research on gene–environment interactions (GEIs), from interest in the 1970s fueled by rapid technological capacities in the 1980s and particularly 1990s (see Shostak 2003 for a brief history), significantly raised interest in mapping out the specificity of the interactions between internal and external risk factors. In post-genomic science, GEI emerged as a major force in explorations of genetically defined susceptibility to a range of exposures and the exposure-mediated regulation of gene expression (Frickel 2004; Shostak 2013). This research was part of a shift from a "nature versus nurture" dichotomy to a focus on complexity and interactions of multiple environments (internal and external to the body) and led to a growing "interactionist consensus" within post-genomic science (Landecker & Panofsky 2013: 349). Yet despite consensus about the importance of considering interactions and even the artificiality of dissociating the impacts of genes and environments, researchers have found it challenging to define, measure, and analyze the environment in genomic science (Darling et al. 2016: 51).

The environment – as a variously defined space in and outside the body – takes on an important role in GEI research and beyond, and in particular in environmental epigenetics research. In these models, individuals and subpopulations with or without "genetic risk factors" are understood as susceptible in particular places and times, creating the necessity of introducing the lives of individuals and communities into analyses of health and disease, drawing together a new canvas of complex factors. In these visions, not only is risk located both outside the body and within the genes (Shostak 2003), but these two domains or levels are in fact merged. As internal and external environments are observed to interact, with external factors exerting an important influence on gene expression, levels of scale generally conceived of as distal and proximal become increasingly difficult to tease apart. The artificiality of their separation becomes apparent. If this approach "emplaces" (Shostak 2003: 2336) genes in environments, how and where are they thought to act? How does the environment get in? What is the "dose" from the environment? Where does it act from, the inside or the outside or both?

Genes and environments in environmental epigenetics research on suicide

Since summer 2013, our team has been carrying out ethnographic research at the McGill Group for Suicide Studies (MGSS), a research unit focused on understanding the development of suicide risk. Altogether, the MGSS is made up of approximately 40 researchers (including principal investigators,[3] trainees, and research staff) with specializations ranging from molecular biology, neuroanatomy, genetics, and epigenetics, to brain imaging, psychology, and anthropology. Our ethnographic research examines all of these subspecialties and their interactions, but focuses most closely on the unit's epigenetic research. Fifty-nine interviews with members of the MGSS have been carried out and a member of our team attended nearly all lab meetings and journal clubs between spring 2013 and summer 2015. The director of the group, Gustavo Turecki, is a central figure in epigenetics research on suicide. In addition to having established an international research profile, he and his colleagues have developed an unusual set of raw materials, in the form of the brains of "suicide completers." These brains are housed in the Douglas–Bell Canada Brain Bank along with the brains of people who have died with other specific disease profiles (e.g., Alzheimer's disease). Indeed, the research being carried out at the MGSS would not be possible without access to these tissues.

Environmental epigenetics and neuroplasticity research at the MGSS uses the brain tissue of "suicide completers" to assess the extent to which life experiences, particularly early childhood abuse, may have caused changes in brains – at the epigenetic, morphological, and even immunological levels – that increased their risk of committing suicide. The characteristics of the brains of "suicide completers" who experienced early abuse are compared to controls classified as not having experienced such adversity, among other comparisons. Information regarding the life experiences of people whose brains have been donated is collected through a "psychological autopsy," in which a psychological profile of an individual is constructed by means of interviews with close loved ones. Researchers at the MGSS are especially interested in identifying whether or not people experienced childhood trauma, as one of the key assumptions underlying much of this research is that early childhood is a "critical window" characterized by a higher level of cellular, neural, and developmental plasticity and that abuse during this time might leave indelible traces in their brains (Ernst et al. 2009; Turecki et al. 2012).

Research at the MGSS is built around the "molecular conduit" model in which the epigenome acts as the "interface" between environmental inputs (in the form of "early life adversity") and genetic pathways linked to risk of suicide. A diagram of hypothesized "contributors to suicidal behavior" from a recent review article by Turecki gives us some idea of how the relationship between distinct mechanisms across different scales is being imagined (see Figure 31.1 below).

The scalar work which this diagram is accomplishing is worth noting. First of all, despite the attempt to simplify the range of "factors" associated with increased suicide risk, these hypothesized mechanisms are not causal in any straightforward way. The factors are distributed, systemic, and deeply complex in their effects: what Hannah Landecker and Aaron Panofsky have called, "new ontologies of outcome" (Landecker & Panofsky 2013: 342). Distal or predisposing factors such as "early life adversity" are thought to activate signaling cascades resulting in the epigenetic regulation of gene pathways linked to stress, neural plasticity, and a range of cognitive processes, themselves associated with behavioral traits (such as impulsive aggression), which are, in turn, associated with increased suicide risk. "Epigenetic factors" thus appear in each of the three scalar domains here, figuring as distal, developmental, and proximal. This is particularly significant because of the ways in which such scalar concepts have long been used in the health sciences, and particularly how they have been associated with "the social" and "biological."

Scales and environments of suicide

Social epidemiologist Nancy Krieger has drawn attention to the ways in which scaling concepts can naturalize certain assumptions in the health and biosciences. Krieger focuses on the key

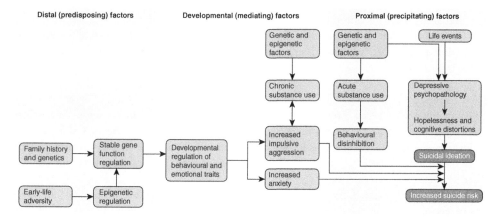

Figure 31.1 Overview of the contributors to suicidal behaviour. Reprinted by permission from Springer Nature: Nature Reviews Neuroscience, Gustavo Turecki, "The Molecular Bases of the Suicidal Brain," 15(12):802–16. Copyright 2014.

terms "proximal" (often glossed as "downstream") and "distal" (often glossed as "upstream"). Both terms, she points out, originated in nineteenth-century biology and were "invented to describe anatomical location and distance, as measured on a spatial scale" (Krieger 2008: 222). Soon the concepts were taken up in other disciplines, such as geology, in which they took on a temporal meaning in addition to their spatial one. By the late nineteenth century the usage of the terms expanded even further, from spatiotemporal scale to "describing levels and causal hierarchies" (ibid.). This was especially clear in the form of nineteenth-century social thought which conceptualized the social or the body politic as a kind of organism. Here "proximal" and "distal" were used to describe hierarchically arranged – and often nested – structural "levels" of societies. By the early twentieth century, biologists also began to use the terms in newly expansive ways with the development of the modern evolutionary synthesis – in which "proximate" or physiological causes were contrasted with "ultimate," "distal" or evolutionary ones. Krieger emphasizes that by the mid-twentieth century, public health researchers had also taken up these terms with several assumptions: 1) that "proximal" factors are either biological or those directing acting on or within the body; 2) that all other factors including "social" ones were thus "distal"; 3) that "causal potency" was linked to "distance" – (the closer, the stronger the effect) and that "distal causes necessarily exerted their influence through successively more proximal factors" (Krieger 2008: 223). These assumptions, Krieger argues, continue to structure much work both in GEI studies and in social determinants of health research. Epigenetic models of scale complicate these long-presumed relationships by being both distal and proximal, and always micro.

In MGSS models, specific forms of differential methylation are seen as being both specific *effects of* early child adversity and as *factors associated with* increased suicide risk. They are material remnants of the former enacting specific effects on the body, in a two-step argument of scale in which a "distal" environment is rendered a molecularized "proximal" risk, and the past exerts influence on the future. Naguib Mechawar, a molecular biologist and principal investigator at the MGSS, described how he sees the first step in this process occurring, the process through which a life experience becomes a bioactive molecular profile:

> What does abuse do? It's psychological stress. It's physical stress. What does that trigger? It triggers stress hormones, adrenalin. It triggers something that should not be triggered as intensely and as frequently [as in the case of] sexual or physical abuse when you're nine [and] your brain is maturing… I think that the consequences of chronic stress on the brain give rise to a lot of changes that can be [long term] – because this is such a vulnerable period during which the brain is much more plastic than in adulthood and *this leaves traces at different levels.*

These molecularized experiences, or "miniaturized environments" (Landecker & Panofsky 2013) are then seen to take on active lives of their own. Naguib and his fellow researchers believe that understanding these miniaturized environments – seen as the material traces of trauma, abuse, or stress and leading to aggression and impulsivity – is key to explaining why, given the same life experiences and clinical profiles (e.g., depression) some people commit suicide whereas others will never do so.

Much as in GEI research, the MGSS's epigenetic findings lead them to believe that no single trait or experience is sufficient to explain suicide. Gustavo, who in addition to being trained in neuroscience and an epigenetics researcher is also a clinical psychiatrist who works in a highly specialized depressive disorders unit, in no way attempts to link suicide solely to early childhood abuse, acknowledging that abuse is "not specific to suicide by any means." Yet he nonetheless sees it as critical factor in being able to predict "maladapted behavioural trajectories," whereby "suicide risk is perhaps the most severe negative end point of those psychopathological conditions that are in turn predicted by these negative trajectories."

This portrait of "epigenetic risk" is highly complex, associating molecularized environments with precise effects, which are in turn thought to be involved in complex and multiple interactions. This reflects sociologist Hannah Landecker's assertion that:

> Of significance is not the fact of delineation of environments as molecular, but that these environments are seen as bioactive with very high specificity – these molecules are investigated in relation to one another, within long chains or nets of causality across space and time that reach in and through the body.
>
> *(2011: 179)*

Moreover, Landecker has argued that the molecularization of the environment as theorized in epigenetics research is distinct from previous eras of the molecularization of the body as it foregrounds "molecular interrelation and critical timing rather than the search for answers in the structural enumeration of the molecules themselves; in epigenetics one sees an understanding of the body's molecules as hung in the same network of interaction as environmental molecules…" (Landecker, 2011: 170). Epigenetic arguments about risk thereby undercut notions of proximal and distal to the extent that the distal – the environment – becomes proximal in terms of space. In these scalar arguments of the development of suicide risk, both innate characteristics of a person, their already acquired internal environments, and the impacts of specific experiences such as abuse are thought to interact within the same space. In this view, experiences that could otherwise be described as distal and proximal, are seen to interact at a molecular level.

Researchers at the MGSS see the environment as capable of being internalized, in a potentially indelible manner, affecting functions within the body as well as what is behaviorally expressed. The embodied past thereby enacts change on the present and future from within the body. These interactions, or relationships, complicate efforts to associate the large with the small (Tsing 2004: 507), or abuse or suicide risk with an epigenetic profile. What could alternately be referred to as the "distal" and the "proximal" are constantly interacting in the same space over time, informed by past experiences and potentially leading to future behaviors. These interactions and "unruly" relationships across time and space complicate researchers' attempts to identify clear relationships between the macro and the micro (and back again) as well as between factors which exert their influence through past experiences and those which work through an integration into present functions. Along the way concepts such as "distal" or "proximal" lose their prior meaning.

Conclusion

The body, environments, and risk take on new meanings in environmental epigenetics research, making exposures visible at a molecular level and collapsing the distal and the proximal into the body. This has expanded the jurisdiction of molecular biology and the potential applications of molecular measures and techniques to encompass a wide range of phenomena previously understood through social or psychological metrics (e.g. social-economic status, stress, acculturation, economic deprivation, perceived racism). At the same time, environmental epigenetics highlights the interpenetration of bodies and places. The body is porous (Niewöhner 2011; Shostak 2003) and continuous with its environment. It becomes a "molecular archeological site revealing the past history of exposures and potential future harms (Proctor, 1995, p. 235), both containing and being transformed by the many places in which it has been situated" (Shostak 2003: 2338). The emerging vision of the body in this postgenomic era is considered to have the potential to produce new assemblages, transcending the reductionisms of apparently bygone eras.

Yet trying to conceive of the relationships between genes and environments remains a challenge for researchers. Even the definition of the environment remains unclear. Katherine Darling and colleagues identified at least three ways in which GEI researchers conceived of the body vis-à-vis the environment: bodies as environments in and of themselves, bodies as permeable, and bodies seen as the molecular materialization of social experiences (Darling et al. 2016: 52). In our research, we have found considerable oscillation among these three perspectives, among others, even by the same individual. A great deal of work remains to be done in order to understand how "environments" (variously conceptualized) react and change and solidify over time and space. What is clear is that in environmental epigenetics research a wide variety of social and material environments are being transformed into phenomena best measured and conceptualized at a molecular level. Paradoxically, this form of analysis turns attention away from the social conditions and contexts associated with the specific environments of interest to epigeneticists, the "causes of causes," that affect people (Darling et al. 2016; Krieger 2011; Phelan et al. 2010). Instead, as environments are embodied, specific trajectories of "causal accountability" (Krieger 2008) are produced in which responsibility for health risks resides within individuals. For instance, researchers focus on early abuse rather than the social, economic, or political conditions that tend to be associated with instances of abuse and the effects of the embodied abuse are traced at an individual level.

Ultimately, this turn to the environment has solidified the tendency in Western society to see illness from an internalizing perspective, with the relevant causes seen as immediate and localized in the body (Young 1976; Kroll-Smith & Floyd 1997). Thus, while Niewöhner (2011) has argued that the increasing appreciation of the molecular mechanisms of epigenetic regulation means that researchers are less "able to ignore the many ties that link the individual body and its molecules to the spatio-temporal contexts" outside the body (2011: 290), the epigenetics researchers we have observed acknowledge those ties, but then proceed to investigate the "environment" of interest, i.e., early childhood adversity, as independent of its original context. In other words, the environment – whose importance in much social science literature is its local nature, its political and social value, its "emplacement" – is studied as separable from the time and space of its origins. Some element of that original exposure is seen as existing independently of its original context, continuing to affect a person well beyond the original experience. The environment, in epigenetics research, becomes something that gets in, and in some cases, cannot get back out, whether due to passive mechanisms (a trace simply left untouched on a post-mitotic neuron) or through a series of complex active mechanisms (via brain-based changes, trait development, subsequent behavioral profiles or lifestyles, all providing feedback to one another). In either case, the original epigenetic profile is implicated as a part of the past, shaping the present and future. These are the specificities of risk from an environmental epigenetics perspective. While the relationship between the body and the environment is opened up and rendered more complex, we see a number of neoreductionisms (Lock 2015) within this research, as the environment becomes molecularized and associated with very specific events and long chains of causality. The body is seen as responsive, but at least in the case of the epigenetic profiles MGSS researchers are attempting to identify and associate with suicide risk, the body is also seen as condemned from early in life.

The MGSS research responds to contemporary imperatives to personalize or individualize risk predictions, public health messages, and biomedical treatments and further responds to demands to "go into the body" to most credibly capture the environment (Darling et al. 2016).

The molecularization of suicide risk enacted through their routine collection of biological specimens and accounts of people's lives provides "molecular credibility" (Kenney & Müller 2017) and allows them to at once open their analytic gaze toward the environment, while at the same time domesticating and containing its unruly complexity.

Notes

1 Methylation is perhaps the most well-studied of epigenetic mechanisms. It involves a chemical modification of DNA in which a methyl molecule is added to the nucleoside cytosine (C) with the effect of altering the expression of the gene.
2 Such critiques of genetic reductionism are aimed not at the heuristic methodological reductionism commonly employed in the biological sciences, but to the extension of such methodological approaches to matters of ontology (Sloan 2000, pp. 16–17; Opitz 2000, p. 446).
3 Lloyd is a principal investigator in the group.

References

Billings, P, Beckwith, J & Alper, J 1992, "The Genetic Analysis of Human Behavior: A New Era?" *Social Science & Medicine* vol. 35, no. 3, pp. 227–238.
Carr, ES & Lempert, M 2016, *Scale: Discourse and Dimensions of Social Life*, University of California Press, Berkeley.
Conrad, P 1999, "A Mirage of Genes," *Sociology of Health & Illness*, vol. 21, no. 2, pp. 228–241.
Cunningham-Burley, S & Kerr, A 1999, "Defining the 'social': Towards an Understanding of Scientific and Medical Discourses on the Social Aspects of the New Human Genetics," *Sociology of Health & Illness*, vol. 21, no. 5, pp. 647–668.
Darling, KW, Ackerman, SL, Hiatt, RH, Lee, S & Shim, J 2016, "Enacting the Molecular Imperative: How Gene-Environment Interaction Research Links Bodies and Environments in the Post-Genomic Age," *Social Science & Medicine*, vol. 155, pp. 51–60.
Douglas, M 1986, *Risk Acceptability According to the Social Sciences*, Russell Sage Foundation, New York.
Draper, E 1991, *Risky Business: Genetic Testing and Exclusionary Practices in the Hazardous Workplace*, Cambridge University Press, Cambridge, UK.
Ernst, C, Mechawar, N & Turecki, G 2009, "Suicide Neurobiology," *Progress in Neurobiology*, vol. 89, no. 4, pp. 315–333.
Franklin, S & Roberts, C 2006, *Born and Made: An Ethnography of Preimplantation Genetic Diagnosis*, Princeton University Press, Princeton, NJ.
Frickel, S 2004, *Chemical Consequences: Environmental Mutagens, Scientist Activism, and the Rise of Genetic Toxicology*, Rutgers University Press, New Brunswick, NJ.
Greely, HT 1992, "Health Insurance, Employment Discrimination, and the Genetics Revolution," in DJ Kevles & L Hood (eds.), *The Code of Codes: Scientific and Social Issues in the Human*, Harvard University Press, Cambridge, MA.
Heath, D, Rapp, R & Taussig, K-S 2008, "Genetic Citizenship." in D Nugent & J Vincent (eds.) *A Companion to the Anthropology of Politics*, Wiley-Blackwell, London.
Ho, M-J 2003, "Migratory Journeys and Tuberculosis Risk," *Medical Anthropology Quarterly*, vol. 17, no. 4, pp. 442–458.
Kenney, M & Müller, R 2017, "Of Rats and Women: Narratives of Motherhood in Environmental Epigenetics," *BioSocieties*, vol. 12, no. 1, pp. 23–46.
Krieger, N 2008, "Proximal, Distal, and the Politics of Causation: What's Level Got to Do With It?" *American Journal of Public Health*, vol. 98, no. 2, pp. 221–230.
Krieger, Nancy 2011. *Epidemiology and the People's Health: Theory and Context*. Oxford University Press.
Kroll-Smith, S & Floyd, HH 1997, *Bodies in Protest: Environmental Illness and the Struggle over Medical Knowledge*, NYU Press, New York.
Landecker, H 2011, "Food as Exposure: Nutritional Epigenetics and the New Metabolism," *BioSocieties*, vol. 6, no. 2, pp. 167–194.
Landecker, H & Panofsky, A 2013, "From Social Structure to Gene Regulation, and Back: A Critical Introduction to Environmental Epigenetics for Sociology," *Annual Review of Sociology*, vol. 39, no. 1, pp. 333–357.

Lippman, A 1991, "Prenatal Genetic Testing and Screening: Constructing Needs and Reinforcing Inequities," *American Journal of Law & Medicine*, vol. 17, nos. 1–2, pp. 15–50.

Lock, M 1998, "Breast Cancer: Reading the Omens," *Anthropology Today*, vol. 14, no. 4, pp. 7–16.

Lock, M 2015, "Comprehending the Body in the Era of the Epigenome," *Current Anthropology*, vol. 56, no. 2, pp. 151–177.

Lock, M, Lloyd, S & Prest, J 2006, "Genetic Susceptibility and Alzheimer's Disease," in L Leibing & L Cohen (eds.), *Thinking about Dementia: Culture, Loss, and the Anthropology of Senility*, Rutgers University Press, New Brunswick, NJ.

Lupton, D 1995, *The Imperative of Health: Public Health and the Regulated Body*, SAGE, London.

Nguyen, V-K & Peschard, K 2003, "Anthropology, Inequality, and Disease: A Review," *Annual Review of Anthropology*, vol. 32, pp. 447–474.

Niewöhner, J 2011, "Epigenetics: Embedded Bodies and the Molecularisation of Biography and Milieu," *BioSocieties*, vol. 6, no. 3, pp. 279–298.

Obermeyer, CM 2000, "Risk, Uncertainty, and Agency: Culture and Safe Motherhood in Morocco," *Medical Anthropology*, vol. 19, no. 2, pp. 173–201.

Opitz, JM, 2000, "The Geneticization of Western Civilization: Blessing or Bane?" in PR Sloan (ed.), *Controlling Our Destinies: Historical, Philosophical, Ethical, and Theological Perspectives on the Human Genome Project*, University of Notre Dame Press, Notre Dame, IN.

Phelan, J, Link, BG & Tehranifar, P 2010, "Social Conditions as Fundamental Causes of Health Inequalities Theory, Evidence, and Policy Implications," *Journal of Health and Social Behavior*, vol. 51, pp. S28–40.

Proctor, RN 1995, *Cancer wars: How politics shapes what we know and don't know about cancer*, Basic Books, New York.

Rapp, R 1999, *Testing Women, Testing the Fetus: The Social Impact of Amniocentesis in America*, Routledge, New York.

Rhodes, T, Singer, M, Bourgois, P, Friedman, SR & Strathdee, SA 2005, "The Social Structural Production of HIV Risk among Injecting Drug Users," *Social Science & Medicine*, vol. 61, no. 5, pp. 1026–1044.

Rothstein, MA 1997, *Genetic Secrets: Protecting Privacy and Confidentiality in the Genetic Era*, Yale University Press, New Haven, CT.

Shostak, S 2003, "Locating Gene–environment Interaction: At the Intersections of Genetics and Public Health," *Social Science & Medicine*, vol. 56, no. 11, pp. 2327–2342.

Shostak, Sara 2013. *Exposed Science: Genes, the Environment, and the Politics of Population Health*. Berkeley: University of California Press.

Sloan, PR 2000, "Completing the Tree of Descartes," in PR Sloan (ed.), *Controlling Our Destinies: Historical, Philosophical, Ethical, and Theological Perspectives on the Human Genome Project*, University of Notre Dame Press, Notre Dame, IN.

Tsing, A 2004, *Friction: An Ethnography of Global Connection*, Princeton University Press, Princeton, NJ.

Turecki, G, Ernst, C, Jollant, F, Labonte, B & Mechawar, N 2012, "The Neurodevelopmental Origins of Suicidal Behavior," *Trends in Neuroscience*, vol. 35, no. 1, pp. 14–23.

Turecki, G 2014, "The Molecular Bases of the Suicidal Brain," *Nature Reviews Neuroscience*, vol. 15, no. 12, pp. 802–816.

Young, A 1976, "Internalizing and Externalizing Medical Belief Systems: An Ethiopian Example," *Social Science & Medicine*, vol. 10, no. 3–4, pp. 147–156.

32

Stem cells: global cells, local cultures

Jennifer Liu

Introduction – stem cells

Stem cell research, and especially human embryonic stem cell research, has generated hope, hype and fears as well as substantial scholarly and political attention. Stem cells are scientific objects but they are also cultural and social objects. Thus, cultural framings shape the situated practices and meanings of stem cells even as they circulate in a global frame, and certain kinds of regulatory reconfigurations facilitate the capitalization of therapeutic potential. With the announcement in November 1998 that James Thomson's team at the University of Wisconsin had derived the first human embryonic stem cell lines, stem cells became objects of enhanced media and political attention, generating imaginations of both transformative therapeutic hope and dystopian fears.

Embryonic stem cells are noteworthy for two main characteristics: immortality and pluripotency. Immortality in stem cells refers to the capacity for self-renewal – ongoing cell division and proliferation without differentiation into specific tissue types. Pluripotency refers to the cellular capacity to differentiate into tissues from all of the body's three germ layers; pluripotent cells can thus theoretically give rise to all of the cell types found in an adult. These cell lines thus represented important new research tools as well as possibilities of unprecedented therapeutic intervention. Thomson's team suggested, "(t)hese cell lines should be useful in human developmental biology, drug discovery, and transplantation medicine" (Thomson et al. 1998). That is, such cells could be used for basic research, for development and testing of new pharmaceuticals, and to make new tissues and organs for transplantation.

Although the first human bone marrow (hematopoietic stem cell) transplant took place in 1968, the derivation and stabilization of human embryonic stem cell lines – immortal and pluripotent – signalled the potential to form any cell type, and the possibility of intervening in various diseases, including degenerative ones such as Parkinson's and Huntington's. Thus, human embryonic stem (hES) cells seemed to offer potentially lifelong and in many cases life-saving therapies to previously incurable diseases (also including diabetes, spinal cord injury, Alzheimer's, etc.). They appeared to provide the foundation for potentially unlimited therapeutic value.

Economic potential in shifting legal contexts

In an era of enhanced capitalization of biological products and even life itself, stem cell research also seemed to carry nearly infinite economic potential. In the United States, two legal decisions in particular changed the economic landscape of the biosciences, generating in new forms of biological value and capitalization. The Bayh-Dole Act of 1980 changed the terrain of intellectual property rights arising from federally funded science. Whereas previously intellectual property rights resulting from such research were held by the federal government, the Bayh-Dole Act not only granted such rights to universities and other agencies, but effectively required them to patent and commercialize findings derived from federally funded research (Hayden 2003: 27). Bayh-Dole also shifted the boundaries of scientific partnerships, prompting new forms of academic–corporate–government partnerships, ushering in new forms of privatization and profit-making in the biosciences, thus enhancing biotech industrialization. Arguments for California's Stem Cell Research and Cures Act, for example, relied heavily on the potential of stem cell research to generate substantial economic as well as therapeutic rewards.

Also in 1980, the U.S. Supreme Court adjudicated the *Diamond v. Chakrabarthy* case which centered around the question of whether living things were subject to patent. The court ruled that Chakrabarthy's oil-decomposing bacterium was indeed patentable, "a live human-made micro-organism is patentable subject matter."[1] Previously, the U.S. Patent Office had denied Chakrabarthy a patent arguing that living things are products of nature and thus not subject to patent. The Chakrabarthy case ruling thus opened up a new class of property and formed the foundation for a massive influx of venture capital into biotech. Given the wide range of potential therapeutic and research applications for stem cells, they became targets for both government R&D and venture capital investment. Specific organisms, cell lines (including hES), and even genetic information became subject to patent and property rights claims.

These U.S. legal decisions shaped bioresearch elsewhere as well, facilitating new forms of value, extraction and stabilization (Franklin and Lock 2003), bodily commodification and fragmentation (Waldby and Mitchell 2006), and the production of biocapital (Sunder-Rajan 2006). New forms of life emerged out of stem cell laboratories – hybrid organisms were made, cell lines immortalized, and profit actualized. Snuppy, the Afghan hound cloned by infamous stem cell scientist Woo Suk Hwang, was the first cloned dog. Dr. Hwang was working at Seoul National University (SNU) at that time, and SNU (for whom the dog is named) holds the patent to the procedure. Snuppy was named *Time Magazine*'s "Most Amazing Invention of 2005." In Taiwan, in 2006, stem cell researchers at National Taiwan University produced fluorescent pigs – with jellyfish genes – that glowed green "from the inside out" under black light (Hogg 2006). In other cases, researchers have sought to patent the DNA of Indigenous peoples, and genetic materials and cell cultures derived from Taiwan's Indigenous peoples are available for purchase from the Coriell Institute for Medical Research in New Jersey (Liu 2017), which also maintains a stem cell biobank.

Ethics

At the turn of the millennium, even as hES cells seemed to hold nearly limitless healing potential, they proved to be problematic objects. Most notably, this was because their origins relied upon the destruction of early human embryos, most of them so-called "surplus" or "leftover" embryos from *in vitro* fertilization (IVF). Human ES cells thus became available because of the proliferation of IVF (itself once a highly controversial technology). Sarah Franklin has written comprehensively on IVF, stem cells, and the IVF-stem cell interface

(Franklin 2013). By now, over 5 million children have been born through IVF and millions of embryos created but not implanted for reproduction. Many such embryos have been destroyed, many linger in liquid nitrogen tanks, some have been donated for research and others for reproduction.

In the United States, the embryonic derivation of hES cells overlaps with fraught abortion politics that center the moral status of the human embryo, and culminated in George W. Bush's 2001 restriction of federal funds for hESC research to only those cell lines already in existence. Subsequently, several states passed funding measures to support stem cell research, most notably California's Stem Cell Research and Cures Act which committed $3 billion over ten years to support SCR including hESCR (Benjamin 2013, Thompson 2013). The federal funding restriction coupled with the perception of the U.S. government as anti-science, or at least anti–stem cell science, prompted a specific period of "brain drain" of U.S.-based scientists to more stem cell science–friendly places.

Such places included especially the U.K. and certain East Asian sites such as Singapore, China, and Taiwan. In contrast to the United States, in the U.K. the Human Fertility and Embryology Authority (HFEA) along with other government bodies made an explicit attempt to provide a comprehensive, forward-looking, and relatively permissive governance structure for new biomedical technologies (Franklin and Roberts 2006). They have approved various controversial techniques including, in 2015, mitochondrial transfer procedures for reproductive purposes (HFEA 2016). Singapore, China, and Taiwan all built robust stem cell and biotech research centers, and recruited scientists to run the programs and often to build international collaborative stem cell networks. While at first, some of these sites sought to gain research advantage through permissive regulatory regimes, by now all have robust regulatory apparatuses regarding human embryo and stem cell research.

Scientists responded to the problem of the human embryo, in part, by designing around it. Somatic cell nuclear transfer (SCNT) was thought to be the way to derive patient specific embryonic stem cells. SCNT entails the insertion of the nucleus removed from a patient's (e.g. skin) cell into an egg that has had the nucleus removed. It is the technique that was used to produce Dolly, the first cloned sheep, in 1996 and Snuppy in 2005; but in stem cell research it is used for therapeutic rather than reproductive cloning. Because of its close relationship to reproductive cloning, and the requisite destruction of the embryo in therapeutic cloning, SCNT was a controversial technique. When I arrived in Taiwan in 2005, a scientist there was working on a new idea. He proposed using a technique from preimplantation genetic diagnosis (PGD) that involved the removal of a single cell of a four cell blastocyst, thereby leaving the embryo intact. Before he could complete his research, however, an American team had established proof of concept and published their work using eight and ten cell embryos (Klimanskaya et al. 2006).

More significantly, in 2007 Shinya Yamanaka's team's induced pluripotent stem (iPS) cells served to bypass the question of the embryo and furthered the reconceptualization of biology in the new millennium (Takahashi et al. 2007, also Yu et al. 2007). Whereas previously, biological cellular development was thought to be unidirectional, iPS cells showed that somatic cells could be reprogrammed to recuperate pluripotency. Human embryos, it seemed, were perhaps no longer necessary for human pluripotent stem cell research, and differentiation and apoptosis were no longer the only cellular options. Although immortalized cell lines had already established that cellular development – differentiation – could be precluded, iPS cells reversed the process. This had been previously thought to be impossible.

The question of the human embryo was, of course, not the only ethical question. Other ethical debates ignited around various concerns. For instance, where would eggs used in stem cell research come from, and ought women to be compensated for them? For some, paying women for eggs brings human body parts into the market in unacceptable ways. Charis Thompson (2007) and Insoo Hyun (2006) argue for compensation for the act of egg donation – rather than payment for eggs – asking, among other things, why women donors are excluded from compensation when everyone else involved stands to profit. Others asked what kinds of hybrid or chimeric organisms are permissible, and what if part of the hybrid is human? Who would have access to therapies and "cures," and who would be used as experimental subjects? What are, or should be, the limits of human intervention? These questions, among many others, were asked and answered, even if not resolved, in various ways.

Global stem cells

While acknowledged internationally as requiring some special consideration, the embryo and its use in stem cell research is not everywhere so fraught. While in the United States, "Snowflake" embryos are regarded and treated as future children to be adopted into new families (Ganchoff 2004), elsewhere embryo adoption is illegal (Clarke 2008). Varying notions of meaningful life and of appropriate kinship are in play and at stake in these negotiations alongside questions of the status of the embryo itself.

Sarah Franklin emphasizes that stem cell technologies do not come into existence outside of social realms, but are infused through and through with various values.

> Stem cell technologies…demonstrate how biological properties are increasingly not only being "discovered", but are here being created, in ways that reveal specific national and economic priorities, moral and civic values, and technoscientific institutional cultures.
>
> *(Franklin 2005: 61)*

This is not an unveiling of "nature" – of biology – but a "making" of the biological infused with the social. Franklin further suggests that stem cells are a "global biological." Even as stem cell research takes place in local spaces, the U.K., as elsewhere, is responding to SCR in a global (or at least transnational) frame, and stem cell research is a "global biological enterprise" (Franklin 2005: 61). Stem cells are also global in the sense of a total system or system-wide approach – a totality (Franklin 2005: 61).

Indeed, stem cells may have "global" impacts by which I mean species level effects that implicate future generations of humans, their interspecies relations (including model and hybrid organisms), and may even render planetary effects. But stem cells might be considered global in many additional senses of the terms. Assumptions of a global morality are often in play as consideration of the human embryo is written into presumably global guidelines and regulations, such as those of the International Society for Stem Cell Research (ISSCR). Global flows of capital underpin the stem cell enterprise. In addition, stem cells are global in that sites of early clinical work involving hES cell implantation simultaneously interact with global regulatory apparatuses, affect patients from diverse global sites, and shape global reputations of scientists, countries and regions.

For instance, Aditya Bharadwaj (2013) has studied a controversial stem cell clinic in New Delhi for over a decade. The doctor's patients come from over 22 countries seeking treatment for spinal cord injury, lyme disease, among other disorders. Even as patients improve, Bharadwaj reflects on the clinic doctor's own realization that her work takes place within hierarchical structures of science. Her scientific and therapeutic work, she says, would be viewed as groundbreaking were she in the West, but scientific breakthroughs are not supposed to come from India. Instead, she is often positioned by most of the scientific and media community as a "maverick" scientist (Bharadwaj 2014). This example shows that stem cells may indeed be a kind of global biological, but science takes place within pronounced global hierarchies.

My own research on stem cells and related biotechnologies in Taiwan was initially motivated by U.S.-based rhetorics of international competition and "wild" or otherwise deficient Asian ethics. After the federal funding restriction in the U.S., many in Asian countries saw stem cell research as a way to "catch up" with the West in this promising (scientifically, therapeutically, economically) field. And, in the United States, many worried that the U.S. would fall behind given the funding restrictions.

But whereas a lack of regulatory constraints may have been viewed initially as promoting stem cell research progress, it quickly became clear that guidelines and regulations were required to oversee this research (Ong 2005, Liu 2016). Countries' reputations were at stake, and the marketability of downstream products and technologies may well depend on a kind of ethical provenance (Franklin 2003). In my own work in Taiwan, while initially the government took a wait and see approach, I found somewhat ironically that the initial lack of a regulatory apparatus had the unintended effect of slowing research progress. There, a leading embryonic stem cell researcher at a top research institution would not work with human embryos in the absence of clear guidelines and public approval.

Global is/as local

Stem cells are not only scientific objects. Indeed, they emerged as objects of significant cultural import. Biologically, they rewrote prior understandings of the limits and irreversibility of cellular processes. Ethically, they gave rise to fraught public debates. Stem cells thus revealed themselves to be important cultural objects calling for analysis beyond the biological and moral questions in which they were embroiled.

Studying stem cells as cultural objects suggests that they are not stable entities or categories. Rather, they are shaped in varied realms of significance, meaning, and value and they are infused with diverse interests that may shift in different contexts and at different scales. They are not ontologically given; they are made in relation to the circumstances of their emergence and use. Stem cells are richly productive in revealing the mutual imbrications of sciences and social orders – their co-production (Jasanoff 2004).

Even as stem cell research is a global – or at least transnational – enterprise, it is practiced everywhere in local settings. A global science is not a uniform science, and stem cell research reveals situated local priorities and values. Anthropologists, long used to paying attention to local contexts, have studied how stem cell sciences, even as they are part of global science, both reveal and take on specific meanings. Here, I briefly discuss a very limited selection of anthropological writing that has emerged on stem cell standardization, in science, in regulatory practice, and translational research and treatment.

In her research on the characterization of stem cell lines, Linda Hogle stays close to both the science and to stem cells as material objects in order to adequately understand their cultural aspects. The impulse to standardize "definitions, testing, and research practices" (2010: 434) is heightened in fields and situations of uncertainty, such as that of SCR, and Hogle tracks difficulties in characterizing and standardizing stem cell lines. Stem cells are recalcitrant objects that resist facile characterization, yet characterization and standardization of cells and practices is a priority in both basic and translational research. Hogle shows that cell characterization – designed to stabilize the field and to control cells' variability – takes on unanticipated social elements including attachment to the laboratories of their derivation, and thus remain linked to specific forms of local and tacit knowledge (2010: 442). Specific techniques and cell lines thus may not travel well. Furthermore, cell lines may be unofficially ascribed personality characteristics and reputations (Hogle 2010), as well as emotional states such as (un)happiness (Franklin 2013, chapter 2).

Achim Rosemann (2014) discusses the standardization of regulatory and clinical protocols in stem cell science. Rosemann follows a case of transnational research collaboration in translational regenerative medicine. The China Spinal Cord Network is the first clinical trial collaboration of its kind, with participants from China, Hong Kong, Taiwan, and the United States. In this spinal cord stem cell clinical trial, Rosemann shows how western standards of regulation and clinical practice are required by the research and trial protocol. However, he shows that in China, the standardization that occurs for the clinical trials does not completely replace other kinds of treatment and research practices. That is, they do not wholly overtake local modes but rather create a zone of specific standardization. Thus, multiple practices coexist in a kind of clinical translational pluralism. Like Hogle's cells, clinical and research practices may also resist standardization.

Salter and Salter (2007), in their comparative assessment of national hESC regulations show how, even as some countries are more permissive than others, they nonetheless fall within a general framework in which the terms of comparison and relative acceptability are standardized. Bharadwaj asks us to move beyond a good–bad dichotomy and to interrogate what underlies such a formulation as he witnesses healing in New Delhi. Priscilla Song, working on and in Chinese stem cell clinics, suggests that "biotech pilgrims" – patients who travel to China seeking stem cell treatments unavailable in their home countries – fit neatly into neither their home country's nor China's stem cell regulatory regimes. These transnational clinical interactions lead to the emergence of hybrid practices of, among other things, biotech governance and entrepreneurship (Song 2011).

Charis Thompson (2010), in her comparison of two stem cell labs, one in South Korea and one in Singapore, shows how within these laboratories specific national and local characteristics are expressed, even as both countries share labels as Asian Tigers (noted for their rapid economic development including Taiwan and Hong Kong). Building upon Thompson's work, I suggest that even within a single national space we find heterogeneous research practices and meanings. That is, "widely variable nationalistic goals and international aspirations often play out in unpredictable and sometimes contradictory ways" (Liu 2012a: 412). A shared label of Asian Tiger also resists standardization in SCR.

Taiwanese stem cells

In Taiwan, stem cell research is a component of a larger government strategy of biotech promotion. Biotech is seen as a growth industry with the potential for economic, scientific, and therapeutic payoffs. It holds the promise of launching Taiwan as a full-fledged member of the knowledge economy on the global stage of science. SCR and regenerative medicine also offer the therapeutic promise of addressing regionally important medical challenges that many feel have been inadequately researched in western laboratories. Regional and ethnic health concerns align with a larger project of creating a place in the global science–capital nexus for Taiwan.

I moved to Taiwan in 2005 to study stem cell research as a cultural form and to see how related ethical questions might be formulated and addressed in a non-western, newly democratic country. By tracking the questions, discussions and topics that arose in "following stem cells around," I became acquainted with the varied ways that stem cells are expressive of uniquely Taiwanese concerns, even as the domain of SCR is global. I found that stem cells "with the genetic characteristics of the Taiwanese" are imagined by some as a way to include Taiwanese in the therapeutic promise of stem cell research while also participating in global scientific progress, and that the derivation of "Taiwanese" hES cell lines overlaps with claims made in a genetic register that distance Taiwanese from (northern Han) Chinese (Liu 2010, 2012b). Thus, in a context in which Taiwan's sovereignty is continually under challenge from China, stem cells and genetics articulate an ethno-political identity with high stakes.

Taiwan achieved full electoral democracy in 1996, having previously endured the world's longest period of continuous martial law.[2] Scientists, some of whom had been blacklisted during the martial law period, returned from careers overseas to usher in a new era of political and scientific progress alongside their more rooted compatriots. Scholars have long examined the democracy–science interface. Robert Merton (1942) argued that good politics and good science share the same foundational principles, and more current work focuses largely on the ways in which science might be practiced in more democratic and inclusive ways (e.g. Douglas 2009, Hilgartner et al. 2015, Sleeboom-Faulkner and Hwang 2012). Thompson (2013) in particular has shown how democratic publics contribute to the making of "good" stem cell science. In Taiwan, this remains an especially salient set of questions as newly democratic publics challenge previously unassailable claims to expertise and authority. An engaged and critical community of scholars and activists insist that biomedical projects receive adequate public input, represent varied publics, and include elements of deliberative democracy (e.g. Rei 2010, Rei and Yeh 2009, Wu 2012, Deng and Wu 2010, Tsai 2010).

In stem cell and human embryo research fields, meetings of experts soon gave way to calls for broader public consultations in order to assess public perspectives on stem cell research in Taiwan, even as public interest remained relatively low. "We're a small country," a legal scholar told me, "so we have to give it one shot, one try." A strong sense of having to do it right pervades much of the stem cell scene, from ethical regulation to standards of best practice including GMP, GCP, and GLP (Good Manufacturing, Clinical, and Laboratory Practice respectively). And while scandals have occurred in neighboring countries (e.g. South Korea's famous 2005 Woo Suk Hwang and Japan's 2014 Haruko Obokata) in which supposedly spectacular scientific achievements quickly transformed into spectacular shame, in Taiwan most biotech scandals relate to conflict of interest or related financial issues. And procurement problems related to extractive, and even duplicitous, approaches to collecting Indigenous biosamples, have been addressed (if not resolved) by newly empowered Indigenous communities and their allies (Liu 2017).

Thus Taiwan remains relatively unexceptional in both stem cell research and related ethical regulation. So far, it is neither a leader of groundbreaking stem cell science nor of great scandal or controversial "maverick" therapies. Analytically, however, Taiwan provides a rich terrain to study the ways in which global science is (re)shaped by local and regional concerns (e.g. political, economic, medical, ethical), and the ways in which, even as regulatory documents conform to global standards, they nonetheless come out of a recognized need to represent local public values (Liu 2016).

Conclusion

The global does not merely map onto local sites, and specific locals are not simply geographical sites of relative difference or sameness. Globalization is uneven and it may skip large regions (Ferguson 2006); its interconnections are often messy (Tsing 2000) and characterized by friction that (re)shapes knowledge in "zones of awkward engagement" (Tsing 2005: xi). Its nodes and assemblages (Ong and Collier 2005) comprise multiple and varied elements that may resist stabilization. Stem cells are global objects, but they are not everywhere the same; they take on local complexities and are brought into a wide variety of both therapies and discourses. Their therapeutic and economic potential is joined with their potential for ethical challenges and scandalous science.

The role of the social analyst is to go beneath the dominant discourses and representations to see what other kinds of meanings, conflicts or stabilizations that cultural objects, like stem cells, might evoke. Our role is to avoid as best as we can what Bharadwaj (2012: 312) calls "anthropology capture," in which the analyst is taken in by the normative discourses on stem cell ethics thereby limiting the questions and categories of analysis. SCR has largely become stabilized or normalized in what Thompson describes as the "end of the beginning of human pluripotent stem cell research" (2013: 19). However, as Franklin suggests – regarding IVF and invoking Alice in Wonderland – even as new biomedical technologies and techniques become normalized, they may become "curiouser and curiouser" (2013: 2) as they give rise to new lived experiences, new relations, and new questions.

Notes

1 35 U.S.C. Section 101 – Diamond v. Chakrabarty https://supreme.justia.com/cases/federal/us/447/303/case.html.
2 1949–1987, 38 years. This has since been exceeded by Syria.

References

Benjamin, Ruha. 2013. *People's science: Bodies and rights on the stem cell frontier*. Stanford, CA: Stanford University Press.

Bharadwaj, Aditya. 2014. Experimental subjectification: The pursuit of human embryonic stem cells in India. *Ethnos* 79(1): 84–107.

Bharadwaj, Aditya. 2013. Subaltern biology? Local biologies, Indian odysses, and the pursuit of human embryonic stem cell therapies. *Medical Anthropology* 32: 359–373.

Bharadwaj, Aditya. 2012. Enculturating Cells: The Anthropology, Science, and Substance of Stem Cells. *Annual Reviews Anthropology* 41: 303–317.

Clarke, Peter. 2008. Embryo Donation/Adoption: Medical, Legal and Ethical Perspectives. *Internet Journal of Law, Healthcare and Ethics* 5(2). http://ispub.com/IJLHE/5/2/11953.

Deng, Chung-Yeh and Chia-Ling Wu. 2010. An innovative participatory method for newly democratic societies: The 'civic groups forum' on national health insurance reform in Taiwan. *Social Science and Medicine* 70: 896–903.

Douglas, Heather. 2009. *Science, policy, and the value-free ideal*. Pittsburgh, PA: University of Pittsburgh Press.

Ferguson, James. 2006. *Global shadows: Africa in the neoliberal world order*. Durham, NC: Duke University Press.

Franklin, Sarah. 2003. Ethical biocapital: New strategies of cell culture. In Sarah Franklin and Margaret Lock, eds., *Remaking life & death: Toward an anthropology of the biosciences*, 97–128. Santa Fe, NM: School of American Research.

Franklin, Sarah. 2005. Stem cells R us: Emergent life forms and the global biological. In Aihwa Ong and Stephen J. Collier, eds., *Global assemblages: Technology, politics, and ethics as anthropological problems*, 59–78. Malden, MA: Blackwell.

Franklin, Sarah. 2013. *Biological relatives: IVF, stem cells, and the future of kinship*. Durham, NC: Duke University Press.

Franklin, Sarah and Margaret Lock, eds. 2003. *Remaking life & death: Toward an anthropology of the biosciences*. Santa Fe, NM: School of American Research.

Franklin, Sarah and Celia Roberts. 2006. *Born and Made: An Ethnography of Preimplantation Genetic Diagnosis*. Princeton, NJ: Princeton University Press.

Ganchoff, Chris. 2004. Regenerating movements: Embryonic stem cells and the politics of potentiality. *Sociology of Health & Illness* 26(6): 757–774.

Hayden, Cori. 2003. *When nature goes public: The making and unmaking of bioprospecting in Mexico*. Princeton, NJ: Princeton University Press.

HFEA – Human Fertilisation and Embryo Authority. 2016. Mitochondria scientific review update: call for evidence. http://hfeaarchive.uksouth.cloudapp.azure.com/www.hfea.gov.uk/docs/Fourth_scientific_review_mitochondria_2016.pdf

Hilgartner, Stephen, Clark A. Miller and Rob Hagendijk, eds. 2015. *Science and democracy: Making knowledge and making power in the biosciences and beyond*. New York: Routledge.

Hogg, Chris. 2006. Taiwan breeds green-glowing pigs. BBC News, 12 January. http://news.bbc.co.uk/2/hi/4605202.stm.

Hogle, Linda. 2010. Characterizing human embryonic stem cells: Biological and social markers of identity. *Medical Anthropology Quarterly* 24(4): 433–450.

Hyun, Insoo. 2006. Fair payment or undue inducement? *Nature* 442(7103): 629–630.

Jasanoff, Sheila. 2004. *States of knowledge: The co-production of science and social order*. New York: Routledge.

Klimanskaya, Irina, Young Chung, Sandy Becker, Shi-Jiang Lu and Robert Lanza. 2006. Human embryonic stem cell lines derived from single blastomeres. *Nature* 444(7118): 481–485.

Liu, Jennifer. 2017. Postcolonial biotech: Taiwanese conundrums and subimperial desires. *East Asian Science, Technology and Society: An International Journal* 11: 1–26.

Liu, Jennifer. 2016. Emerging science, emerging democracy: Stem cell research and policy in Taiwan. *Perspectives on Science* 24(5): 609–636.

Liu, Jennifer 2012a. Asian regeneration? Technohybridity in Taiwan's biotech? *East Asian Science, Technology and Society: An International Journal* 6(3): 401–414.

Liu, Jennifer. 2012b. Aboriginal fractions: Enumerating identity in Taiwan. *Medical Anthropology* 31: 329–346.

Liu, Jennifer. 2010. Making Taiwanese (stem cells): Identity, genetics, and purity. In Aihwa Ong and Nancy Chen, eds., *Asian Biotechnology: Ethics and Communities of Fate*, 239–262. Durham, NC: Duke University Press.

Merton, Robert K. 1942. Science and technology in a democratic order. *Journal of Legal & Political Sociology* 1: 15–26.

Ong, Aihwa. 2005. Ecologies of expertise: Assembling flows, managing citizenship. In Aihwa Ong and Stephen J. Collier, eds. *Global assemblages: Technology, politics, and ethics as anthropological problems*, 337–353. Malden, MA: Blackwell.

Ong, Aihwa and Stephen J. Collier, eds. 2005. Global Assemblages: Technology, Politics and Ethics as Anthropological Problems. Malden, MA: Blackwell.

Rei, Wenmay and Jiunn-Rong Yeh. 2009. Promises and pitfalls of using national bioethics commissions as an institution to facilitate deliberative democracy: Lessons from the policy-making of human embryonic stem cell research. *National Taiwan University Law Review* 4(2): 69–105.

Rei, Wenmay. 2010. Toward a governance structure beyond informed consent: A critical analysis of the popularity of private cord blood banking in Taiwan. *East Asian Science, Technology and Society: an International Journal* 4(1): 53–75.

Rosemann, Achim. 2014. Standardization as situation-specific achievement: Regulatory diversity and the production of value in intercontinental collaborations in stem cell medicine. *Social Science & Medicine* 122: 72–80.

Salter, Brian and Charlotte Salter. 2007. Bioethics and the global moral economy: The cultural politics of human embryonic stem cell science. *Science, Technology, & Human Values* 32(5): 554–581.

Sleeboom-Faulkner, Margaret and Seyoung Hwang. 2012. Governance of stem cell research: Public participation and decision-making in China, Japan, South Korea and Taiwan. *Social Studies of Science* 42(5): 684–708.

Song, Priscilla. 2011. The proliferation of stem cell therapies in post-Mao China: Problematizing ethical regulation. *New Genetics and Society* 30(2): 141–153.

Sunder Rajan, Kaushik. 2006. *Biocapital: The constitution of postgenomic life*. Durham, NC: Duke University Press.

Takahashi, Kazutoshi, Koji Tanabe, Mari Ohnuki, Megumi Narita, Tomoko Ichisaka, Kiichiro Tomoda, and Shinya Yamanaka. 2007. Induction of pluripotent stem cells from adult human fibroblasts by defined factors. *Cell* 131(5): 861–872.

Thompson, Charis. 2013. *Good science: The ethical choreography of stem cell research*. Cambridge, MA: The MIT Press.

Thompson, Charis 2010. Asian regeneration? Nationalism and internationalism in stem cell research in South Korea and Singapore. In Aihwa Ong and Nancy Chen, eds., *Asian Biotechnology: Ethics and Communities of Fate*, 95–117. Durham, NC: Duke University Press.

Thompson, Charis. 2007. Why we should, in fact, pay for egg donation. *Regenerative Medicine* 2(2): 203–209.

Thomson, James A., Joseph Itskovitz-Eldor, Sander S. Shapiro, Michelle A. Waknitz, Jennifer J. Swiergiel, Vivienne S. Marshall, Jeffrey M. Jones. 1998. Embryonic stem cell lines derived from human blastocysts. *Science* 282(5391): 1145–1147.

Tsai, Yu-yueh. 2010. Geneticizing ethnicity: A study on the 'Taiwan Bio-Bank'. *East Asian Science, Technology and Society: An International Journal* 4(3): 433–455.

Tsing, Anna L. 2000. The global situation. *Cultural Anthropology* 15(3): 327–360.

Tsing, Anna L. 2005. *Friction: An ethnography of global connection*. Princeton, NJ: Princeton University Press.

Waldby, Catherine and Robert Mitchell. 2006. *Tissue economies: Blood, organs, and cell lines in late capitalism*. Durham, NC: Duke University Press.

Wu, Chia-Ling. 2012. IVF policy and global/local politics: The making of multiple-embryo transfer regulation in Taiwan. *Social Science & Medicine* 75: 725–732.

Yu, Junying, Maxim A. Vodyanik, Kim Smuga-Otto, Jessica Antosiewicz-Bourget, Jennifer L. Frane, Shulan Tian, Jeff Nie, Gudrun A. Jonsdottir, Victor Ruotti, Ron Stewart and James A. Thompson. 2007. Induced pluripotent stem cell lines derived from human somatic cells. *Science* 318(5858): 1917–1920.

33

Co-producing animal models and genetic science

Carrie Friese

This chapter reviews the role of animal models in genetic research, both historically and in the contemporary moment. Consistent with the social science and historical literatures on animal models, I argue that genetic science and animal models have 'co-produced' (Jasanoff, 2004; Reardon, 2001) one another across the twentieth and into the twenty-first centuries (Clarke & Fujimura, 1992; Lederman & Burian, 1993). This means that genetic knowledge and the animal used have shaped one another in an iterative and open-ended manner rather than in a one way and deterministic manner. As Hans-Jörg Rheinberger (1997, 2000, 2010) has argued, nature does not come ready made into the laboratory; nature (here laboratory animals) must instead be remade as research material, and rendered meaningful through laboratory practices that occur within the histories of scientific disciplines. But just as science reshapes the biologies of animals upon entering the lab, the biologies of animals also act back in meaningful ways that shape what can be known about genetics and genomics.

The chapter begins with a brief conceptual introduction to models and modelling in the life sciences. This provides a framework for exploring some of the key non-human species used to produce genetic knowledge since the start of the twentieth century. This review will provide a basis for then discussing the standardization of nonhuman animal bodies within the laboratory, and the corresponding idea of 'generalization' – that species conserve biological traits through evolution so that one species can stand for another in establishing universal biological facts. This allows nonhuman animal species to act as models of humans in the production of biomedical knowledge. However, the twin processes of standardization and generalization have been problematized by recent developments in genetic science itself. Rather ironically, genetically uniform animals demonstrate the limits of genetic determinism as not only proofs of epigenetics (e.g., the agouti mouse); animals with the same genomes also show how nonhuman animals develop in different ways over time and in the contexts of laboratory science.

Models and modelling in the life sciences

Models are generally understood as physical representations of something else, serving as physical or material sites of practice that make future scenarios possible. Mary S. Morgan and Margaret Morrison (1999) develop the notion of models as 'mediators', or things that are patched together with theory in order to produce knowledge. By looking at the genetic sciences, Evelyn Fox Keller (2000) extends this idea of models as mediators by distinguishing between 'models of' something else, as in Morgan and Morrison's delineation, from 'models for' doing certain things in practice. Building on the idea of models for future action, Angela Creager (2002: 4–5) uses the tobacco mosaic virus as a case study to show how some organisms come to stand as not only exemplars of other things or species but also as exemplars for conducting research. Creager denotes that model organisms become 'model systems' in such cases.

Other distinctions have been usefully developed when analyzing how models work, particularly in the life sciences. For example, Jessica Bolker (2009) distinguishes between 'surrogate models' where one species stands for another species (such as dogs standing for humans in biomedical research) and 'exemplary models' where one species stands for a more basic process that exists across a whole range of species. The representational credibility of both surrogate and exemplary models is constantly in question, however. For example, Nicole Nelson (2012) develops the idea of an 'epistemic scaffold' to show how the case for using surrogate models (i.e., mice as models of humans) in biomedical research is built up over time by scientists. However, this scaffolding is precarious and constantly at risk of being dismantled if claims are over generalized.

Meanwhile, the limits of what counts as a 'model organism' is another question. Rachel Ankeny and Sabina Leonelli (2011) distinguish between 'model organisms' and 'experimental organisms', a difference that they argue is important as 'model organisms' have become ubiquitous in life science research due to funding and publishing pressures to use them. Ankeny and Leonelli argue that model organisms are extensively studied species, with significant human communities and technical apparatuses surrounding them, which are used to understand a broad range of organisms as part of interdisciplinary and comparative work that is integrative (i.e., exemplary models). In contrast, experimental organisms are used to study more specific phenomenon or processes from a particular discipline or perspective. Certainly some of the animal models of genetic science have become model organisms in this sense (e.g., *Drosophila* and mouse). But genetic science also uses experimental organisms in its research, particularly in biomedical research.

Across these delineations – and incremental developments in sociological, historical and philosophical approaches to modelling – there is consensus that the ability of models to surprise scientists makes them good to think and work with; models are thus crucial for generating new knowledge (Ankeny & Leonelli, 2011: 315; Creager, 2002; Keller, 2000; Morgan & Morrison, 1999; Rheinberger, 1997, 2010). Model organisms and experimental organisms have played a key role in developing genetic knowledge often because of the unanticipated findings generated by working with them.

The animal models of genetic sciences

The story of animal models in genetic science is embedded in the transition from natural history, where a small number of many different kinds of animal species were needed for the purpose of classification and comparison, to experimental science, where large numbers of selects animals were required for the purpose of hypothesis testing (Clarke, 1995; Kohler, 1994). Over the twentieth century specific organisms have played crucial roles in the experimental approaches that have been central to genetic knowledge production, including maize, mice, rats, *Drosophila, C. elegans, E. coli* and *Neurospora*. In this context, there have been a select number of monographs and research papers that tell the history of genetic science by tracing these species. I briefly review this literature with some depth in order to provide an introduction to the role of specific species in the history of genetics, and the history of genetics in creating two of the 'model organisms' (Ankeny & Leonelli, 2011) par excellence: *Drosophila* and mouse.

Robert Kohler (1994) provides a now canonical historical study of early twentieth-century genetic science by following a specific nonhuman animal species. Kohler follows *Drosophila* across T.H. Morgan's laboratory, and through the careers of the scientists who worked in his lab, in order to understand how genetic knowledge was produced in practice.[1] What emerges is neither a story of biological nor social constructionism, but rather a historical narrative regarding the ways in which *Drosophila* were created in order to address certain questions, and in turn delimit other types of questions from being asked. This is a story of ironic twists, as *Drosophila* came into Morgan's lab in order to ask questions about evolution (Kohler, 1994: 42). However, Kohler shows how, as *Drosophila* took over Morgan's laboratory, a focus on experimental evolution gave way to research on neo-Mendelian experimental heredity and genetic mapping. Indeed, the fecundity of flies was bound up in the moral economy of Morgan's laboratory and early *Drosophila* exchange networks based on communal sharing and collaboration. And this made it difficult to use *Drosophila* for evolution and development research. Indeed, George Beadle turned to *Neurospora* as a better model organism for studying biochemical genetics, or what some people were starting to call 'molecular biology' (Kohler, 1994: 243). And Theodosius Dobzhansky required wild type flies in order to conduct his research in evolution. *Drosophila* eventually fell out of favour amongst molecular biologists by the middle of the twentieth century as new questions came to the fore.

During this same time period, mice were also becoming a prominent model organism in genetic science (Rader, 2004: 16–18), with a specifically medical focus (Gaudillière, 2004). Karen Rader explores the ways in which mice were industrialized as part of scientific and biomedical research. She focuses specifically on the creation of Jackson Laboratories in the United States, which continues to fund its genetic research in part through its extensive sale of mice to other labs. Rader shows how C.C. Little increased the availability of mice for genetic research (Rader, 2004: 29–30) through his links with cancer research (Rader, 2004: 46). But maintaining a mouse colony was considerable when compared to *Drosophila*, which led Little to found the Jackson Laboratory. In this context, mice were commercialized as another kind of biomedical supply for a full range of scientific and biomedical disciplines (Rader, 2004: 99) in order to facilitate genetic research with applications to human medicine (Gaudillière, 2004). As Jean-Paul Gaudillière (2004) notes, in this context mouse geneticists at Jackson Laboratory were not only doing research on genes, but were also and crucially modelling human pathologies. He argues: 'The mapping of mice did not follow the mapping of flies, because mice were constructed as human beings' (Gaudillière, 2004: 200). Gaudillère contends that this meant genetic research using mice differed significantly from genetic research using *Drosophila*, such that the cultures and practices of genetic science varied significantly in the twentieth century not simply because the biologies of model organisms differed but also because the goals of the science differed.

Animals like *Drosophila* and mice lost prominence in much of genetic science during the middle of the twentieth century, when bacteria and phages increasingly became the preferred organisms to work with (Creager, 2005; Kohler, 1994). However, both mice and *Drosophila* regained prominence in molecular biology during the 1980s and 1990s (Davies, 2013b: 3; 2013c; Weber, 2007). In addition, other species like *C. elegans* (Leonelli & Ankeny, 2012) became important genetic tools around this time. The re-emergence of *Drosophila* and mice as model organisms has been understood as the logical next step in gene mapping (Davies, 2013b, 2013c; Leonelli & Ankeny, 2012). But the re-emergence of these species in genomics has also occurred because these species were standardized across the twentieth century (Weber, 2007), allowing for genetic sequencing to be used in order to answer new questions (Leonelli &

Ankeny, 2012: 30). The favoured species of genetic science have become 'model organisms' (Ankeny & Leonelli, 2011) par excellence because there is so much knowledge and infrastructure surrounding them, and so these species are unlikely to ever truly go extinct from the lab (Kohler, 1994).

Standardization and generalization: the genetics of animal models

The sociological, historical and philosophical literature surrounding the use of nonhuman animal species as models in genetic science centres largely around two interrelated concepts: standardization and generalization. Standardization in particular carries with it a range of definitions and meanings including: 1) being genetically inbred; 2) being ubiquitous; 3) being a standard-bearer; and 4) uniting what was previously ad hoc. The object of standardization is similarly diffuse, including: 1) the animal species itself; 2) the language about the animal; and 3) the environment of the animal. This heterogeneity, and the aim of controlling it, is interlinked with the goal of producing generalizable knowledge that can both extend beyond the local laboratory setting and beyond the species of animal used in the experiment, whether it be as a surrogate or exemplary model, or a model or experimental organism.

Standardization

Standardization is a key theme across the literature on animal models in genetic science. Kohler (1994) focuses on how *Drosophila* had to be standardized in order to become tools for chromosome mapping, which focused on the order and distance between genes using quantitative methods. The flies in Morgan's laboratory were literally reconstructed genetically through selection and inbreeding, causing mutants that simply could not be understood according to neo-Mendelian formulas. Kohler emphasizes that *Drosophila*, and laboratory animals more generally, become instruments through the accretion of craft knowledge and skill gained by tinkering with these animals (Kohler, 1994: 6–7). For Morgan's lab, standardization and genetic mapping were part and parcel of one another.

Rader emphasizes that mice were standardized during the early twentieth century as well, also through selection and inbreeding to create mutants. She carefully defines standardization as both ubiquitous and conspicuous, or as representing something that everyone uses and as a marker for and a rallying point. Because mice had been standardized at the locus of the gene they are standard-bearers for genetic research and its salience for medicine, while also being the most frequently used 'sentient' nonhuman animal used in science today (Rader, 2004: 16–18).

Ankeny and Leonelli emphasize that standardization is a crucial component in research involving model organisms. These species are often chosen because they have characteristics amenable to genetic science, including: small physical and genetic sizes, short generation times, short life cycles, high fertility rates and high mutation rates (Ankeny & Leonelli, 2011: 316). These species have also been reworked in the laboratory, at the genetic level, through a range of techniques. As such, the status of a species as a model organism has come to depend upon the development of a genetic account of the organism in terms of gene sequence, gene function and phenotype (Ankeny & Leonelli, 2011: 316). However, Ankeny and Leonelli emphasize that genetic standardization alone is not enough; model organisms also require a standardized nomenclature as well as a standardized experimental setting – standardizing language and space is a crucial aspect of making model organisms work in producing generalizable knowledge (Leonelli & Ankeny, 2012: 31–33).

Robert Kirk's (2010) work on the standardization of laboratory animals is instructive in this regard. Kirk explores the creation of national standards for laboratory animal production in the United Kingdom across the middle of the twentieth century. He argues that the Medical Research Council transformed ad hoc relations between local animal dealers and individual laboratories into the laboratory animal industry we see today. Accreditation and corresponding standards for laboratory animal production – including housing, feeding and hygiene – were the primary technologies used in this move toward centralization (Kirk, 2010: 70). This was supported by a communication technology aimed at linking laboratory animal producers and consumers: a bimonthly newsletter (which would have been familiar to 'fly people' (Kohler, 1994) and 'mouse people' (Rader, 2004)). But to truly standardize, Kirk argues that a research base for laboratory animal breeding and management had to be created in order to support the discipline of laboratory animal science (Kirk, 2010: 75). Kirk (2010: 79) emphasizes that pathogenic standardization was of far greater concern in the UK in this context (see also Kirk, 2008, 2012). In other words, if the standardization of mice in the United States was linked with making genetic research a standard-bearer according to Rader, the standardization of laboratory animals in the UK was linked to the politics of 'socialism as democracy' in post-war Britain, according to Kirk. Centralization was therefore crucial to standardization in this context (Kirk, 2010: 93).[2]

Transgenics has been a key site where the processes of standardization have been both extended and transformed. As Donna Haraway (1997: 57) notes, transgenic organisms are 'at once completely ordinary and the stuff of science fiction'. Laboratory animals have been remade through transgenics, directly at the molecular level as opposed to through selection and inbreeding. Here genes are inserted in order to either deactivate or introduce a genetic sequence that is of interest. The OncoMouse – developed by Du Pont Corporation in the 1980s – is one of the most famous examples of this. Joan Fujimura (1996: 7) shows this genetically modified strain of mouse further standardized cancer research. By incorporating a specific oncogene, *ras*, the strain was advertised and promoted for its ability to reliably develop cancer tumours within a specified time period: one month. OncoMouse was at once an ordinary commodity not unlike the mice from Jackson Labs, but also a significant shift. As Donna Haraway (1997: 79) notes, OncoMouse was the first patented animal in the world; transgenics made this laboratory animal into a human invention that could be patented in a manner that was not true of selection and inbreeding, which 'mimics' 'natural' processes.

Generalization

Kirk (2010: 66) emphasizes that standardization discourses arose in conjunction with the goal of creating universal knowledge. In other words, standardization is linked with the need and desire to generalize scientific findings beyond the laboratory setting. In the context of using nonhuman animals as models, generalization has relied upon the idea that species conserve biological processes through evolution. This makes it possible for one species to stand for another (Burian, 1993; Nalbandov, 1976; Rader, 2004; Rheinberger, 2010).

The idea of species generality unevenly solidified across the twentieth century (Logan, 2001, 2002). Historian of science Cheryl Logan (1999, 2001, 2002, 2005) explores when and how the presumption that species conserve biological forms came into being. Through a number of case studies, Logan has shown that the extent of physiological similarity and difference across species was not assumed at the end of the nineteenth century, but was instead an empirical question. It was not until after 1900 that species similarity became an a priori assumption in life science research. In this context, some species became 'exemplary models'

(Bolker & Raff, 1997), whose bodies could stand as universal representations of biological forms and processes across a full range of other species. Logan (2001, 2002) attributes the solidification of 'generality' to the increasing availability of standardized animals as research materials through commodification.

Rachel Ankeny (2007) has contested the idea that generality took hold as strongly as either Logan or Bolker contend. In her analysis of *C. elegans* as a model organism in molecular biology, Ankeny (2007) has noted that much modelling work implicitly assumes that findings about one species can be generalized to another. Nonetheless, she argues that comparison is one of the key features of modelling practices, and so there is a feedback loop between knowledge about the index case (the well described model) and the subject being modelled (Ankeny, 2007: 55). Ankeny argues that the work involved in making comparisons delimits the assumption of species conservation, thereby calling into question the extent to which generalization stabilized across the twentieth century. This critique is further developed with the definition of 'model organisms' (Ankeny & Leonelli, 2011) as entities capable of bringing a range of different disciplines together in order to *compare* species more holistically.

Troubling standardization and generality

Can a mouse stand for a human? Can a mouse in a laboratory stand for anything outside the laboratory? Gail Davies (2010, 2011, 2012a, 2012b, 2013a, 2013b, 2013c) explores these questions at length in her transnational study of the endeavour to make and use humanised mice. The idea here is that a humanised mouse will serve as a better surrogate model of humans because it will have greater 'corporeal equivalence' to humans (Davies, 2012b). As another site of transgenics, the humanised mouse is another attempt at greater standardization with the aim of improving generality. But Davies shows that genetics alone cannot solve the problems of generality, as the very different spaces in which humans and animals live thwart translational efforts. In the process, both standardization and generality come into question.

As an example, Davies (2012a) points to the tragic case of preclinical safety trials at Northwick Park in the U.K. This trial tested a humanized monoclonal antibody to treat leukaemia and rheumatoid arthritis. No adverse effects were found in laboratory animals, but human immune systems overreacted in a manner that led to systemic organ failure in all trial participants. Davies states: 'The mice used for safety testing, kept in sterile laboratories, had not had the same immune challenges as the human trial subjects. Their inexperienced immune systems did not react. In humans, the memory of past infections contributed to the detrimental response. The species, but also the spaces, were not representative' (Davies, 2012b: 132–33). Davies's argument parallels beautifully Canguilhem's reminder that: 'We must not forget that the laboratory itself constitutes a new environment in which life certainly establishes norms … Whole extrapolation does not work without risk when removed from the conditions to which these norms relate' (Gaudilliere & Rheinberger, 2005: 2–3).

Postgenomics and post-standardization: genetics, model organisms and the environment

Both standardization and generality are idealized goals that – while surely productive of knowledge practices (Kirk, 2010: 65) – are nonetheless always already partial and incomplete. Today both standardization and generalization are increasingly being questioned in the context of postgenomics. This raises new questions about the role of the environment in genetic expression as well as the ability to generalize research findings beyond the laboratory and beyond the animal species or strain in question.

Davies (2013b: 7) emphasizes that the intertwined goals of standardization and generalization are based on a particular notion of genetic reductionism, wherein research is focused on specific gene foci in animals so as to strategically blackbox biological complexity. She contends that this notion is linked to the notion of biological generality, wherein the mice in the experiment represent mice in general and/or humans in general. But the idea of a standardized mouse assumes a kind of genetic reductionism that the Human Genome Project rather ironically destabilized, ushering in postgenomic research that explores gene–gene, gene–epigenetic and gene–environment relations. This has raised questions about the role of animals as models in genetic science and translational research more generally (Davies, 2012b). Davies (2013a, 2013b, 2013c) has compellingly argued that this also destabilizes the focus on 'standardization' in historical and social science accounts of laboratory animals, as genetic science itself has become interested in the production of biological difference.

Davies notes that the shifting terrain surrounding standardization and generalization in the context of postgenomics also fundamentally disrupts the idea of replication that is so central to scientific research (Davies, 2013b: 17). In this context, Davies argues that standardization has shifted away from a single, inbred strain of mouse toward a well-characterized genotype that accounts for the life experiences of the individual animal (Davies, 2013a: 426), and therefore takes 'the environment' more fully into account. Indeed, we could say that the agouti mouse stands as a contemporary standard bearer of epigenetics as postgenomics (see for example Landecker, 2011); here genetically identical mice exposed to different food in utero have radically different phenotypes. The environment of the laboratory becomes a crucial data point in this context.

One solution has been to focus greater attention to animal care within the laboratory – specially housing, food and enrichment – so that the entire experimental system of one lab can be reproduced elsewhere (Davies, 2010; 2013b: 8). Vinciane Despret (2004) has offered a theorization of this focus on animal care in science, showing how the model organisms of science come to do what scientists care for them to do. I have ethnographically charted how scientific practices change when animal care comes to fore in rather uncanny ways (Friese, 2013). Kirk (2010, 2014), Tone Druglitrø (2017) and Beth Greenhough and Emma Roe (2011) have explored the rise of laboratory animal science in this context, emphasizing the ways in which animal care has become central to experimental research. And Nicole Nelson (2013) has shown how science and welfare concerns merge when gene–environment interactions are of research interest. The argument here is that genetic standardization is not the problem for many areas of research, but rather standardizing the environment (Ankeny et al., 2014; Nelson, Forthcoming).

Conclusion

This chapter has reviewed the roles of animal models in genetic research, showing how the genetics of animal models and the production of genetic knowledge have co-produced one another since the beginning of the twentieth century. Species like *Drosophila* and mice have been remade in the laboratory as they have become model organisms par excellence through the work of genetic and other medical/scientific disciplines. This work has rested upon twin beliefs in and processes of standardization and generalization. However, genetically uniform animals demonstrate the limits of genetic determinism, as not only proofs of epigenetics (e.g., the agouti mouse) but also in terms of the consequences of care for the ways animals develop over time and in science. The environment is becoming an important source of new genetic knowledge, and an important site for renewed efforts in standardization, in the shifting landscapes of postgenomic research.

Notes

1 That said, Evelyn Fox Keller's (1983) biography of Barbara McClintock very much weaves the story of this scientist's career with the story of maize.
2 See also Kirk (2008) on the links between standard experimental animals in Britain and the political contexts for this endeavour.

References

Ankeny, R. A. (2007). Wormy logic: Model organisms as case-based reasoning. In A. N. H. Creager, E. Lunbeck, & M. N. Wise (Eds.), *Science Without Laws: Model Systems, Cases, Exemplary Narratives* (pp. 46–58). Durham, NC: Duke University Press.

Ankeny, R. A., & Leonelli, S. (2011). What's so special about model organisms? *Studies in History and Philosophy of Science*, 42, 313–323.

Ankeny, R. A., Leonelli, S., Nelson, N. C., & Ramsden, E. (2014). Making organisms model human behavior: Situated models in North-American alcohol research, since 1950. *Science in Context*, 27(3), 485–509.

Bolker, J. A. (2009). Exemplary and surrogate models: Two modes of representation in biology. *Perspectives in Biology and Medicine*, 52(4), 485–499.

Bolker, J. A., & Raff, R. A. (1997). Beyond worms, flies and mice: It's time to widen the scope of developmental biology. *The Journal of NIH Research*, 9, 35–39.

Burian, R. (1993). How the choice of experimental organism matters: Epistemological reflections on an aspect of biological practice. *Journal of the History of Biology*, 26(2), 351–368.

Clarke, A. E. (1995). Research Materials and Reproductive Science in the United States, 1910–1940 with Epilogue: Research Materials Revisited. In S. L. Star (Ed.), *Ecologies of Knowledge: New Directions in Sociology of Science and Technology* (pp. 183–225). Albany: State University of New York Press.

Clarke, A. E., & Fujimura, J. H. (1992). *The Right Tools for the Job: At Work in Twentieth-Century Life Sciences*. Princeton: Princeton University Press.

Creager, A. N. H. (2002). *The Life of a Virus: Tobacco Mosaic Virus as an Experimental Model, 1930–1965*. Chicago: University of Chicago Press.

Creager, A. N. H. (2005). Mapping genes in microorganisms. In J.-P. Gaudillière & H.-J. Rheinberger (Eds.), *From Molecular Genetics to Genomics: The Mapping Cultures of Twentieth-Century Genetics* (pp. 9–41). London: Routledge.

Davies, G. F. (2010). Captivating behaviour: Mouse models, experimental genetics and reductionist returns in the neurosciences. *The Sociological Review*, 58, 53–72.

Davies, G. F. (2011). Playing dice with mice: Building experimental futures in Singapore. *New Genetics and Society*, 30(4), 433–441.

Davies, G. F. (2012a). Caring for the multiple and the multitude: Assembling animal welfare and enabling ethical critique. *Environment and Planning D: Society and Space*, 30, 623–638.

Davies, G. F. (2012b). What is a humanized mouse? Remaking the species and spaces of translational medicine. *Body & Society*, 18, 126–155.

Davies, G. F. (2013a). Arguably big biology: Sociology, spatiality and the knockout mouse project. *Bio-Societies*, 4(8), 417–431.

Davies, G. F. (2013b). Mobilizing experimental life: Spaces of becoming with mutant mice. *Theory, Culture & Society*, 30, 129–153.

Davies, G. F. (2013c). Writing biology with mutant mice: The monstrous potential of post genomic life. *Geoforum*, 48, 268–278.

Despret, V. (2004). The body we care for: Figures of anthropo-zoo-genesis. *Body & Society*, 10(2), 111–134.

Druglitrø, T. (2017). 'Skilled care' and the making of good science. *Science, Technology and Human Values, Online First*. doi:10.1177/01622439/6688093.

Friese, C. (2013). Realizing the potential of translational medicine: The uncanny emergence of care as science. *Current Anthropology*, 54(7), S129–S138.

Fujimura, J. H. (1996). *Crafting Science: A Sociohistory of the Quest for the Genetics of Cancer*. Cambridge, MA: Harvard University Press.

Gaudilliére, J.-P. (2004). Mapping as technology: Genes, mutant mice, and biomedical research (1910–1965). In H.-J. Rheinberger & J.-P. Gaudilliére (Eds.), *Classical Genetic Research and its Legacy: The Mapping Cultures of Twentieth-Century Genetics* (pp. 173–203). London: Routledge.

Gaudilliére, J.-P., & Rheinberger, H.-J. (2005). Introduction. In J. -P. Gaudilliére & H. -J. Rheinberger (Eds.), *From Molecular Genetics to Genomics: The Mapping Cultures of Twentieth-Century Genetics* (pp. 1–6). London: Routledge.

Greenhough, B., & Roe, E. (2011). Ethics, space, and somatic sensibilities: Comparing relationships between scientific researchers and their human and animal experimental subjects. *Environment and Planning D: Society and Space*, 29, 47–66.

Haraway, D. J. (1997). *Modest_Witness@Second_Millennium.FemaleMan©_Meets_OncoMouse^{TM}: Feminism and Technoscience*. New York: Routledge.

Jasanoff, S. (2004). *States of Knowledge: The Co-Production of Science and Social Order*. London: Routledge.

Keller, E. F. (1983). *A Feeling for the Organism: The Life and Work of Barbara McClintock*. New York: Holt Paperbacks.

Keller, E. F. (2000). Models of and Models for: Theory and Practice in Contemporary Biology. *Philosophy of Science*, 67(Supplement), S72–S86.

Kirk, R. G. W. (2008). 'Wanted – standard guinea pigs': standardisation and the experimental animal market in Britain ca. 1919–1947. *Studies in the History and Philosophy of Biology and Biomedical Sciences*, 39(3), 280–291.

Kirk, R. G. W. (2010). A brave new animal for a brave new world: The British Laboratory Animals Bureau and the constitution of international standards of laboratory animal production and use, circa 1947–1968. *Isis*, 101, 62–94.

Kirk, R. G. W. (2012). Standardization through Mechanization: Germ Free Life and the Engineering of the Ideal Laboratory Animal. *Technology and Culture*, 53(1), 61–93.

Kirk, R. G. W. (2014). The invention of the 'stressed animal' and the development of a science of animal welfare, 1947–1986. *Stress, Shock, and Adaptation in the Twentieth Century* (pp. 241–263). Rochester, NY: University of Rochester Press.

Kohler, R. E. (1994). *Lords of the Fly: Drosophila Genetics and the Experimental Life*. Chicago: Chicago University Press.

Landecker, H. (2011). Food as exposure: Nutritional epigenetics and the new metabolism. *BioSocieties*, 6(2), 167–194.

Lederman, M., & Burian, R. M. (1993). Introduction. *Journal of the History of Biology*, 26(2), 235–237.

Leonelli, S., & Ankeny, R. A. (2012). Re-thinking organisms: The impact of databases on model organism biology. *Studies in History and Philosophy of Biological and Biomedical Science*, 43, 29–36.

Logan, C. A. (1999). The altered rationale behind the choice of a standard animal in experimental psychology. Henry H. Donaldson, Adolf Meyer, and 'the albino rat'. *History of Psychology*, 2(1), 3–24.

Logan, C. A. (2001). '[A]re Norway rats … things?' Diversity versus generality in the use of albino rats in experiments on development and sexuality. *Journal of the History of Biology*, 34(2), 287–314.

Logan, C. A. (2002). Before there were standards: The role of test animals in the production empirical generality in physiology. *History of the Journal of Biology*, 35(2), 329–363.

Logan, C. A. (2005). The legacy of Adolf Meyer's comparative approach: Worcester rats and the strange birth of the animal model. *Integrative Physiological and Behavioral Science*, 40(4), 169–181.

Morgan, M. S., & Morrison, M. (1999). *Models as Mediators: Perspectives on Natural and Social Science*. Cambridge: Cambridge University Press.

Nalbandov, A. (1976). *Reproductive Physiology of Mammals and Birds: A Comparative Physiology of Domestic and Laboratory Animals and Man* (3rd ed.). San Francisco: W.H. Freeman.

Nelson, N. (2012). Modeling mouse, human and discipline: Epistemic scaffolds in animal behavior genetics. *Social Studies of Science*, 43(1), 3–29.

Nelson, N. C. (Forthcoming). *Model Behavior: Animal Experiments and the Genetics of Human Behavior*. Chicago: University of Chicago Press.

Rader, K. (2004). *Making Mice: Standardizing Animals for American Biomedical Research, 1900–1955*. Princeton: Princeton University Press.

Reardon, J. (2001). The Human Genome Diversity Project: A case study in coproduction. *Social Studies of Science*, 31, 357–388.

Rheinberger, H.-J. (1997). *Toward a History of Epistemic things: Synthesizing Proteins in the Test Tube*. Stanford: Stanford University Press.

Rheinberger, H.-J. (2000). Beyond nature and culture: Modes of reasoning in the age of molecular biology and medicine. In M. Lock, Allan Young and Alberto Cambrosio (Ed.), *Living and Working with the New Medical Technologies: Intersections of Inquiry*. Cambridge: Cambridge University Press.

Rheinberger, H.-J. (2010). *An Epistemology of the Concrete: Twentieth-Century Histories of Life*. Durham: Duke University Press.

Weber, M. (2007). Redesigning the fruit fly: The molecularization of Drosophila. In A. N. H. Creager, E. Lunbeck, & M. N. Wise (Eds.), *Science Without Laws: Models Systems, Cases, Exemplary Narratives* (pp. 23–45). Durham: Duke University Press.

34

Making microbiomes

Amber Benezra

Microbiomics

In the study of human microbial ecology, *microbiota* are the microorganisms that make up the ecological communities on the human body, and *microbiome* refers to the microbial genes in these communities. It is estimated there are at least as many microbial cells in our bodies as human cells, and the number of genes represented in our indigenous microbial communities likely matches our 23,000 *Homo sapiens* genes with eight million microbial genes. From the biological science point of view, humans are supraorganisms (a system of multiple organisms functioning as one) or holobionts (the singular ecological unit made up of symbiotic assemblages) – composites of human and microbial selves. We have evolved from, and with these microbes. Human microbiomes inextricably entangle biological processes with social practices; microbial populations – affected by how and where we are born, what food we eat, who we live with and love – are digesting our food, training our immune systems, interacting with our states of health and illness, moods, and behavior. This chapter addresses three concerns: how the microbiome is produced as an experimental object in the context of genomic knowledge, what social science approaches to microbiome science have emerged, and how knowledge about microbiota is crucially bound to the highly specific technologies used to sequence microbial genomes.

Metagenomic technologies propel future research, raise bioethical and privacy issues, disarrange logics of funding and regulation, and complicate categories of species, community, and self. As human microbial ecologists seek to explain interrelationships between microbes in the gut, mouth, vagina, and skin to autism, drug efficacy, postnatal health, and cancer (to name a few), a microbial focus materializes for sociology, anthropology, philosophy, and public health concerning issues of environment, genomics, and molecular biomedicalizations. Interdisciplinary insights have the potential to facilitate the application of social science tools in the design of observational and interventional microbiome studies. As colleagues and I have written elsewhere, "Bringing anthropology and human microbial ecology into a meaningful dialogue allows for new modes of collaborative research. It should create a symbiosis that enables both fields to co-develop in ways that encourage a more profound view of our 'humanness' – transforming our categories of 'community,' 'individual,' and 'life,' and in the process helping to address major global health inequities" (Benezra et al. 2012, 6380). Microbiome research creates spaces for social and biological scientists to jointly reckon with the sociomaterial conditions and "local microbiologies" (Koch 2011) in which microbiomes are enacted.

Though microbes have been on earth 3.49 billion years longer than *Homo sapiens*, there is no history of microbes separable from a history of microbiology. Most reductively, microbes are microscopic organisms, that is, they require a microscope to be seen by the human eye. Thus "microbes" (and subsequently, "microbiomes") always come into being for humans by way of scientific interlocutors, and in this way have been enacted throughout history in tandem with the technologies developed to see them.

In the first two decades of the twenty-first century, the development of metagenomics and associated data technologies completely transformed how scientists study microbes. By way of microscopy, laboratory cultures, and molecular biology, advances in high-throughput sequencing technology have revealed a previously unseen, massive microbial diversity on and in the human body, ushering in a new era for the study of microbial communities in environments and hosts. "Seeing" microbes through their genomes has made visible a tremendous number of formerly unknown organisms, and has moved research away from conventional hypothesis driven science. Metagenomics, also referred to as environmental genomics or community genomics, pools and studies the genomes of all of the organisms in a community, and all of the functions encoded in the community's DNA (Handelsman 2007). Metagenomics allows for vast communities of microbes, previously unculturable outside their habitats, and functionally inseparable from their communities, to be identified through their DNA; it addresses a more ecological approach to understanding microbes within their environment, how they function, and their intricate evolutionary relationships. The methods of metagenomics include a series of experimental and computational approaches centered around shotgun sequencing, massively parallel sequencing, and bioinformatics. Genomic technologies and high-throughput sequencing such as those used in metagenomics have transformed what it means to be a biological scientist. Big data innovations have changed what sort of knowledge is both knowable and valued (Leonelli 2015), how scientists conduct research and consequently, how they understand biology (Leonelli 2014).

The study of microbiomes pivots on the union of these new metagenomic technologies with gnotobiotics,[1] the scientific practice of creating and using "germ-free" experimental animals. Human microbial ecology depends heavily on mouse models – the standard experimental protocol is to transplant the microbes from human samples into germ-free mice. Gnotobiotic technology has become a standard experimental tool, as a way of isolating communities of microbes, studying interrelationships of microbes and their hosts, and monitoring external effects such as diet and environment. These experiments are meant to capture an individual's microbial community at one moment in time, and then replicate it in multiple mice in order to make generalized human conclusions, and determine the degree to which phenotypes can be transmitted via microbiota.

According to a survey done by 14 scientific government agencies in 2016, the majority of microbiome research in the US (37 percent) is done on human microbiota, followed by non-human lab studies at 29 percent. Studies on soil, water, and atmospheric microbes collectively make up only 22 percent of all research (Stulberg et al. 2016). Undoubtedly, how microbes make up (and work within) human bodies is the primary scientific concern.

The social lives of microbes

An emerging literature in anthropology, sociology, and philosophy of science is also taking particular notice of microbial life in human bodies. These texts explore a microbial ethos, and to varying degrees call for a microbe-down reorientation of the hierarchy of life forms. Social scientists have joined the conversation about the commensal biosocialities of human microbial ecology, and lines are being drawn between the "perils and promises" of thinking with microbial communities as model ecosystems (Paxson and Helmreich 2014). There are many social science microbial perspectives rooted in the work of Lynn Margulis, revolutionary biologist whose concept of symbiogenesis[2] emphasizes the genetically historical interdependence of all life forms on earth, and places microbes squarely at the originary center of evolution, society, and life. Donna Haraway embraces Margulis' thinking, first in *When Species Meet* (2008) and later in *Staying with the Trouble* (2016), conceiving of microbes as companion species, messmates, tiny makers of our material and discursive selves that we become-with, or not at all (2016, 4). Anthropologist Gisli Palsson continues that train of thought in his volume with Tim Ingold, *Biosocial Becomings: Integrating Social and Biological Anthropology*. For Palsson, humans (by way of microbiomes) are aggregates of life forms and the outcomes of ensembles of biosocial relations, "Humans may usefully be regarded as fluid beings, with flexible, porous boundaries; they are necessarily embedded in relations, neither purely biological nor purely social, which may be called biosocial; and their essence is best rendered as something constantly in the making and not as a fixed, context-independent species-being" (2013, 39). Palsson goes on to argue that the human–microbe, biological–social assemblages affect the very way social scientists can conduct research, and thus, "a radical separation between social and biological anthropology seems theoretically indefensible" (2013, 39).

Philosophers of science argue that taking more notice of microbes as the dominant life form on the planet, both now and throughout evolutionary history, will transform some of the philosophy of biology's standard ideas on ontology, evolution, taxonomy and biodiversity. Further, metagenomics also impacts how social scientists make sense of communities, interactions and environments, shifting analyses of life from entities to process (O'Malley and Dupré 2007). Many social science approaches to microbiota are philosophical in nature, such as sociologist Myra Hird's work on microontologies: "beginning to think through the parameters of bringing the microcosmos to bear on our approach to social scientific topics. Microontologies refers to a microbial ethics, or, if you will, an ethics that engages seriously with the microcosmos" (2009, 1) and for Hird (like Palsson), theorizing this ethics is crucial to the future of the social sciences (2010). Anthropologists like Alex Nading attempt to translate microbial thinking into methodology, raising questions about how to do a social study of the microbiome, "in places where it does not (yet) exist as a category of expert practice or public discourse" (2016, 560). Nading compares microbiome work of US scientists and Nicaraguan hygienists suggesting different ways in which the cultural/interpretive evidence of paraethnography interfaces symbiotically with the quantitative/statistical evidence of bioscience, in order "to suggest that critical science studies can anticipate, rather than simply await, the emergence of global categories of action and inquiry" (2016, 578).

Social studies of microbiomes respond to the call to account for the political and social lives of nonhumans found in multispecies ethnography (Kirksey and Helmreich 2010). Microbiomes follow multispecies trajectories through the work of those like anthropologist John Hartigan, who theorizes culture and society across species lines, attempting to reverse the trajectory by which social theory is generated, eschewing the assumption that society and culture are solely about people (2014a, 2014b). For Hartigan, cultural analyses of microbiomics are not just about how humans think about nonhumans—they allow us to think bioculturally. Human geographer Jamie Lorimer's work on microbiome therapies suggest new ways of thinking of companionship and human immunity as more-than-human (2017, 2016). Work by scholars in human geography engages with the growing scientific, popular, and policy interest in the microbiome. Academic research from groups like the ESRC-funded "Good Germs/Bad Germs" project at Oxford, the Canadian Institute for Advanced Research's Humans and the Microbiome Program, and the University of California Humanities Research Institute Microbiosocial Working Group explores the transformative potential of recent developments in metagenomics for developing new public understandings of the microbiome.

However, there are those turning a wary eye to the social science and philosophical potentialities of microbiome studies. Anthropologists Heather Paxson and Stefan Helmreich worry about the focus on the materiality of microbes (especially in terms of microbial agency) and warn against "new reductionisms" and return to scientific determinism that they see emerging from taking microbiome science for granted. Instead they focus on the model ecosystems that microbes represent, "The microbial realm, shared across scales and contexts, variously and simultaneously universal, ubiquitous and unique, has become a fresh court of appeal for those who would model new modes of living with and within biological nature. The question is not simply 'what is life?' but rather, 'what forms of life do we wish to insist upon?'" (2014, 185). Helmreich develops these ideas further in *Sounding the Limits of Life: Essays in the Anthropology of Biology and Beyond*, constructing the figure of the microbial human, *homo microbis*. Again, taking an anti-new-materialist position, Helmreich describes a "microbiomania" amongst science studies scholars, "The microbiome is a novel kind of object or figure in biology, to be sure, but its multiple meanings do not themselves follow from the fact that microbiomes are composed of a multiplicity of organisms" (2015, 65). Concerned that biology must not be seen as "speaking for itself," Helmreich redraws the same lines between biology and culture, human and nonhuman, that microbial ecology has the potential to blur. This work illustrates the important concomitance of microbiome science with debates in the social sciences concerning materiality, biology, and nature.

Taking this critique as a starting point, perhaps we can use microbiome research to follow Elizabeth A. Wilson's suggestion that a better, more generous engagement with biology will reanimate and transform the foundations of social theory (Wilson 2015). The technologies and practical futures of microbiome research may address concerns about disarticulating biology and culture and those about rematerializing matter, and move toward finally developing a conceptual toolkit in which we can "take biomedical data seriously but not literally" (Wilson, 2015, 12). In this space, social scientists may be able to analyze and speak with human microbial ecology.

What technology begets

Like many working in the fields of genomics and personalized medicine, human microbial ecologists participate in "promissory science" (Hedgecoe 2004), building expectations and future contexts for a technology-reliant discipline based on the "anticipation" (Adams et al. 2009) of a predictably uncertain technoscientific and biomedical future, one that so far exists mostly in speculations and promises. The ways in which technological visions are deployed can both direct the paths of research and shape our hopes for this science. As the microbial terrain of the human body is translated into colossal data sets for algorithmic probing, there emerges a governance of health that is increasingly understood through metagenomics. Studies of microbiota are attempting to reach beyond DNA to characterize the expressed functions of microbe communities by using genome-scale platforms. The large amounts of data these platforms produce can be seen to be making what Adams et al. (2009, 247) call speculative forecasts, defining the present of microbiome science through its anticipated future, and creating real, material trajectories where health states and treatments will be diagnosed and prescribed based on the functional genomics[3] of one's personal microbiota (Benezra 2016).

The analytic tools used for deciphering the immense data sets from human gut microbiota make implicit assumptions about what the data says about genes, microbes and human health, influencing research trajectories toward the molecular basis of disease. Responsibility for health becomes focused outward on places, practices and exposures, and inward on genes and genomes (Shostak 2003). Microbiota make up a significant part of a human's cells, genes, and metabolic function – they are us and their compliance (Whitmarsh 2009) or non-compliance in healthy behavior is our biomedical fate. This calls up what Shostak and Landecker have described the "environmental genetic body" or "epigenetic metabolic body," one susceptible to social and biological exposures, simultaneously the site of the past (genetic histories) and future (epigenetic risks) (2003, 2011 respectively). Both scholars query whether genetic technologies can end in public health interventions that are not molecularly myopic, reifying social, economic, and cultural difference on the molecular level. As shown in the following "Microbiosocial Futures" section, studies of microbiomes are at risk for playing out in these same ways. Microbes are studied and understood through their collective genomes and metabolic processes, and human health states are evaluated and diagnosed through a microview of a person's microbiota. But microbes interestingly complexify what is understood about the molecularized body and the molecularized environment, becoming a potential constant between the two. Instead of looking at how environments are turned into the biology of the body, human microbial ecology looks more widely and across individual development at how microbial populations are constituted by worldly environments. Then, microbes come to constitute the corporeal environment within the human body; food, medicine, pathogens, all affect our joint human–microbe biology through our microbial partners. Microbes are in and out, human and nonhuman, simultaneously environment and body.

Studies of human microbiota not only require new scientific skills from geneticists and biologists, they produce a very specific kind of information, and create critically and potentially conflicting bioinformatic considerations familiar in the post-genomic era: subject privacy and rapid, public data release (Gevers et al. 2012). The ways in which the microbiome is understood metagenomically translates into the way microbes are studied. As Clarke et al. describe biomedicalization, "the increasingly complex, multisited, multidirectional processes of medicalization, both extended and reconstituted through the new social forms of highly technoscientific biomedicine" (2003, 162), microbiome experiments cannot rely on *in vitro* testing, instead, the *in vivo* of germ-free mice is turned into the *in silico* of shotgun and 16S data.[4] This type of vision begets a specific type of intervention, one that focuses on addressing human health from a microbial perspective; a dream of an eventual "personalized gnotobiotics" that would customize medicine based on metabolomics, and genetic testing. Interpreting all of this data is a means to seeing people as human–microbe composites or as "polygenomic organisms" (Dupré 2012) and is based on a supposition that the actions of our microbial partners are accurately deciphered through metagenomics. Yet such precision microbiome manipulation is still only aspirational in terms of new technologies and new actions; these methods are still subject to all the instability to which such dreams have succumbed in deriving many straightforward health interventions from mapping human genomes.

Microbiosocial futures

The new tools and creative experimentalism of microbiome research open up discussions of the limits as well as the horizons of possibility; there are growing ethical, legal and social challenges resulting from studies of human microbiota. Besides worries that not enough is understood about host–microbial and other interspecies interactions to avoid unforeseen adverse outcomes, microbiome scientists are extremely careful about making ambitious microbiota/health claims and urge caution against any prescriptive recommendations. Microbiologists and epidemiologists also warn against an overinvestment in the microbiome as the next biological panacea for all humanity's ills. Next-generation sequencing technology can enable the discernment of "metabolic networks" and reveal biochemical reactions but, "genomes are littered with clues both true and false, such as 'hypothetical proteins' and genes that are understood poorly or not at all, but could make for important differences in what metabolic networks do" (Hanage 2014, 248). The vastness and complexity of microbiome networks and their interactions (including social and environmental ones) need to be characterized before causation can be determined. Professor of Medicine (with a focus on the epidemiology of risk) J. Dennis Fortenberry calls for special attention to be paid to the sociopolitical nature of race in biomedical research so that microbiome work can, "contribute to reduction of health disparities while avoiding attribution of causal inference to specific race and ethnicity categories" (2013, 165). This raises further concerns about the implications of using microbiota to profile, or forensically identify individuals or populations – microbiome science ultimately reifying categories it had endeavored to disrupt.

Medical ethics and public health scholars are concerned with dietary supplements resulting from human microbiome research, which open the door to what they call *commercialized intervention*, "the proliferation of commercial products that claim maintenance or restoration of good health, and prevention of disease or sickness with the use of good bacteria" (Slashinski et al. 2012, 6). Slashinski et al. argue that human microbial ecology makes claims about the "biological vitality" of human microbiomes in ways that instrumentalizes knowledge about microbes to create commodities to be sold under the guise of contributing to human health (2012). This research direction raises questions of profit and unequal access to and distribution of technologies, as well as the vulnerability of publics to "molecularized marketing" (Landecker 2011).

Microbiome research seeks to find how microbes can be utilized, genetically altered, or even prescribed to effect durable changes in environment, food, and bodies. As the biovalue of human microbiome research is in danger of being increasingly assimilated into capital value, public health interests, especially in the case of traditionally vulnerable populations (where disease burden and structural violence are greatest, and a history of research exploitation persists) need to be carefully balanced with the scientific research, funder goals, and the industry marketplace.

As human microbial ecologists stand on the precipice of defining human health through relationships with our co-evolved microbiota, work also emerges for scholars in the social sciences, public health, and bioethics, investigating the proprietary ownership of microbes and new necessities of biomolecular privacy required. We must think about how our genomic intertwining with our microbial partners changes how we understand relatedness and selfhood. To reconsider humans in the context of the microbial forces a multispecies perspective on a reluctant anthropocene. Most importantly, we can use social science methodologies to elucidate the ethical and material conditions in which microbiomes come into being, to strive toward forging health intervention strategies that account for different relational ontologies between people and microbes.

Notes

1 Gnotobiotics and germ-free are not technically the same, but are talked about interchangeably. Gnotobiotic refers to an experimental animal in which the only microorganisms present are known (*gnostos*: known, *bios*: life) while germ-free means completely without microbes. Specific Pathogen Free (SPF) animals, which have become a lab standard can be seen as a universally-used type of

gnotobiology – they have the same undefined commensal microbes as conventional mice, but have a discrete set of pathogens filtered out.

2 Symbiogenesis is the emergence of a new organism (phenotype, organ, tissue or organelle) from the interconnectivity of two separate organisms – symbiosis over time, a relationship between organisms that results in the evolutionary change of both (Margulis and Sagan 1997). Or as Haraway puts it, evolution through the "long-lasting intimacy of strangers" (2016, 60).

3 While early microbiome research attempted to identify the specific members of the community (and became quickly overwhelmed because such a small percentage of microbial life is known and cataloged, and because it was determined that every person has microbiota as unique as a fingerprint) a focus on functional genomics developed, attempting to describe which biological tasks the members of a community carry out – *what* they do, rather than *who* is there.

4 16S ribosomal RNA is a type of RNA that is involved in the production of proteins. All bacteria have 16S rRNA, and it is conserved within a species, so it is used to identify organisms whose 16S rRNA signature is known (or to identify unknown organisms if the signature if not known). The 16S rRNA gene is amplified and sequenced, allowing the taxa of bacteria present in a community to be identified.

References

Adams, V., M. Murphy, and A. Clarke (2009) "Anticipation: Technoscience, life, affect, temporality." *Subjectivity* 28: 246–265.

Benezra, A. (2016) "Datafying microbes: Malnutrition at the intersection of genomics and global health." *BioSocieties* 11(3): 334–351.

Benezra, A., J. DeStefano, and J. I. Gordon (2012) "Anthropology of microbes." *PNAS* 109(17): 6378–6381.

Clarke, A. *et al.* (2003) "Biomedicalization: Technoscientific Transformations of Health, Illness and US Biomedicine." *American Sociological Review* 68(2): 161–194.

Dupré, J. (2012) *Processes of Life: essays in the philosophy of biology*. Oxford: Oxford University Press.

Fortenberry, J. D. (2013) "The uses of race and ethnicity in human microbiome research." *Trends in Microbiology* 21(4):165–166.

Gevers, D. *et al.* (2012) "Bioinformatics for the Human Microbiome Project." *Public Library of Science, Computational Biology* 8: 1–8.

Hanage, W. P. (2014) "Microbiology: Microbiome science needs a healthy dose of skepticism." *Nature* 512(7514): 247–248.

Handelsman, J. (2007) "Metagenomics and microbial communities." In *Encyclopedia of the Life Sciences*. Chichester: John Wiley and Sons, Ltd.

Haraway, D. (2008) *When Species Meet*. Minneapolis: University of Minnesota Press.

Haraway, D. (2016) *Staying with the Trouble: Making Kin in the Chthulucene*. Durham: Duke University Press.

Hartigan, J. (2014a) "Multispecies vs Anthropocene." *Somatosphere*. http://somatosphere.net/2014/12/multispecies-vs-anthropocene.html (accessed December 12, 2014).

Hartigan, J. (2014b) *Aesop's Anthropology: A Multispecies Approach*. Minneapolis: University of Minnesota Press.

Hedgecoe, A. (2004) *Politics of Personalised Medicine: Pharmacogenetics in the Clinic*. Cambridge: Cambridge University Press.

Helmreich, S. (2015) *Sounding the Limits of Life: Essays in the Anthropology of Biology and Beyond*. Princeton: Princeton University Press.

Hird, M. J. (2009) *The Origins of Sociable Life: Evolution After Science Studies*. New York: Palgrave Macmillan.

Hird, M. J. (2010) "Symbiosis, Microbes, Coevolution and Sociology," *Ecological Economics*, 69(4): 737–742.

Ingold, T. and G. Palsson eds. (2013) *Biosocial Becomings: Integrating Social and Biological Anthropology*. Cambridge: Cambridge University Press.

Kirksey, S. E. and S. Helmreich (2010) "The Emergence of Multispecies Ethnography." *Cultural Anthropology* 25(4): 545–576.

Koch, E. (2011) "Local microbiologies of tuberculosis: Insights from the Republic of Georgia." *Medical Anthropology* 30(1): 81–101.

Landecker, H. (2011) "Food as exposure: Nutritional epigenetics and the new metabolism." *BioSocieties* 6(2): 167–194.

Leonelli, S. (2014) "What difference does quantity make? On the epistemology of big data in biology." *Big Data & Society* 1(1): 1–11.

Leonelli, S. (2015) "What counts as scientific data? A relational framework." *Philosophy of Science* 82(5): 810–821.

Lorimer, J. (2017) "Probiotic Environmentalities: Rewilding with Wolves and Worms." *Theory, Culture and Society* 34(4): 27–48.

Lorimer, J. (2016) "Gut Buddies: Multispecies Studies and the Microbiome." *Environmental Humanities*, 8(1): 57–76.

Margulis, L. and D. Sagan (1997) *Microcosmos: Four Billion Years of Evolution from Our Microbial Ancestors.* Berkeley: University of California Press.

O'Malley, M. A. and J. Dupré (2007) "Size doesn't matter: towards a more inclusive philosophy of biology." *Biology and Philosophy* 22: 155–191.

Nading, A. (2016) "Evidentiary Symbiosis: On Paraethnography in Human–Microbe Relations." *Science As Culture* 25(4): 560–581.

Paxson, H. and S. Helmreich (2014) "The perils and promises of microbial abundance: Novel natures and model ecosystems, from artisanal cheese to alien seas." *Social Studies of Science* 44(2): 165–193.

Shostak, S. (2003) "Locating gene-environment interaction: at the intersections of genetics and public health." *Social Science and Medicine* 56: 2327–2342.

Slashinski, M. J., S. A. McCurdy, L. S. Achenbaum, S. N. Whitney, and A. L. McGuire, (2012) "Snake-oil," "quack medicine," and "industrially cultured organisms:" Biovalue and the commercialization of human microbiome research. *BioMed Central Medical Ethics* 13(28): 1–8.

Stulberg, E. *et al.* (2016) "An assessment of US microbiome research." *Nature Microbiology* 1(1):15015.

Whitmarsh, I. (2009) Medical Schismogenics: Compliance and "Culture" in Caribbean Biomedicine. *Anthropological Quarterly* 82(2): 447–475.

Whitmarsh, I. (2013) "The Ascetic Subject of Compliance: The Turn to Chronic Diseases in Global Health." In: J. Biehl and A. Petryna, (eds) *When People Come First: Critical Studies in Global Health.* Princeton: Princeton University Press, pp.302–324.

Wilson, E. A. (2015) *Gut Feminism.* Durham: Duke University Press.

35

Behavior genetics: boundary crossings and epistemic cultures

Nicole C. Nelson and Aaron Panofsky

Introduction

Behavior genetics asks some of science's most provocative questions: why are individuals and groups different in their degrees of success, levels of intelligence, risks of addiction, and propensities to violence? As a field, behavior genetics straddles intellectual boundaries between biology and social science, and practitioners' attempts to work across these boundaries have repeatedly generated controversy. Critics continue to question whether behavior geneticists have struck the right balance in emphasizing inherited and environmental contributions to behavior, or whether they should be using biological tools to examine "social" problems at all.

The term "behavior genetics" came into use in the late 1950s, and it was a phrase self-consciously coined by practitioners hoping to found a new field free from the dark legacies of eugenics and scientific racism (Panofsky 2014). The aspiration was to create a new biosocial science that would draw multi-disciplinary research teams to address trans-disciplinary research questions using humans and animal models, and psychosocial and biogenetic research tools (Panofsky 2016). But those outside the field have repeatedly disagreed with the way that behavior geneticists have drawn the boundaries of their discipline. Some have pushed back on the distinction practitioners have made between early twentieth-century eugenics and the contemporary field of behavior genetics (e.g. Paul 1998); others have argued more forcefully that the field is fundamentally ideological rather than scientific (e.g. Lewontin et al. 1984, Lewontin 1992). Some attempt to place behavior genetics firmly on the side of biology to prevent it from seeping into the social sciences, while behavior geneticists themselves have sought to ground their work in a universal social science.

Science and Technology Studies (STS) scholars have been both participants in and analysts of disputes about the appropriate relationship between genetic analysis and the behavioral sciences. Some have intervened in methodological discussions, offering conceptual clarifications about the capacities of genetic tools. Evelyn Fox Keller (2010) and Peter Taylor (2014) have both written detailed critiques of heritability studies, pointing out problems with and common misinterpretations of these studies. Many others have focused on the potential harms of framing public problems in genetic terms, such as the potential for genetic approaches to divert attention and resources away from the structural causes of these problems and reinforce discrimination against marginalized people (e.g. Lippman 1992, Nelkin and Lindee 1995, Duster 2003). While most STS scholars have been critical of what they see as the incursion of genetic approaches into the social and behavioral sciences, some have called for greater integration. Jeremy Freese and Sara Shostak (2009) have argued, for example, that social science might improve behavior genetics by offering a better understanding of the social dynamics that shape the distribution of genetic disease. Finally, some have focused on the dynamics of the disputes themselves. James Tabery (2014), for example, uses historical and philosophical tools to explain why behavior geneticists have been unable to resolve disputes about gene–environment interaction effects.

This chapter uses disputes along the bio/social boundary to explain the emergence of multiple distinctive epistemic cultures within the field of behavior genetics. We draw on Karin Knorr Cetina's (1999) description of "epistemic cultures" as distinctive technical, social, and symbolic systems for making knowledge. Knorr-Cetina's analysis shows the diversity of knowledge-making cultures within the sciences; cultures that do not always map onto other categories such as "discipline" or "field." We use the term "field" in Bourdieu's (2004) double sense: a site of struggle that is simultaneously social and intellectual, where scientists compete for status and to define what counts as good science. While all scientific fields are to some extent heterogeneous, the terrain of behavior genetics is especially varied. Practitioners disagree on whether they should exert control over the field's tools or give them away, attend different annual meetings and publish in different journals, subscribe to different visions of what it means to be a responsible scientist—and yet, they all see themselves as behavior geneticists. Faced with this heterogeneity it may be tempting to conclude that behavior genetics is simply not a field at all, but we argue that analyzing behavior genetics as a site of collective struggle is productive because of this heterogeneity, not in spite of it. Considering behavior genetics as a field, its interactions with neighboring fields, and individual practitioners' memberships in multiple fields makes visible the niches where the two epistemic cultures we describe here have taken hold. We draw on a variety of data sources, both broad and deep; including historical, interview-based, and ethnographic research.

Fracturing the field

Behavior genetics was founded in the 1960s by social scientists, biologists, psychologists, geneticists, and animal behaviorists. This intellectually diverse group was interested in creating an integrated transdiscipline devoted to studying the heredity of behavior but free from the divisive and discredited ideas of racist eugenics (Panofsky 2014). This dream of synergistic, intellectual biosocial cooperation was interrupted when psychologist Arthur Jensen wrote his now infamous article "How much can we boost IQ and achievement?" in 1969. Drawing extensively from nascent behavior genetics research, Jensen argued that IQ, racial achievement gaps, and social inequality were largely genetically determined and thus not amenable to social intervention. Appearing in the late years of the civil rights movement, such claims were taken as a stunning provocation and sparked wide-ranging reactions, from public debate to activist disruptions of behavior genetics meetings. Most vociferous among academic responses were radicals and Marxists like Leon Kamin (1974), Richard Lewontin (1970), and Stephen J. Gould (1981), who charged behavior genetics with fraudulent data, elitist conceptions of intelligence and behavior, and ideologically reactionary understandings of biology as determining social life.

While a minority of behavior genetics welcomed Jensen's work as a brave biological grounding of social science with a willingness to speak politically uncomfortable truths, most field members were dismayed that Jensen had dragged squarely into their field matters they had carefully pushed out of bounds. Those who were not initially supportive of Jensen adopted different strategies to deal with the controversy his work provoked. A small but vocal group of behavior geneticists wanted to save the field by forcefully expelling Jensen and his ideas. They joined in the chorus of outside critics who denounced Jensen's methods and conclusions as fundamentally flawed. But most members sought to litigate the details of Jensen's claims and to place behavior genetics in a middle ground between Jensen and his radical critics. The majority followed this route because they saw the critics' attacks as an existential threat to the field: they felt the only option was to circle the wagons and defend Jensen's academic freedom, and by extension their own.

This controversy effectively ended behavior geneticists' hope that their field would be able to peacefully straddle the biology/social science boundary. If that arrangement was no longer viable, how would behavior genetics bridge that gap in the future? In the aftermath of the Jensen controversy, behavior geneticists were faced with a crisis of credibility. Mainstream geneticists had largely written off the field as politically dangerous and scientifically flawed (Lewontin 1975, Provine 1986), and behavioral scientists saw it as a revival of genetic determinism and scientific racism (SPSSI 1969, see also Eckberg, et al. 1977). The interdisciplinary goodwill that had characterized behavior genetics research in the 1960s was gone. Many fled the field, and the researchers who remained – mainly human psychologists and animal behaviorists – faced the daunting question of how to rebuild its scientific reputation. The different strategies that practitioners pursued in rebuilding behavior genetics have resulted in a field that is today "less a soccer field and more an archipelago," as Panofsky (2014: 32) puts it. The contemporary field of behavior genetics consists of a set of relatively isolated "islands" of practitioners that only loosely cohere under the banner of behavior genetics. We examine how two of these clusters of practices and practitioners formed through the process of managing disputes along the bio/social boundary.

Giving the field away

One strategy for drawing attention to the marginalized field, pursued by many behavior geneticists especially through the early 1990s, was to aggressively confront many sacred cows of the socialization paradigm in social science. For example, practitioners used studies noting similarities between monozygotic twins reared separately and comparing psychological traits between adoptees and their birth and adoptive parents to attack the idea that forces of socialization shaped early childhood development. Adherents of this approach were fond of arguing that apart from situations of true deprivation or abuse, schools and parents "don't much matter" for making children who they are – pre-social genetic potential mostly establishes a person's intelligence and personality (Scarr 1992, Rowe 1994).

Though excellent for grabbing scientific and media attention, this approach tended to limit the reach and appeal of behavior genetics. Consequently, another strategy emerged: practitioners sought to establish partnerships with psychologists, psychiatrists, and social scientists and help them bring genetically sensitive designs to their own research, a strategy that a leading behavior geneticist described as "giving the field away." Researchers who pursued this approach argued that genetic heterogeneity might be confounding other researchers' claims about how social environment shapes individual outcomes, and offered their expertise to these researchers in order to tease out the causes. "Giving the field away" would not require a researcher to re-train. This style of research was "generous" in Nikolas Rose's (1992) sense, because it didn't seek to denigrate a paradigm, colonize a field, or convert counterparts. Rather, it put behavior geneticists at their partners' service: practitioners would bring the tools of genetic analysis and often the valuable twin and adoptive research subjects, while their partners would bring the substantive expertise, precise theoretical questions, and connections to other research communities.

"Giving the field away" produced a characteristic style of research practice. Behavior geneticists focused on experimental systems – twin and family studies and later genetic association studies – that were portable and could be applied to any substantive problem for which a paper and pencil behavioral assessment existed. The result was a saltating agenda, hopping from topic to topic, where the net effect was the massive accumulation of heritability estimates (the genetic proportion of variance in a trait for a particular population) and genetic associations (correlations between traits and genetic markers). With this more generous style of research came a more conciliatory tone. Though this accumulation of findings aimed to demonstrate the "first law of behavior genetics" (Turkheimer 2000) – that everything is heritable – talk turned away from the limited power of environmental forces and toward an appreciation for "gene–environment interplay" (Rutter 2006). Yet controversy over behavior genetics never abated as critics continued to charge the field with crude conceptions of genes and environment that tended not to reliably replicate and to exaggerate the determining force of genetics (e.g., Charney 2012).

The Behavior Genetics Association (BGA) has historically been the main institutional home for these researchers. Founded in 1970 as the society representing the new multidisciplinary field, during the Jensen years the BGA became a mutual support group for the embattled scientists who remained associated with the science. Members counted on each other to defend behavior genetics against its critics and not to criticize each other's science aggressively. They adhered to a strong ethic of scientific freedom – their version of "scientific responsibility" was to defend each other's rights to pursue questions no matter how provocative to social morals or other fields' scientific standards. From this relatively safe community, behavior geneticists could launch their efforts to breach the genetic science/behavioral science boundary, and later to bring social and psychological scientists across it by giving the field away.

Cultivating cautious claiming

Another way that practitioners addressed the field's credibility problems was by cultivating practices aimed at carefully managing the scientific claims made using behavior genetics tools. This approach emerged out of the small but vocal group of dissenters who responded to Jensen's 1969 paper by arguing alongside critics that Jensen had misused behavior genetics' analytical techniques and made conclusions that were unwarranted on the basis of his data. Animal behavior geneticist Douglas Wahlsten is an example of a practitioner who embodied this approach. In the course of his career, Wahlsten has authored numerous opinion pieces on topics such as the limits of heritability analyses (Wahlsten 1994a) and the dangers of ignoring histories of eugenics in discussions about the ethics of behavior genetics (Wahlsten 2003). In contrast to those behavior geneticists who responded to the heredity, race, and intelligence controversies of the 1970s by holding tight to the value of academic freedom, these same events convinced Wahlsten that he had a responsibility as a behavior geneticist to engage in these disputes and prevent the spread of incorrect or actively misleading ideas (Wahlsten 1994b).

This strategy put practitioners in conflict with the majority of their fellow behavior geneticists, who sought to defend Jensen rather than ejecting him from the field. In response to these internal tensions, the strategy evolved over time into one that focused more on managing claims rather than identifying problematic practitioners. Instead of managing boundary disputes by trying to banish problematic practitioners from the field (or conversely, relaxing boundaries to encourage outsiders to take up behavior genetics), cautious claimers attempted to police the scientific claims made using behavior genetics techniques irrespective of who was using them or where they were being used.

Practitioners employed a variety of techniques to cultivate cautious claiming amongst users of behavior genetics tools. Authoring articles on methodological confounds and using the peer review process to advocate for more moderate conclusions in publications allowed concerned researchers to engage with what they saw as problematic claims. The sheer volume of studies published on tools such as the "elevated plus" maze (a rodent model of anxiety) shows how some behavior geneticists persisted in making methodological issues visible even concerning established research tools (Nelson 2013). Although the elevated plus maze has been in routine use since the 1980s, new methodological studies continue to appear in the literature, investigating topics such as whether building the maze out of plastic or metal affects experimental outcomes (Hagenbuch et al. 2006), proposing new maze designs (Fraser et al. 2010), or questioning whether the measurements taken in the maze can really be said to measure "anxiety" (Milner and Crabbe 2008).

This strategy of encouraging methodological discussion, however, conflicted with the "bunker mentality" (Panofsky 2014) of other behavior geneticists, who felt that those within the field should refrain from making the kinds of critiques that outsiders regularly made of their work. Many cautious claimers thus employed a subtler strategy of insisting on particular linguistic formulations for describing the tools and findings of the field, formulations that signaled the need for methodological caution without being overtly critical (Nelson 2018). Rather than referring to the elevated plus maze as a "test of anxiety," for example, some insisted that it should be referred to as a "test of anxiety-like behavior." Such cautious terminology was important, argued one prominent practitioner, because it allowed the field to avoid the impression that "it [was] possible to create a comprehensive mouse model of a human mental illness" (Crawley, 2007: 261–262).

While those who sought to "give the field away" focused their efforts on building alliances in the social sciences, the cautious claimers were more preoccupied with managing disputes with those in molecular biology and neuroscience. The "molecular revolution" of the 1990s brought molecular biologists into greater contact with behavior geneticists, and the new techniques developed in this period created possibilities for claims about genes and behavior with an unprecedented degree of specificity (Nelson 2015). Behavior geneticists, particularly those with backgrounds in psychology, responded to these new combinations of biological techniques and behavioral claims by pointing out the numerous confounds in molecular biologists' study designs and by critiquing molecular biologists' naïve claims to have identified "genes for" particular behaviors. In the view of many veteran behavior geneticists, molecular biology techniques such as "knocking out" specific genes in the mouse genome created more problems than they solved when applied to the study of behavior. In one particularly strident letter to the editor in a 1995 issue of *Nature,* a psychology-trained behavior geneticist suggested that the entire field was being "led into a technological cul-de-sac" by the increasing adoption of knockout techniques that were "wholly inappropriate for resolving the issues for which [they were] intended" (Routtenberg 1995, 314; see also Gerlai 1996). The letter writer argued that such drastic alterations to the genome would surely result in compensatory changes in the expression of other genes, making it difficult to discern what role any of these genes played in the behavior of normal animals.

Even after debates about specific methodological issues had come to closure, cautious claimers continued to retell stories about these events to impart lessons to their fellow practitioners about how to formulate claims about genes and behavior. Stories about the molecular revolution made excellent fodder for cautionary tales both because of who had made those claims and because their claims looked audacious in retrospect (Nelson 2018). Researchers' retellings of these events resembled a narrative "status degradation ceremony" (Garfinkel 1956), where molecular biologists started out as confident, powerful entrants onto the behavioral research scene and ended up humbled by the complexity of behavior – and by the knowledge of veteran animal behavior geneticists. While these stories did create distinctions between insiders and outsiders to the field of behavior genetics, the more crucial distinction they made was between practitioners who were willing to adjust their claims and practices, and those who continued to speak and behave as though "genes for" behaviors existed. These stories and linguistic conventions were part of an alternative professional ethics for behavior genetics, one that imbued particular ways of speaking and experimenting with moral significance. By attempting to apply the brakes on "speed genomics" (Fortun 1999), researchers saw themselves as upholding the virtues of disinterestedness and organized skepticism, and preventing scientifically dubious and socially dangerous claims from taking root in public discussions.

The International Behavioral and Neural Genetics Society (IBANGS) became an institutional home for practitioners who favored this approach. Formed in 1996, this new society advocated for a return to a "big tent" approach to behavior genetics, encouraging interaction between animal and human research, and European and North American practitioners. The society's founding coincided with another crucial event in the field's history – a speech made at the 1995 BGA annual meeting by outgoing president Glayde Whitney. At the closing banquet celebrating the twenty-fifth anniversary of the society, Whitney argued that it was time for the field to move toward a new agenda of studying differences between human racial groups. He presented data on the murder rates in countries and cities with different "racial compositions" in their populations, and he argued that it was a reasonable hypothesis to investigate whether these differences in the murder rates could be attributed to genetic differences between races. For those who saw his remarks as irresponsible speculation rather than defensible scientific inquiry, the newly formed IBANGS offered a place to start over and establish new norms of professional conduct.

An intersectional field analysis

The two epistemic cultures we have sketched out map imperfectly onto a division in the field between animal and human research: psychologists doing heritability studies on human populations have tended to employ the "give the field away" approach, while animal researchers have adopted the cautious claiming strategy. An intersectional analysis examining researchers' positions within multiple fields helps explain this pattern.

For human researchers, disseminating their research tools in larger fields such as psychology and psychiatry has helped them gain new allies and allay the concerns of potential critics. By inviting outside researchers to collaborate with them or to take up the tools of genetic analysis on their own, human behavior geneticists have increased the number of practitioners who see these approaches as scientifically valuable. Cultivating relationships with practitioners in more established fields has also helped these behavior geneticists increase their scientific standing by association. The external-facing orientation of human researchers has resulted in an ironic situation where those who have the most power within the field of behavior genetics are the most ambivalent in their commitment to it. While animal research was central to the newly established field of behavior genetics at mid-century, human studies came to dominate venues such as the BGA meetings in the latter half of the twentieth century. The ethic of defending academic freedom also prevails at contemporary BGA meetings, despite the fact that those who are likely to hold such views are also likely to view their participation in the BGA as simply a means toward other scientific ends.

Animal behavior geneticists, in contrast, have found themselves in a relatively weak position in many of the fields in which they participated post-Jensen. Their research programs and ethical sensibilities are not well represented in the venues most closely associated with behavior genetics today, nor have they benefitted from alliances with broader fields such as neuroscience in the same way that human researchers have. For example, while funding from the National Institute of Mental Health for genetic studies in animal models increased in the late 1990s, most of this investment went into knockout mouse models and large-scale mouse mutagenesis projects (Hyman 2006) and not the selective breeding techniques used by veteran animal behavior geneticists. The uneven terms on which animal behaviorists felt they were competing intensified rather than weakened their desire to maintain control over their experimental techniques.

Animal researchers found it difficult to exert control over the tools of the field, however, because they saw the barriers to entry to behavioral testing as low. Unlike molecular biology, where the equipment needed was expensive and it was quite difficult to get new techniques to work, behavioral tests were inexpensive and seemingly easy to run. A researcher interested in testing mice for anxiety could buy an elevated plus maze relatively cheaply, and since the test involves simply measuring how much time a mouse spends in various areas of the maze, there was no way not to produce data (barring a mouse who jumps out of the maze). Animal researchers argued that the material culture of animal research created an asymmetry where outsiders were free to take up behavioral techniques without consultation, whereas they themselves needed to seek out experienced collaborators if they wanted to take up molecular techniques. Rather than happily "giving away" the tools of behavioral testing, then, animal researchers tended to feel as though those tools were being taken away from them and used inappropriately. While they believed they lacked the ability to physically maintain control over inexpensive behavioral equipment, they believed they could exert control over the interpretation of those experiments and broader narratives about the field's future (Nelson 2018).

Considering individual practitioners' positions in multiple fields, then, explains why they adopted different strategies for rebuilding the field post-Jensen and belong to different epistemic cultures today. Human behavior geneticists' control over their home field of behavior genetics provided them with a strong position from which to share their research tools with other fields. In contrast, animal behavior geneticists' weak position in multiple fields meant that they could neither prevent their colleagues from giving away the tools of behavior genetics or molecular biologists from taking those tools, and so were forced to rely on other approaches.

Conclusion

"Epistemic cultures" (Knorr Cetina 1999) and "boundary work" (Gieryn 1999) are foundational concepts in STS, and in this chapter we have attempted to bring these two lines of thinking into greater contact through an analysis of the field of behavior genetics. In some senses behavior genetics is an exceptional case, but the intensity of boundary-making activities and the epistemic diversity of this unique field provide opportunities for identifying patterns that may hold true in other cases. Boundary work contributes to the development of epistemic cultures in some obvious ways: for example, by creating and maintaining divisions between fields that allow distinctive epistemic cultures to evolve. But as we have seen in this case, work along the outer edges of a field can also have consequences for the internal structure of a field. Practitioners' memberships in multiple fields and engagements in multiple boundary disputes can contribute to the further diversification of knowledge-making practices within shared arenas, allowing distinctive epistemic cultures to arise within scientific fields.

This case study also demonstrates the analytical value in employing units such as "the field," even in a historical moment when vaguely defined, interdisciplinary spaces are proliferating in the postgenomic sciences. Our analysis shows that far from being a poor candidate for a field analysis because of its heterogeneity, considering behavior genetics as field helps explain its diversity. The various institutions and practices of contemporary behavior genetics can only be understood by setting them in relation to one another and in relation to the multiple fields that have contributed to their emergence. Analysts faced with similarly unruly scientific spaces may also find a consideration of shared spaces of social and intellectual struggle useful for making sense of heterogeneity.

References

Bourdieu, Pierre. 2004. *Science of Science and Reflexivity*. Translated by R. Nice. Chicago: University of Chicago Press.

Charney, Evan. 2012. "Behavior Genetics and Postgenomics." *Behavioral and Brain Sciences* 35(6): 1–80. doi:10.1017/S0140525X11002226.

Crawley, Jacqueline. 2007. *What's Wrong With My Mouse: Behavioral Phenotyping of Transgenic and Knockout Mice*. 2nd edition. New York: Wiley-Liss.

Duster, Troy. 2003. *Backdoor to Eugenics*. 2nd edition. New York: Routledge.

Eckberg, Douglas Lee, David F. Haas, Glenn M. Vernon, Lee Ellis, Robert Bierstedt, Dan Clawson, Meryl Fingrutd, Frank Sirianni, Mary Ann Clawson, Diana Powers, Don Palmer, Ed Royce, Linda Hartig, Mark Mizruchi, Michele Ethier and Steve Buechler. 1977. "Sociobiology and the Death of Sociology: An Analytic Reply to Ellis [with Rejoinder]." *American Sociologist* 12(4): 191–200.

Fortun, Michael. 1999. "Projecting Speed Genomics." In *The Practices of Human Genetics*, eds. Michael Fortun and Everett Mendelsohn, 25–48. Dordrecht (Netherlands): Springer.

Fox Keller, Evelyn. 2010. *The Mirage of a Space between Nature and Nurture*. Durham, NC: Duke University Press.

Fraser, Leanne, Richard E. Brown, Ahmed Hussin, Mara Fontana, Ashley Whittaker, Timothy P. O'Leary, Lauren Lederle, Andrew Holmes, and André Ramos. 2010. "Measuring anxiety- and locomotion-related behaviours in mice: a new way of using old tests." *Psychopharmacology* 211(1): 99–112.

Freese, Jeremy and Sara Shostak. 2009. "Genetics and Social Inquiry." *Annual Review of Sociology* 35: 107–128.

Garfinkel, Harold. 1956. "Conditions of Successful Degradation Ceremonies." *American Journal of Sociology* 61(5): 420–424. doi:10.1086/221800.

Gerlai, Robert. 1996. "Gene-Targeting Studies of Mammalian Behavior: Is It the Mutation or the Background Genotype?" *Trends in Neurosciences* 19(5): 177–181. doi:10.1016/S0166-2236(96)20020-7.

Gieryn, Thomas F. 1999. *Cultural Boundaries of Science: Credibility on the Line*. Chicago: University of Chicago Press.

Gould, Stephen Jay. 1981. *The Mismeasure of Man*. New York: Norton.

Hagenbuch, Niels, Joram Feldon and Benjamin K. Yee. 2006. "Use of the Elevated Plus-Maze Test with Opaque or Transparent Walls in the Detection of Mouse Strain Differences and the Anxiolytic Effects of Diazepam." *Behavioural Pharmacology* 17(1): 31–41.

Hyman, Steven. 2006. "Using Genetics to Understand Human Behavior: Promises and Risks." In *Wrestling with Behavioral Genetics: Science, Ethics and Public Conversation*, eds Erik Parens, Audrey R. Chapman, and Nancy Press, 109–130. Baltimore, MD: Johns Hopkins University Press.

Kamin, Leon J. 1974. *The Science and Politics of I.Q.* Potomac, MD: L. Erlbaum Associates.

Knorr-Cetina, Karin. 1999. *Epistemic Cultures: How the Sciences Make Knowledge*. Cambridge, MA: Harvard University Press.

Lewontin, Richard C. 1970. "Race and Intelligence." *Bulletin of the Atomic Scientists* 26: 2–8.

Lewontin, Richard C. 1975. "Genetic Aspects of Intelligence." *Annual Review of Genetics* 9: 387–405.

Lewontin, Richard C. 1992. *Biology as Ideology: The Doctrine of DNA*. New York, NY: HarperPerennial.

Lewontin, Richard C., Steven P. R. Rose and Leon J. Kamin. 1984. *Not in Our Genes: Biology, Ideology, and Human Nature*. New York: Pantheon Books.

Lippman, Abby. 1992. "Led (Astray) by Genetic Maps: The Cartography of the Human Genome and Health Care." *Social Science & Medicine* 35(12): 1469–1476. doi:10.1016/0277-9536(92)90049-V.

Milner, Lauren C. and John C. Crabbe. 2008. "Three Murine Anxiety Models: Results from Multiple Inbred Strain Comparisons." *Genes, Brain and Behavior* 7(4): 496–505.

Nelkin, Dorothy and M. Susan Lindee. 1995. *The DNA Mystique: The Gene as a Cultural Icon*. New York: Freeman.

Nelson, Nicole C. 2013. "Modeling Mouse, Human, and Discipline: Epistemic Scaffolds in Animal Behavior Genetics." *Social Studies of Science* 43(1): 3–29.

Nelson, Nicole C. 2015. "A Knockout Experiment: Disciplinary Divides and Experimental Skill in Animal Behaviour Genetics." *Medical History* 59(3): 465–485.

Nelson, Nicole C. 2018. *Model Behavior: Animal Experiments, Complexity, and the Genetics of Psychiatric Disorders*. Chicago: University of Chicago Press.

Panofsky, Aaron. 2014. *Misbehaving Science: Controversy and the Development of Behavior Genetics*. eds. Scott Frickel, Mathieu Albert, and Barbara Prainsack, 107–126.

Panofsky, Aaron. 2016. "Some Dark Sides of Interdisciplinarity: The Case of Behavior Genetics." in *Investigating Interdisciplinary Collaboration: Theory and Practice across Disciplines*, edited by S. Frickel, M. Albert and B. Prainsack. New Brunswick, NJ: Rutgers University Press.

Paul, Diane B. 1998. *The Politics of Heredity: Essays on Eugenics, Biomedicine, and the Nature–Nurture Debate*. Albany: State University of New York Press.

Provine, William. 1986. "Geneticists and Race." *American Zoologist* 26: 857–887.

Rose, Nikolas. 1992. "Engineering the Human Soul: Analyzing Psychological Expertise." *Science in Context* 5(2):351–369.

Routtenberg, Aryeh. 1995. "Knockout Mouse Fault Lines." *Nature* 374(6520): 314.

Rowe, David C. 1994. *The Limits of Family Influence: Genes, Experience, and Behavior*. New York: Guilford.

Rutter, Michael. 2006. *Genes and Behavior: Nature–Nurture Interplay Explained*. Malden, MA: Blackwell.

Scarr, Sandra. 1992. "Developmental Theories for the 1990s: Development and Individual Differences." *Child Development* 63: 1–19.

SPSSI. 1969. "Statement on the Current IQ Controversy: Heredity Versus Environment." *American Psychologist* 24(11): 1039–1040.

Tabery, James. 2014. *Beyond Versus: The Struggle to Understand the Interaction of Nature and Nurture*. Cambridge, MA: MIT Press.

Taylor, Peter J. 2014. *Nature–Nurture? No: Moving the Sciences of Variation and Heredity Beyond the Gaps*. Cambridge, MA: The Pumping Station.

Turkheimer, Eric. 2000. "Three Laws of Behavioral Genetics and What They Mean." *Current Directions in Psychological Science* 9(5): 160–164.

Wahlsten, Douglas. 1994a. "The Intelligence of Heritability." *Canadian Psychology* 35(3): 244–260.

Wahlsten, Douglas. 1994b. "Nascent Doubts May Presage Conceptual Clarity: Reply to Surbey." *Canadian Psychology* 35(3): 265–267.

Wahlsten, Douglas. 2003. "Airbrushing Heritability." *Genes, Brain, and Behavior* 2(6): 327–329.

36

Synthetic biology

Deborah Scott, Dominic Berry and Jane Calvert

Introduction

New and emerging sciences and technologies present particular analytical problems for their investigators. This chapter addresses some of these problems while unpacking the example of synthetic biology. The latter can be defined in numerous and sometimes contradictory ways, but at its most general, synthetic biology is an effort to re-imagine biological science and technology in light of engineering. A further defining feature of synthetic biology has been the conspicuously rapid rise of research interest from scholars in the social sciences and humanities, even leading some to coin the term 'para-synthetic biologist' (Campos 2012: 138).

Webster (2007) has argued that a distinctive feature of emerging technologies is that they provide openings for social scientific engagement since they create 'new relations of nature, culture and technology' (p. 463). They are indeterminate, uncertain, and their future paths are not set, meaning that social scientists might yet have a greater influence on their development than in more established fields. But by having more influence, these social scientists also become more implicated and involved. As scholars in the field of science and technology studies (STS), we are aware that by agreeing to write this chapter on synthetic biology we are reinforcing its legitimacy as an appropriate topic for social scientific investigation. This is potentially problematic since we observe the term 'synthetic biology' declining in importance precisely at the moment that synthetic biology is becoming a required chapter heading in a Handbook such as this. For example, the flagship 'Synthetic Biology Engineering Research Center' (SynBERC) in the USA has rebranded as the 'Engineering Biology Research Consortium'.[1] In Europe, UK synthetic biologists are increasingly being awarded funding under the heading of 'Industrial Biotechnology', and the European Commission's new working definition of synthetic biology is so broad that it arguably includes all of biotechnology (EC 2014). Is this merely a decline in the popularity of yet another buzzword (Bensaude-Vincent 2014; Hilgartner 2015), or does it reflect changes in priorities for engineers, scientists, and policy makers? Is it significant that the number of newcomers publishing in the field of synthetic biology has declined since 2010 (Raimbault et al. 2016)? Does a focus on synthetic biology become less relevant in the context of the rapid uptake of interventionist techniques such as 'gene editing' and 'gene drives' across the life sciences? More broadly, is synthetic biology just a name given to some of the latest developments in biotechnology?

With these questions in mind, we have decided that, rather than discussing synthetic biology *per se*, in this chapter we will focus on what synthetic biology can deliver to researchers in the social sciences and humanities on topics that are of broader interest, namely: engineering knowledge, moral economy, governance, publics, and collaboration. These topics cut across 'that thing sometimes called synthetic biology' in ways that allow us to keep the ambiguity of its constitution central to our analysis.

Engineering knowledge

Historians and philosophers of science have been characteristically quick to ask whether synthetic biology constitutes a new means of knowledge production. For example, O'Malley et al. (2007) argue that if one looks at practices and goals, there are at least three different categories of synthetic biologist. The first aim to build standardised interchangeable biological 'parts' (i.e. DNA sequences) that have expected functions when inserted into recipient organisms. The second group work at the level of whole genomes, sometimes with the intention of reducing a genome to its minimum possible size, other times to redesign existing genomes. The third group seeks to create 'protocells' (simple living systems) from non-living components. All three approaches attempt to engineer with life.

Because they aspire to engineer, in what ways are these groups producing knowledge? In other words, what do we mean by knowledge in the context of engineering? Synthetic biologists often argue that growth in knowledge is shown by subsequent improvements in the organism or construct's functioning, that is, they learn how to *make* better. But philosophers such as Fox Keller (2009) are sceptical about whether all the making in synthetic biology contributes new biological knowledge, on the grounds that making really is distinct from knowing or understanding.

One of the ways in which synthetic biology becomes valuable to historical, philosophical, and social studies of science is that it provokes a deeper exploration of what engineering knowledge entails. Recognising this opportunity, Schyfter (2013) has used arguments developed by Vincenti (1990) in the context of aeronautical engineering to analyse cases in biological engineering. In an article addressing the improvement of a collection of genetic constructs, Schyfter highlights the ways in which synthetic biologists can be understood as engaging in 'parameter variation', a method employed by engineers to arrive at an optimal solution in an experimental context lacking a clear theoretical framework. Parameter variation then appears to be a strong candidate for a practice particularly characteristic of engineering knowledge. At the same time it is also resonant with (and potentially contributory to) 'exploratory experimentation', a form of experimentation recently recognised and developed by philosophers of science. In an exploratory experiment, experiments are conducted to see what emerges, rather than for hypothesis testing or demonstration purposes (Franklin 2004). We find ourselves then understanding engineering better, but in ways that might also diminish epistemological boundaries between science and engineering. This opportunity for productive interaction between sociological and philosophical investigation would be left unexplored if we failed to take synthetic biology's engineering aspirations seriously.

The moral economy of the biosciences

Many scientists and engineers were initially drawn to synthetic biology out of a dissatisfaction with the contemporary biosciences, be it how they are organised, funded, or owned. As well as building technologies, some synthetic biologists saw themselves as building a community with distinctive values. The concept of a moral economy (Thompson 1971), which identifies the moral principles and practices of social groups, helps interpret these aspects because it entwines economic, social, and moral goals. Moral economies can be hard to see, their power in part derived from remaining unspoken. They become particularly visible, however, when accepted principles come under threat and require defence. Desired readjustments to the limits of good or acceptable practice in bioscience and biotechnology are readily understood on Thompson's terms (Frow 2013).

A prime area of activity for synthetic biology in the moral economy of the biosciences is intellectual property (IP). Synthetic biology exists at a time when the functions, roles, and purposes of IP in the biosciences are under considerable scrutiny. The most widely cited example of recent flux is that of the 2013 US Supreme Court decision in *Association for Molecular Pathology v. Myriad Genetics, Inc.* in which it was decided that DNA sequences found in nature are not appropriate subject matter for a patent claim (in contradiction to decades of patenting convention), because they fall under the category 'product of nature' (Kevles 2015). Views on the issue of patenting differ considerably within synthetic biology (Nelson, 2014), but what makes the community particularly worthy of attention is that many synthetic biologists are vocal in their objection to existing patenting practices within biotechnology (see Campos 2012). For them, keeping gene sequences non-proprietary achieves not only a scientific and economic good (by 'de-risking' the IP landscape for those attempting to create novelties), but also a social and moral one.

The *Myriad* decision has been widely celebrated as a victory by those lawyers, scholars, and campaigners who think biotechnology needs to leave patenting behind. This same decision, however, can also be read as pushing *against* those pursuing a non-proprietary IP agenda within biotechnology. As Ghosh (2014) has emphasised, the Supreme Court's decision placed considerable importance on the synthesised (as opposed to natural) origins of patentable genetic material, offering a signal to the biotechnology industry as a whole that synthesising DNA is one of the most appropriate ways to demonstrate that claimed sequences are *not* products of nature. This leaves open the question of how different a synthetic sequence must be from a natural one in order to claim it as a patentable novelty. Such questions link the moral economy to the broader governance of synthetic biology by recognising that 'being intellectual property' is only one aspect of an innovation's identity, to be set alongside its economic worth, values, and the meanings it embodies. This is particularly true in relation to fair and equitable use of genetic resources, discussed below.

Governance

For scholars of the governance of emerging science and technology, synthetic biology provides opportunities to theorise and experiment with decision-making. For example, it has invigorated debate around the legacy of the 1975 Asilomar conference on recombinant DNA technologies (Berg et al. 1975). Asilomar is often invoked as a positive model for self-regulation of synthetic biology. Organisers of the original Asilomar meeting have encouraged the application of its template – a voluntary temporary moratorium by scientists and practitioners, followed by expert discussion and measures agreed by the community – to issues such as human germline gene modification (Baltimore et al. 2015). STS scholars, however, argue that Asilomar is not an appropriate model for decision-making, particularly because participation at the meeting was restricted to technical experts and its scope to technical considerations. Instead, these scholars call for deliberative processes involving a broad range of experts and publics, with a wider, more inclusive framing of what is at stake (Hurlbut et al. 2015).

When they were setting out to establish a new field, synthetic biologists needed to make claims of novelty. At the same time, they did not want to attract new regulatory requirements, and thus emphasised 'continuity with the past' to regulators (Tait 2009: 150). National and international advisory bodies have largely accepted this framing, concluding that current and near term synthetic biology is not sufficiently novel to challenge regulatory and ethical frameworks, although future synthetic biology advances might (see PCSBI 2010; EC 2014).

As more techniques and products enter industrial use and commercialisation, policy discussions of governance seem to be shifting away from the generic term 'synthetic biology' and focusing on specific tools, trends, and products. This shift is providing openings to acknowledge novel aspects of these techniques and to try new forms of oversight. For instance, with CRISPR-Cas9 and other gene-editing techniques, US regulators have already determined that CRISPR-edited mushrooms fall outside the American regulatory regime for GMOs (Waltz 2016). Discussions under the heading of 'synthetic biology' at the UN Convention on Biological Diversity in late 2016 focused on whether the Nagoya Protocol (an international legal agreement to ensure fair and equitable access and benefit-sharing arrangements for genetic resources) was undermined in light of unrestricted movement of digital sequence information. Developing country delegations freely acknowledged that these developments were not unique to synthetic biology, but used the agenda item as an opening to demand action on what they termed a new form of biopiracy (Tsioumani et al. 2016). And while discussions on synthetic biology have been stymied over 'precautionary' vs 'proactive' approaches, more specific applications such as gene drives are generating fresh ideas of how process requirements could enable research responsive to its genetic, ecological, and social contexts (Kaebnick et al. 2016).

The many publics of synthetic biology

Synthetic biology is particularly promissory and future-oriented (Schyfter & Calvert 2015). The promises and envisioned futures of synthetic biology do not just rest on the technical possibilities of an engineering approach to biology, but also rely on visions of particular communities – of those who need synthetic biology, those who use it, and those who would stand in its way.

Take the highly publicised project to use synthetic biology to engineer yeast to produce the precursor of a key compound in an anti-malarial drug. Developing countries relying on artemisinin-based drugs to combat malaria have been framed as vulnerable to climate and market-related supply of the plant *Artemisia*, and thus in need of a more predictable, industrial biotechnology-enabled supply of the drug (Ro et al. 2006). This version of vulnerability has been countered by a focus on the livelihoods of the thousands of small-scale farmers of *Artemisia* in Asia and Africa (ETC 2013). Marris (2015) has intervened in this debate with the STS insight that an increased supply of artemisinin alone will not bring health benefits without attention to relevant political, economic, and social factors.

The increasing accessibility of biotechnology tools is often described as a *democratisation* of biology (Ledford 2015). This political framing seems to centre discussion at two poles: 1) how to police the citizenry of synthetic biology, particularly those on the outskirts of formal scientific communities such as Do-It-Yourself (DIY) amateurs and potential bio-terrorists; and 2) celebrating nascent citizens, such as the undergraduates in the annual International Genetically Engineered Machines (iGEM) competition, who stand poised to innovate. Social scientists have called for reorienting discussion away from these poles. Jefferson et al. (2014) challenge the claim that synthetic biology is actually 'deskilling' the field (specialised infrastructure and local knowledge continue to be vital), and also challenge the corresponding 'policy gaze' on rogue outsiders and terrorists rather than on legitimate insiders and state-sponsored bioweapons programmes (also Marris et al. 2014). Frow (2015) describes how the dominant framing of democratisation endows iGEM competitors and others with a right to innovate, which she then places in tension with discussions of governance and the public good.

Marris (2015) argues that the imagined vision of an ill-informed public rejecting synthetic biology (a vision widely held in the scientific community) has motivated both synthetic biologists and policy actors, directing work on 'Ethical, Legal, and Social Issues' (ELSI) towards an ever-elusive public understanding and acceptance of synthetic biology. Innovative models for decision-making processes such as anticipatory governance (Guston 2013) and Responsible Research and Innovation (Owen et al. 2012, discussed below) reject this deficit framing. These models for decision-making rely on their own visions of publics who are willing to take part in dialogues, be engaged, and collectively grapple with questions of the trajectory and purpose of scientific research and innovation.

Collaboration with the social sciences and humanities

Some of the earliest funding schemes and initiatives that went under the heading of 'synthetic biology', particularly in the UK, called for the involvement of scholars from the social sciences and humanities, and this has continued to the present day. Scholars have taken advantage of these opportunities to experiment with new forms of interdisciplinary collaboration. But there is also a danger of social scientists being drawn into a service role, with pressures to facilitate public acceptance of the technology and help it go to market (Barry et al. 2008).

Collaborations between natural and social scientists are not new, of course. In the 1990s, 3–5 per cent of the total funding for the Human Genome Project was directed towards 'Ethical, Legal and Social Issues' (Fisher 2005). ELSI programmes have since been associated with fields like nanotechnology, stem cell research, and neuroscience. In the European context, public controversies over genetically modified crops have led to calls for the involvement of social scientists in contentious technological areas. A dominant and reductive framing of synthetic biology is that it has the potential 'to be the next GM' (Marris 2015).

Social scientists who have been closely engaged in synthetic biology have had a range of experiences. Anthropologists Rabinow and Bennett (2012), for example, started working with leading US synthetic biologists hoping for genuinely collaborative research, but ended up confronting issues of power and divergent expectations. In contrast, one of us, as part of the 'Synthetic Aesthetics' project, found three-way collaborations between synthetic biologists, social scientists, and artists and designers to be unexpectedly productive (Calvert and Schyfter 2017). Balmer et al. (2015) attempt to capture some of the diversity of these experiences by discussing the roles that social scientists play in synthetic biology: from 'representative of the public', to trophy wife, to critic, to co-producer of knowledge.

These particular social scientists have been dissatisfied with the 'ELSI' framing of their work, because it assumes a conceptual separation of the scientific work and its downstream 'implications'. They have attempted to shift to a 'post-ELSI' mode, where there is an explicit attempt to co-produce knowledge across disciplinary boundaries. Rabinow and Bennett (2012) called their attempt to elicit mutual flourishing 'Human Practices' (a term that was taken up by the iGEM competition, albeit largely stripped of its original connotations). As noted above, we have recently seen the emergence of 'Responsible Research and Innovation' or 'RRI', which, according to some accounts, brings together diverse groups to imagine technological futures differently (Owen et al. 2012). The inclusion of RRI has become a funding requirement for many synthetic biology grants. Definitions of RRI are currently contested, and while some seem to regard it as synonymous with biosafety or public outreach, others hope it will allow for new forms of experimental collaboration. Social scientists working in synthetic biology are currently in the process of simultaneously investigating, creating, and often attempting to deliver RRI.

Conclusions: beyond synthetic biology

As noted in our introduction, the term synthetic biology has recently started to decline in prominence. This decline coincides with other changes in the networks and communities that have been built around synthetic biology. To some there seems to have been a shift in the field's orientation towards industrialisation and militarisation, marked by a significant rise in funding from sources such as the Defense Advanced Research Projects Agency (DARPA) in the USA (WWICS 2015). Although defence funding and commercial ambition have been features of synthetic biology from its beginning (Knight 2002), there are now increasing pressures on synthetic biologists to deliver on their economic promises. Do these changes necessitate different approaches for social scientific study?

One of the main features of STS work on emerging sciences and technologies is the attempt to promote reflexivity among scientists and engineers (and STS researchers themselves). Reflexivity can be defined as 'the process of identifying, and critically examining (and thus rendering open to change), the basic, pre-analytic assumptions that frame knowledge-commitments' (Wynne 1993: 324). The aim is to make all actors involved, 'more self-aware of their own taken-for-granted expectations, visions, and imaginations of the ultimate ends of knowledge' (Macnaghten et al. 2005: 278). In our experience, it is much easier to promote reflexivity when fields are still emerging, because their possible futures are uncertain and it is harder to argue that things cannot be otherwise. The early days of synthetic biology in the mid-2000s were marked by an emphasis on 'openness', both in terms of ownership and in terms of who could consider themselves to be doing synthetic biology – disciplinarily and institutionally. Synthetic biologists recognised themselves as part of a social and political endeavour as well as a technical one, and for this reason welcomed somewhat experimental collaboration with social scientists, humanities scholars, artists, designers, and other groups (Balmer et al. 2016).

It may well prove that there is less scope to engage in collaboration in a field where the parameters of success are being narrowed to the industrial, commercial, and military, if only because of the kinds of research sites that will increasingly come to matter. Commercial enterprises can be hard to gain access to, and when it comes to laboratories run by defence agencies, a social scientist may be in a stronger position to analyse and critique from a greater distance than has characterised social scientist/synthetic biologist relations thus far. Here we have highlighted the earlier kinds of discipline-building and collaboration that occurred in synthetic biology when it began to take shape, and drawn attention to broader themes that the study of synthetic biology illuminates. As social sciences and humanities researchers invested in making biotechnologies diverse, plural, and reflexive, we may come to find our goals are best served by moving beyond synthetic biology.

Acknowledgments

The authors are grateful for funding from the European Research Council (ERC 616510 – Engineering Life).

Note

1 www.ebrc.org accessed 30 January 2017.

References

Balmer, A. *et al.* (2015) Taking Roles in Interdisciplinary Collaborations: Reflections on working in Post-ELSI Spaces in the UK Synthetic Biology Community. *Science and Technology Studies* 28(3): 3–25.

Balmer, A., Bulpin, K., & Molyneux-Hodgson, S. (2016) *Synthetic Biology: A Sociology of Changing Practices* Basingstoke: Palgrave Macmillan.

Baltimore, D. *et al.* (2015) A prudent path forward for genomic engineering and germline gene modification. *Science* 348(6230): 36–38.

Barry, A. *et al.* (2008) Logics of Interdisciplinarity. *Economy & Society* 37: 20–49.

Bensaude-Vincent, B. (2014) The politics of buzzwords at the interface of technoscience, market and society: The case of 'public engagement in science'. *Public Understanding of Science* 23(3): 238–253.

Berg, P. *et al.* (1975) Asilomar Conference on Recombinant DNA Molecules. *Science* 188: 991–994.

Calvert, J. and Schyfter, P. (2017) 'What can Science and Technology Studies learn from art and design? Reflections on Synthetic Aesthetics' *Social Studies of Science* 47(2): 195–215.

Campos, L. (2012) The BioBrick™ road. *BioSocieties* 7(2): 115–139.

ETC Group (ETC). (2013) Potential Impacts of Synthetic Biology on Livelihoods and Biodiversity: Eight Case Studies on Commodity Replacement. www.cbd.int/doc/emerging-issues/emergingissue s-2013-07-ETCGroup%281%29-en.pdf.

European Commission's Scientific Committee on Health and Environmental Risks, Scientific Committee on Emerging and Newly Identified Health Risks, & Scientific Committee on Consumer Safety (EC). (2014) *Opinion on synthetic biology I: Definition.* http://ec.europa.eu/health/scientific_committees/em erging/docs/scenihr_o_044.pdf.

Fisher, E. (2005) Lessons learned from the Ethical, Legal and Social Implications programme (ELSI): Planning societal implications research for the National Nanotechnology Programme. *Technology in Society*, 27, 321–328.

Fox Keller, E. (2009) What does synthetic biology have to do with biology? *BioSocieties* 4: 291–302.

Franklin, L. R. (2004) Exploratory Experiments. *Philosophy of Science* 72(5): 888–899.

Frow, E. (2013) Making big promises come true? Articulating and realizing value in synthetic biology. *BioSocieties* 8(4): 432–448.

Frow, E. (2015) Rhetorics and practices of democratization in synthetic biology. In *Knowing New Biotechnology* (Matthias Wienroth & Eugenia Rodrigues eds.), 174–187.

Ghosh, S. (2014) Nature, nurture and DNA sequences. *Pharmaceutical Patent Analyst* 3(1): 5–7.

Guston, D. H. (2013) Understanding 'anticipatory governance.' *Social Studies of Science* 44(2): 218–242.

Hilgartner, S. (2015) Capturing the Imaginary: Vanguards, visions, and the synthetic biology revolution. In Hilgartner, S. Miller, C., and Hagendijk, R. (eds.) *Science & Democracy: Making knowledge and making power in the Biosciences and Beyond.* New York: Routledge: 33–55.

Hurlbut, J. B., Saha, K., & Jasanoff, S. (2015) CRISPR Democracy: Gene Editing and the Need for Inclusive Deliberation. *Issues in Science and Technology* 32(1).

Jefferson, C., Lentzos, F. & Marris, C. (2014) Synthetic biology and biosecurity: Challenging the 'myths.' *Frontiers in Public Health* 2(115): 21–35.

Kaebnick, G. E. *et al.* (2016) Precaution and governance of emerging technologies. *Science* 354(6313): 710–711.

Kevles, D. J. (2015) Inventions, Yes: Nature, No: The Products-of-Nature Doctrine from the American colonies to the U.S. Courts. *Perspectives on Science* 23(1): 13–34.

Knight, T. F. (2002) DARPA BioComp Plasmid Distribution 1.00 of Standard Biobrick Components. https://dspace.mit.edu/handle/1721.1/21167#files-area.

Ledford, H. (2015) CRISPR, the disruptor. *Nature* 522: 20–24.

Macnaghten, P., Kearnes, M. B., & Wynne, B. (2005) Nanotechnology, governance and public deliberation: What role for the social sciences? *Science Communication*, 27, (2), 268–287.

Marris, C. (2015) The Construction of Imaginaries of the Public as a Threat to Synthetic Biology. *Science as Culture* 24(1): 83–98.

Marris, C., Jefferson, C. & Lentzos, F. (2014) Negotiating the dynamics of uncomfortable knowledge: The case of dual use and synthetic biology. *BioSocieties* 9(4): 393–420.

Nelson, B. (2014) Cultural divide. *Nature* 509(8 May): 152–154.

O'Malley, M. A., Powell, A., Davies, J. F. & Calvert, J. (2007) Knowledge-Making Distinctions in Synthetic Biology. *BioEssays* 30: 57–65.

Owen, R., Macnaghten, P. & Stilgoe, J. (2012) Responsible research and innovation: From science in society to science for society, with society. *Science and Public Policy* 39: 751–760.

Presidential Commission for the Study of Bioethical Issues (PCSBI) (2010) *New directions: The ethics of synthetic biology and emerging technologies*. Washington, DC: PCSBI.

Rabinow, P. & Bennett, G. (2012) *Designing Human Practices*. Chicago: University of Chicago Press.

Raimbault, B., Cointet, J. P. & Joly, P. B. (2016) Mapping the Emergence of Synthetic Biology. *PLoS ONE* 11(9): 1–19.

Ro, D-K., *et al.* (2006) Production of the antimalarial drug precursor artemisinic acid in engineered yeast. *Nature* 440: 940–943.

Schyfter, P. (2013) Propellers and promoters: emerging engineering knowledge in aeronautics and synthetic biology. *Engineering Studies*, 5(1): 6–25.

Schyfter, P. & Calvert, J. (2015) Intentions, Expectations and Institutions: Engineering the Future of Synthetic Biology in the USA and the UK. *Science as Culture* 24(4): 359–383.

Tait, J. (2009) Governing Synthetic Biology: Processes and Outcomes, in *Synthetic Biology: The Technoscience and Its Societal Consequences*, M. Schmidt *et al.* (eds). New York: Springer: 141–154.

Thompson, E. P. (1971) The Moral Economy of the English Crowd in the Eighteenth Century. *Past & Present* 50: 76–136.

Tsioumani, E., *et al.* (2016) Summary of the UN Biodiversity Conference. *Earth Negotiations Bulletin* 9(678).

Vincenti, W. (1990) *What Engineers Know and How They Know It*. Baltimore, MD: Johns Hopkins University Press.

Waltz, E. (2016) Gene-edited CRISPR mushroom escapes US regulation. *Nature* 532: 293.

Webster, A. (2007) Crossing boundaries: Social science in the policy room. *Science, Technology & Human Values*, 32(4): 458–478.

Woodrow Wilson International Center for Scholars (WWICS) (2015) *U.S. Trends in Synthetic Biology Research Funding*. Synthetic Biology Project: Washington DC. www.wilsoncenter.org/sites/default/files/final_web_print_sept2015_0.pdf.

Wynne, B. (1993) Public uptake of science: a case for institutional reflexivity. *Public Understanding of Science* 2(4): 321–337.

Index

For Product Safety Concerns and Information please contact our EU
representative GPSR@taylorandfrancis.com Taylor & Francis Verlag GmbH,
Kaufingerstraße 24, 80331 München, Germany

Printed and bound by CPI Group (UK) Ltd, Croydon, CR0 4YY
01/05/2025
01858576-0001